ENERGY ECONOMICS
Modeling and Empirical Analysis in China

ENERGY ECONOMICS

Modeling and Empirical Analysis in China

Edited by
Yi-Ming Wei
Ying Fan
Zhi-Yong Han
Gang Wu

CRC Press is an imprint of the
Taylor & Francis Group, an **informa** business

CRC Press
Taylor & Francis Group
6000 Broken Sound Parkway NW, Suite 300
Boca Raton, FL 33487-2742

First issued in paperback 2019

© 2010 by Taylor & Francis Group, LLC
CRC Press is an imprint of Taylor & Francis Group, an Informa business

No claim to original U.S. Government works

ISBN-13: 978-1-4398-1121-4 (hbk)
ISBN-13: 978-0-367-38526-2 (pbk)

This book contains information obtained from authentic and highly regarded sources. Reasonable efforts have been made to publish reliable data and information, but the author and publisher cannot assume responsibility for the validity of all materials or the consequences of their use. The authors and publishers have attempted to trace the copyright holders of all material reproduced in this publication and apologize to copyright holders if permission to publish in this form has not been obtained. If any copyright material has not been acknowledged please write and let us know so we may rectify in any future reprint.

Except as permitted under U.S. Copyright Law, no part of this book may be reprinted, reproduced, transmitted, or utilized in any form by any electronic, mechanical, or other means, now known or hereafter invented, including photocopying, microfilming, and recording, or in any information storage or retrieval system, without written permission from the publishers.

For permission to photocopy or use material electronically from this work, please access www.copyright.com (http://www.copyright.com/) or contact the Copyright Clearance Center, Inc. (CCC), 222 Rosewood Drive, Danvers, MA 01923, 978-750-8400. CCC is a not-for-profit organization that provides licenses and registration for a variety of users. For organizations that have been granted a photocopy license by the CCC, a separate system of payment has been arranged.

Trademark Notice: Product or corporate names may be trademarks or registered trademarks, and are used only for identification and explanation without intent to infringe.

Library of Congress Cataloging-in-Publication Data

Energy economics : modeling and empirical analysis in China / Yi-Ming Wei ... [et al.] editors.
   p. cm.
  Includes bibliographical references and index.
  ISBN 978-1-4398-1121-4 (alk. paper)
  1. Energy policy--China. 2. Energy consumption--China. 3. China--Economic policy--1976-2000. 4. China--Economic policy--2000- I. Wei, Yi-Ming. II. Title.

HD9502.C6E55 2010
333.7901'5195--dc22
                                        2009022576

Visit the Taylor & Francis Web site at
http://www.taylorandfrancis.com

and the CRC Press Web site at
http://www.crcpress.com

Contents

Preface xi

Chapter 1 Review of Energy Development in China 1
  1.1 World Energy Development 1
      1.1.1 World Energy Reserves and Distribution 1
      1.1.2 World Economic Growth and Energy Consumption 3
      1.1.3 World Energy Market 4
      1.1.4 International Energy and Environmental Cooperation 6
  1.2 Energy Production in China 7
      1.2.1 Energy Reserves and Distribution in China 7
      1.2.2 China Energy Production and Structure 10
      1.2.3 Production and Development of Renewable Energy in China 12
  1.3 Energy Consumption in China 13
      1.3.1 Energy Consumption Quantity and Structure in China 13
      1.3.2 Energy Consumption in Industrial Sectors 16
      1.3.3 Energy Efficiency and Conservation 17
  1.4 Energy Consumption, Resources and Economy in Regions 18
      1.4.1 Data and Processing 18
      1.4.2 Regional Energy Consumption 19
      1.4.3 Regional Differences of Energy Resources and Consumption 23
      1.4.4 Regional Differences of Energy Consumption, Resources Distribution and Economic Development 26
  1.5 Summary 29

Chapter 2 Structural Relationship between Chinese Energy and Economic Growth 31
  2.1 The Cointegration and Causality between Chinese GDP and Energy Consumption 31
      2.1.1 Method 32
      2.1.2 Cointegration and Causality in Chinese Energy Economy 34
      2.1.3 Cointegration and Causality in Three Industries' Energy Economy 37
      2.1.4 Conclusions and Policy Implications 41
  2.2 Energy Intensity and Economic Structure of China 42
      2.2.1 Decomposition of Energy Intensity Based on SDA 43
      2.2.2 Impact of Chinese Economic Structure Change on Energy Intensity ... 45
      2.2.3 Impact of the Secondary Industry Structure on Energy Intensity 49
      2.2.4 Conclusions 55
  2.3 Energy Structure and Energy Efficiency in China 57
      2.3.1 China's Energy Economy and Energy Efficiency 58
      2.3.2 Empirical Study on the Impact of China's Energy Structure on Energy Efficiency 60

		2.3.3	ME and MRS in Chinese Energy Economy · 63

      2.3.4   Conclusions · 66

  2.4   Summary · 67

Chapter 3   **Analysis and Forecasting of Energy Supply and Demands in China** · 71

  3.1   Forecasting of Energy Requirements in 2020 · 71

      3.1.1   Present Research on Energy Requirement · 72

      3.1.2   Scenario Analysis and Input–Output Analysis · 72

      3.1.3   China Energy Demand Analysis System–CEDAS · 73

      3.1.4   Scenario Analysis of Energy Requirement in 2020 · · · · · · · · · · · · · · · · · · · 73

      3.1.5   Results · 78

      3.1.6   Policy Implications · 85

  3.2   Economic Analysis of Coal Supply System · 86

      3.2.1   System Dynamics Model and Data for Coal Production and Supply · · · · · · · · · · · 87

      3.2.2   Investment Influence on the Capacity of Coal Supply · · · · · · · · · · · · · · · · · 91

      3.2.3   Conclusions and Remarks · 91

  3.3   Summary · 92

Chapter 4   **Fluctuation in Oil Markets and Policy Study** · 95

  4.1   The Characteristics of International Petroleum Price Fluctuation · · · · · · · · · · · · · · · · 95

      4.1.1   International Petroleum Prices and Their Impact on the Global Economy · · · · · · · 95

      4.1.2   Oil Price Level Characteristics of Three Oil Crises · 96

      4.1.3   The Oil Price Level Characteristics in 2005 · 100

      4.1.4   The Long-Term Tendency of International Crude Oil Prices · · · · · · · · · · · · · · · · · 101

      4.1.5   The Short-Term Fluctuations of International and Domestic Crude Oil Prices · · 103

      4.1.6   Concluding Remarks and Policy Suggestions · 105

  4.2   Analysis of the Co-Movement between Chinese and International Crude Oil Prices · 107

      4.2.1   The Long-Term Relationship between International and Domestic Crude Oil Prices · 107

      4.2.2   The Short-Term Relationship between International and Domestic Crude Oil Prices · 108

      4.2.3   The Dynamic Impact of International Crude Oil Prices on Their Chinese Counterpart · 110

      4.2.4   Conclusions · 112

  4.3   Analyses of the Features of the Changes in China's Crude Oil and Its Products · 114

      4.3.1   The Price Correlation Analysis of Oil Products and Crude Oil · · · · · · · · · · · · · · 115

      4.3.2   The Ratio (Logarithmic Difference) Analysis of Oil Products and Crude Oil Prices · 116

      4.3.3   The Dynamic Response Relation between Oil Products and Crude Oil Prices · · · 119

      4.3.4   Concluding Remarks and Policy Suggestions · 123

  4.4   The Impact of Rising International Crude Oil Price on China's Economy · · · · · · · 123

      4.4.1   Computable General Equilibrium (CGE) Model of Oil Prices Fluctuation · · · · · · 124

      4.4.2   Scenarios Analysis of Oil Price Fluctuations · 128

      4.4.3   Concluding Remarks and Main Policy Suggestions · 140

4.5 Forecast of International Oil Price ··· 141
  4.5.1 Changeable International Oil Price ·· 142
  4.5.2 Long-Term Forecast for Oil Price Based on Wavelet Analysis ················· 145
  4.5.3 Mid-Term and Short-Term Forecast Based on Pattern Matching ············· 149
  4.5.4 Mid-Term and Short-Term Forecast Adjusted by Futures ···················· 153
  4.5.5 Conclusions and Policy Suggestions ·· 155
4.6 Study of Chinese Oil Pricing Mechanism ··· 157
  4.6.1 Overview of Chinese Oil Pricing Mechanism Development ·················· 157
  4.6.2 Problems in Chinese Oil Pricing Mechanism ·································· 160
  4.6.3 Suggestions on Reform of Chinese Oil Pricing Mechanism ··················· 161
4.7 Summary ··· 168

Chapter 5 Energy, Environment and CO_2 Abatement in China ············· 171
5.1 Challenges and Opportunities in Kyoto Era ·· 171
  5.1.1 CO_2 Abatement and Global Climate Change ································ 172
  5.1.2 CO_2 Abatement and Economic Growth ······································ 172
  5.1.3 Contemporary Status of Chinese CO_2 Emissions ··························· 173
  5.1.4 Challenges and Opportunities Faced by Chinese Energy and Environment ······ 175
5.2 Characteristics of Carbon Emissions Trend in China ···························· 177
  5.2.1 Trend of Chinese Carbon Emissions Intensity ································· 178
  5.2.2 Comparison of Carbon Emissions Intensity between China and Developed World ·· 179
  5.2.3 Methodology and Data ··· 179
  5.2.4 Empirical Results and Discussion ·· 182
  5.2.5 Conclusions ·· 184
5.3 Impact of Population, Economic Growth and Technology on CO_2 Emissions ··· 185
  5.3.1 Impact Factors of CO_2 Emissions ·· 185
  5.3.2 STIRPAT Model ··· 186
  5.3.3 Trends of Population, Economic Growth, Technology and CO_2 Emissions at Different Income Levels ··· 187
  5.3.4 Results Analysis and Discussions ·· 190
  5.3.5 Conclusions ·· 195
5.4 Impact of Lifestyle on Energy Use and CO_2 Emissions ························ 196
  5.4.1 Relationship of Lifestyle, Energy Use and CO_2 Emissions ················ 197
  5.4.2 Data and CLA Method ·· 198
  5.4.3 Direct and Indirect Impact of Lifestyle on CO_2 Emissions ················ 199
  5.4.4 Conclusions ·· 206
  5.4.5 Policy Implications ··· 207
5.5 Forecasting China's CO_2 Emissions in 2020 ···································· 208
  5.5.1 Energy Consumption and CO_2 Emissions ·································· 208
  5.5.2 Forecasting CO_2 Emissions Based on Energy Consumption ············· 209
  5.5.3 Forecasting CO_2 Emissions under Different Growth Paths ··············· 210
  5.5.4 Policy Implications ··· 215
5.6 Summary ··· 217

Chapter 6 Strategic Petroleum Reserves and National Energy Security ···· 219
6.1 China's Energy Security ·· 219
  6.1.1 Definition and Connotation of Energy Security ······························· 219
  6.1.2 Contemporary Status of China's Energy Security ···························· 221

  6.1.3 Hidden Troubles in China's Energy Security ·· 222
 6.2 Study on Optimal Scale of Chinese Strategic Petroleum Reserves ··············· 223
  6.2.1 Contemporary Status of Chinese Strategic Petroleum Reserves ················ 223
  6.2.2 Optimal Petroleum Reserves Model Based on Decision Tree ···················· 225
  6.2.3 Discussion on Different Reserve Scales ·· 229
  6.2.4 Results ·· 231
 6.3 Risk Assessment of Oil Imports and Countermeasures ································· 231
  6.3.1 Overview of World Oil Trade ··· 232
  6.3.2 Risk Evaluation of Crude Oil Import–Based on HHA Method ················ 234
  6.3.3 Weight Coefficient of Oil Import Risk–Based on AHP Method ·············· 234
  6.3.4 Risk Assessment of Oil Imports in Some Major Oil Importing Countries ······· 236
  6.3.5 Results ·· 243
 6.4 Policy Suggestions on China's Energy Security ·· 244
  6.4.1 Energy Diplomacy Policy ·· 244
  6.4.2 Oil Import Policy ··· 244
  6.4.3 Strategic Petroleum Reserve Policy ·· 245
  6.4.4 Conservation and Renewable Energy Policy ·· 246
  6.4.5 Off-Shore Oil Policy ·· 247
 6.5 Summary ·· 247

Chapter 7 Energy Technology and Its Policy ··· 249

 7.1 Paradigm Transitions of Energy Technological Change ······························· 249
  7.1.1 Phase I: Natural Transitions (before 1859) ··· 250
  7.1.2 Phase II: Hydrocarbon Lock-in and Induced Innovation (1859–1992) ··········· 251
  7.1.3 Phase III: Transition toward Clean and Sustainable Energy System (1992–) ····· 255
 7.2 Oil Shocks and Energy R&D Expenditures Reaction Patterns ····················· 256
  7.2.1 Adjustment from Demand–Side and Supply–Side to Oil Shocks ············· 258
  7.2.2 Energy R&D Reaction Patterns to Oil Shocks ······································· 260
 7.3 Energy R&D Expenditures in Technology Dimension ································· 263
  7.3.1 Entropy Statistics ·· 263
  7.3.2 Energy R&D Expenditures in Technology Dimension ·························· 265
  7.3.3 Energy R&D Expenditures in Country Dimension ································ 267
 7.4 The Substitution Routes of Energy in China and the Policy Analysis on Renewable Energy ··· 270
  7.4.1 Alternatives to Fossil Energy ·· 270
  7.4.2 Case Study: Liquidation Technology of Coal in China ·························· 272
  7.4.3 Policy Analysis on Renewable Energy Technology ······························· 274
 7.5 Policy Suggestions for China's Energy Technology ····································· 277
 7.6 Summary ·· 279

Chapter 8 China Energy Outlook ·· 281

 8.1 Total Energy Consumption: Large Amount, Rapid Growth, Large Regional Differences ·· 281
 8.2 Oil Imports Become Diversified, Imports Risk Reduces Gradually, and SPR Scale Increases Incrementally ·· 282
 8.3 Coal Supply and Demand Are Basically Balanced, Clean Energy Is Developed, and Consumption Patterns Are Diversified ·· 282

- 8.4 Energy Efficiency Is Improved Steadily, Energy Conservation Potential Is Tremendous, and Technological Progress Is the Key 284
- 8.5 Total Emissions Continue to Expand, and the Industrial Structure Goes toward a Carbon-Intensive Trend. The Impact of Consumer Behavior Should Not Be Underestimated 285
- 8.6 Energy Strategies and Policies Should Highlight International Cooperation, Diversification, and Sustainability 287

References 291

Index 307

Preface

Energy, just like labor and capital, is universally acknowledged to be the fundamental production factor and strategic resource of an industrial society. Both energy supply shortages and structural imbalances will directly and significantly impact a country's economic and social development. During past decades, China has carried out a coal-oriented principle in its energy strategy. This has basically guaranteed social development and economic growth, although it also caused China's energy consumption to lag behind the international advanced level. Since the appearance of China's reform and its opening up in 1978, several factors, such as institutional and management innovation, the introduction and proliferation of advanced technology, and so forth, have driven its energy efficiency improvement, and maintained its energy consumption growth at a relatively low pace.

Since the 1990s, China has sustained rapid economic growth. This has resulted in China's ever-growing energy import dependency, especially in oil imports. Meanwhile, world energy markets have fluctuated dramatically, and international crude oil prices have risen sharply, even hitting the historical-high price of 70 U.S. dollars per barrel in 2006. The rise in international oil prices has greatly increased the energy costs of China's economic growth, and national energy security has now become a hot and strategic issue for China's government and society.

We focus on some hot issues concerning China's current energy strategies and policies and analyze the scenarios of different policies, so as to provide a reference for China's energy policy making. To sum up, the issues studied in this volume mainly include:

(1) Gross and structural features of China's energy economy.

Since the reforms and opening up, China has maintained rapid growth of its economy, yet with a slow growth in energy consumption. This has attracted broad attention to the sustainability of China's energy-economy.

Can the relationship between China's rapid economic growth and reduction of energy consumption be maintained in the 21st century? How will Chinese economic growth increase its energy demand? What kind of causal relationship exists between China's energy and economy? Will the control of energy consumption impact China's targeted economic growth? How will China's economic restructuring influence its energy intensity? Can the downward trend of China's energy intensity continue? To what extent have energy inputs supported China's economic growth? What are the differences between the roles of alternative energy sources in supporting economic growth? How can it reach the double target of sustaining economic growth *and* reducing energy consumption through adjusting structure? These are some basic issues in China's energy-economy research as well as the most important issues for China's energy strategy and its policy makers.

(2) Forecasting of China's energy supply and demand.

At present, China's socialist market economic system has been established. However, China's energy industry is still in the process of transformation from a centrally planned system to the market system. As a result, energy strategy and policy formulation methods yield fundamental change, i.e., energy demand and supply can no longer be controlled through state planning. Under such circumstances, scientific analysis and accurate forecasting of energy supply and demand become important for driving China's energy strategy and policy. In particular, during a period of building an affluent society by 2020, how much energy does China need? In which way can China control the rapid growth of energy use? How might it improve the regional balance of energy resources effectively? Can the coal supply meet China's demand during the building of an affluent society by 2020? How might it resolve the overload on production in the state-owned coal mines?

(3) Fluctuations in the international oil market and China's counter-measures.

With the growth of China's oil imports, dependency on international oil markets has significantly increased. China's oil market and the national economy have become increasingly vulnerable to international price changes. What, then, is the rule governing international oil price fluctuations? What kind of interaction is there between international and domestic crude oil prices? What relationship exists between domestic crude oil prices and refined oil prices? How will fluctuations in international oil prices impact China's economy? What is the future movement trend in oil prices? Which kind of oil pricing mechanism should China establish to resist market risks and to better protect oil security? What petroleum strategy should China develop to face the increasingly fierce competition in the world oil market? The study of these problems will provide a scientific basis for China's petroleum strategy.

(4) Energy-environment problems and reduction in CO_2 emissions.

On February 16, 2005, the *Kyoto Protocol* formally took effect. As the world's second largest emitter of CO_2, China's energy-environment problems face both opportunities and challenges. Negotiations in reduction of greenhouse gas emissions have begun for the post-Kyoto era, and mitigating greenhouse gas emissions in developing countries will become one of the key issues in the negotiations. Consequently, China's reduction in CO_2 emissions attracts worldwide attention and several critical issues emerge. For instance, which factors restrict CO_2 emissions? How will these factors impact the reduction at different stages? Currently, with economic development, China's carbon emission intensity has experienced a downward trend, which is contrary to the carbon emission intensity change trajectory of developed countries over the same period. So, what factors have contributed to the trend of rapid decline of China's carbon emission intensity? Could this trend be maintained in the future? What policy measures can be adopted to promote emissions to be further reduced? With China's sustained and rapid economic development in the future, how will its CO_2 emissions, trend change? Empirical analyses of these problems will provide decision support for the greenhouse gas emission reduction strategy and the related energy-environmental policy.

(5) Strategic petroleum reserves and national energy security.

As one of the major oil-importing countries, China has launched its strategic petroleum reserve (SPR). However, such a reserve will still take time to be adequate. Therefore, we should not be overly optimistic about the issue of national energy security. What threats does China's energy security face? Will the 90 days SPR standard set by the International Energy Agency be suitable for China's economic development and energy security? What is the optimal size of SPR for China from 2010 to 2020? How will the diversified index of China's oil imports change? Compared to other major oil-importing countries, what are the sensible ways and means for China to reduce its oil import risk index? Focusing on the issues above, we integrate some qualitative and quantitative methods to develop a series of mathematical models, conduct a comprehensive and systematical analysis, and present several policy recommendations to safeguard China's energy security.

(6) Energy technology progress and change.

During the course of keeping sustainable development of the economy, energy and environment, any progress in energy technology will play a vital role. This requires energy strategy and policy makers to have adequate and comprehensive understanding of energy technology changes—such as the changing impetus, speed and direction of innovation. What stages has the world energy technology-change experienced? What is the impetus of technical change at each stage? What kinds of characteristics do international energy technology R&D input appear to have? What is the trend of its technique portfolio and national distribution? What can we learn from international experiences to improve China's energy policy formulation? Which route should be followed by China's future energy technique substitution? What are the features and shortcomings of China's renewable energy technique policy? Through empirical research on the world's energy technology systems, and analysis on China's energy policy, we propose viewpoints on these critical issues.

As the first volume of a series of China Energy Reports, this book is a collection of the research results on energy strategy and policy issues investigated by the Center for Energy & Environmental Policy Research (CEEP), Institute of Policy and Management (IPM), and Chinese Academy of Sciences (CAS). Based on the changes in the international and domestic energy-economy, different themes will be selected in the China Energy Reports, together with targeted research. The series of China Energy Reports will provide a scientific basis and decision support for Chinese energy strategists and policy makers.

We anticipate that this volume will provide support for decision-makers, and promote the exchange of our findings with energy policy research peers. Therefore, the series of China Energy Reports not only analyzes the policies of specific issues, but also briefly discusses econometric models and methodologies, data sources and pretreatment, and present empirical result analyses and discussions as well as opportunities for further study.

In this volume, Yi-Ming Wei is responsible for compilation, designing, and organizing. Chapter 1 is completed by Hua Liao, Gang Wu, Yu Zhang, Cai-Hua Xu, Xiao-Wei Ma, Ling-Yun He and Ya-Wen Fan; Chapter 2 by Zhi-Yong Han and Yi-Ming Wei; Chapter 3 by Yi-Ming Wei, Qiao-Mei Liang, Rui-Guang Yang and Ying Fan; Chapter 4 by Ying Fan, Qiang Liang, Jian-Ling Jiao and Yi-Ming Wei; Chapter 5 by Lan-Cui Liu, Qiao-Mei Liang, Ying Fan and Yi-Ming Wei; Chapter 6 by Gang Wu and Yi-Ming Wei; Chapter 7 by Jiu-Tian Zhang, Yi-Ming Wei, Ying Fan, Le-Le Zou and Zhi-Yong Han; and Chapter 8 by Yi-Ming Wei, Ying Fan and Gang Wu.

I especially thank the researchers and authors of the chapters for their spirited and patient participation in many rounds of presenting and reworking research, and ultimately revising reports and journal articles into chapters for this volume.

I gratefully acknowledge the financial support from the National Natural Science Foundation of China (NSFC) under Grant Nos. 70425001, 70733005, 70701032 and the National Key Projects from the Ministry of Science and Technology of China (2006-BAB08B01). Also, we received sustained supports from the following experts: Da-Kuang Han, Xu-Peng Chen, Wei-Dou Ni, Wei-Xuan Xu, Hong Sun, Bao-Guo Tian, Jian-Zhong Shen, Ji-Sheng Yan, Mao-Ming Li, Xiao-Tian Chen, Jing-Yuan Yu, Jia-Pei Wu, Shan-Tong Li, Ji-Zhong Zhou,

Shou-Yang Wang, Wei Zhang, Ji-Kun Huang, Xu-Yan Tu, Qin-Lin Gang, Chao-Liang Fang, Lie-Xun Yang, Yong-Gang Zhang, Jing-Ming Li, Lei Ji, Chen Cai, Zhi-Jie Li, Jian-Guo Song, Jia-Li Ge, Ting-Bin Wang, Kang Zhang, San-Li Feng, Ai-Mei Hu, Zhou-Ying Jin, and Xiu-Yuan Liu. To them, I would finally like to express my sincere gratitude and lofty respect.

Yi-Ming Wei
Center for Energy and Environmental Policy Research (CEEP)

CHAPTER 1

Review of Energy Development in China

Energy is one of the essential resources supporting human survival, economic development and social progress. All countries regard their energy strategy as an essential component of their national development strategy. China is the largest developing country in the world, and the second largest energy producer and consumer. Energy issues have become China's developmental bottleneck during its rapid industrialization. When studying China's energy strategy and policy issues, it is necessary to appreciate the international background, history, and circumstances around China's energy growth. Therefore, this chapter addresses the following questions:
- What are the world energy resource reserves and distribution?
- What is the relationship between world economic growth and energy consumption?
- What are the changes in the world energy market?
- What is the progress in international energy and environmental cooperation?
- What are China's energy resource reserves and distribution?
- What are the changes in China's energy production and structure?
- What is the development of China's renewable energy?
- What are the changes of China's energy consumption and structure?
- What is the energy consumption of China's industry?
- What is the situation of China's energy efficiency?
- What are the relationships between China's regional energy resource, consumption, and economic growth?

1.1 World Energy Development

1.1.1 World Energy Reserves and Distribution

1.1.1.1 World Proven Oil Reserves and Distribution

Oil is one of the most important energy sources for the economy and society's development. With the development of the economy and urbanization, the oil demand of every country has rapidly increased. BP Energy Statistical Review (BP, 2005) indicates that world proven oil reserves have continually increased in the last twenty years. Fig. 1-1 shows that world proven oil reserves were about 761.6 billion barrels in 1984, and increased to 1017.5 billion barrels in 1994, and 1188.6 billion barrels in 2004. From 1984–2004, world proven oil reserves increased 56.1%, i.e., 427.0 billion barrels. The growth rate of world proven oil reserves was extremely high during the period 1984–1994, and the proven reserves increased 33.6% that in fact were mainly located in the Middle East. The growth was relatively slow during the period 1994–2004, and the proven reserves only increased 16.8%, located mainly in Russia, Africa and Middle South America.

Proven oil reserves in the Middle East were approximately 733.9 billion barrels at the end of 2004, which was about 61.7% of total world proven oil reserves. This period is thus known as the "World oil storeroom". Europe (although mainly Russia) is the second large region of world oil reserves, whose proven reserve was 139.2 billion barrels, with the share being about 11.7%. The proven oil reserves and share of Africa, Middle South America, and Pacific Asia were 112.2 billion barrels and 9.4%, 101.2 billion barrels and 8.5%, 61.0

billion barrels and 5.1%, 41.1 billion barrels and 3.5%, respectively. Generally speaking, the distribution of world oil resources is regional and irregular.

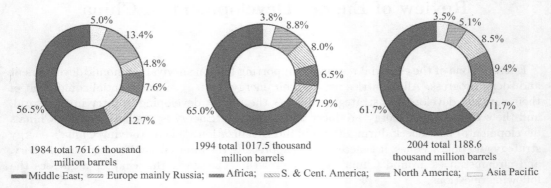

Fig. 1-1 World proven oil reserves and distribution (Source: BP, 2005)

1.1.1.2 World Proven Natural Gas Reserves and Distribution

Natural gas is a clean and highly efficient energy, which has many advantages over alternative sources of power. For example, it has a higher conversion efficiency, lower environmental cost, and lower investment. Consequently, developing natural gas has become an important tide of the world energy industry. Fig. 1-2 shows that world proven natural gas reserves increased rapidly since the 1980s. The proven reserves increased from 96.39 trillion cubic metres in 1984 to 142.89 trillion cubic metres in 1994. World proven natural gas reserves were about 179.53 trillion cubic metres by the end of 2004, which have increased by 83.14 trillion cubic metres, 86.25%, in the last twenty years.

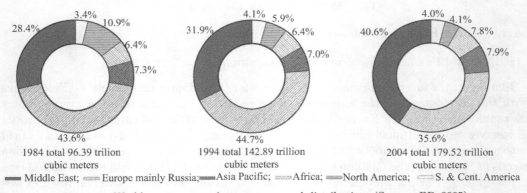

Fig. 1-2 World proven natural gas reserves and distributions (Source: BP, 2005)

The growth rate of world proven natural gas reserves has been slow in recent (last ten) years. However, the growth has quickened for Middle East proven reserves. Fig. 1-2 shows that the proven natural gas reserves of the Middle East increased from the 27.37 trillion cubic metres in 1984 to 72.83 trillion cubic metres in 2004, and the share in total world proven natural gas reserves was about 40.6%. Although natural gas is not principally sourced from the Middle East, just as the distribution of oil, its supply is regional and disproportional. Besides the Middle East and Europe (mainly Russia), the proven natural gas reserves in other regions is less, with only 42.69 trillion cubic metres, corresponding to only 23.7%.

1.1 World Energy Development

1.1.1.3 World Proven Coal Reserves and Distribution

Coal is the most abundant and widest distributed fossil energy on the earth, and which is known as "industry food". So far, coal remains an important raw material and fuel for numerous industry sectors, such as iron and steel, and power generation sectors. Coal resources are distributed in nearly 80 countries and regions, and more than 60 countries have extensively exploited their coal resources. World proven coal reserves were 909.1 billion tons by the end of 2004 (BP, 2005). There is an imbalance in its distribution, to mainly Asia-Pacific, Europe, and North America. As is shown in Table 1-1, coal resource reserves in United States, Russia, China and India are relatively abundant, accounting for 67% of the world total reserves. In China, good quality coal resource reserves are plentiful, but the reserves per capita are still lower than the world average. The world's ratio of workable reserve to annual output is 164 years in 2004, while China's is 59 years.

Table 1-1 Proven coal reserves in the world and main countries in 2004

Country	U.S.	Russia	China	India	Australia	World
Proven coal reserves/Billion tons	246.6	157.0	114.5	92.4	78.5	909.1
Share of the world/%	27.1	17.3	12.6	10.2	8.6	100
Ratio of workable reserved to annual output/Year	245	>500	59	229	215	164

Source: BP (2005).

1.1.1.4 World Renewable Energy Reserves

Because of the exhaustibility of fossil energy (such as coal, oil, and natural gas) and the negative effects of environmental pollution and greenhouse gas emissions caused by burning fossil fuel, countries all over the world are devoted to developing clean and sustainable renewable energy. The world renewable energy resource reserves are abundant. Among which, the reserves of solar energy (power density to the ground) is 1 kWh/m^2, exploitable biomass energy 6.5 billion tce, technically exploitable hydro energy resources 6,960 billion kWh, technically exploitable wind energy resources 9.6 billion kWh, and technically exploitable geothermal resources 50 billion tce (Wang, 2001). Beside hydro energy, the exploitation and employment of wind and solar energy have progressed rapidly over recent years, especially in Germany and Denmark.

1.1.2 World Economic Growth and Energy Consumption

1.1.2.1 Economic Growth and Energy Consumption in the World and Major Countries

Energy is an essential production factor in economic growth. Since the Industrial Revolution, the world economy and energy consumption have increased rapidly. As Table 1-2 shows, the world gross domestic production (GDP) rose from $11,549 billion in 1970 to $31,749 billion in 2004, increasing by 3.0% annually; energy consumption rose from 5,020 million tons of oil equivalent (toe) to 10,220 million toe, growing by 2.1% annually; energy intensity (measured by energy consumption per unit of GDP) declined from 43 toe per thousand U.S. dollars in 1970 to 32 toe per thousand dollars in 2004.

The United States is the largest developed country and energy consumer in the world. As is shown in Table 1-2, its GDP rose from $3,037 billion to $8,785 billion over 1970–2004, and its energy consumption grew from 1,650 million toe to 2,330 million toe, the annual rate being 3.2% and 1.0%, respectively. The first and second world oil crises had a serious effect on the United States economy, resulting in transient economic stagnation and recession. Over 1970–2004, the economy and energy consumption in Japan rose annually by 3.2% and 1.8%, respectively, while the energy consumption growth in several European industrialized countries slowed down in the same period. For example, the GDP annual growth rate in Germany was 2.1% over 1970–2004, while that of energy consumption was 0.2%. China, as a developing country, is rapidly industrializing, therefore its economic scale and energy consumption increases tremendously. In 2004, China's GDP and energy consumption climbed to $1,821 billion and 1,570 Mtoe, respectively.

Table 1-2 GDP and energy consumption in various countries
[GDP: $1000 billion (1990 constant US$); EC indicates energy consumption: billion toe]

Country		Year								
		1970	1975	1980	1985	1990	1995	2000	2002	2004
U.S.	GDP	3.037	3.49	4.185	4.905	5.757	6.506	7.969	8.18	8.785
	EC	1.65	1.69	1.81	1.77	1.97	2.12	2.31	2.29	2.33
Japan	GDP	0.883	1.157	1.407	1.641	2.011	2.277	2.421	2.483	2.538
	EC	0.28	0.33	0.36	0.37	0.44	0.49	0.52	0.51	0.51
Germany	GDP	1.013	1.133	1.331	1.411	1.671	1.849	2.022	2.041	2.07
	EC	0.31	0.32	0.36	0.36	0.35	0.33	0.33	0.33	0.33
Britain	GDP	0.628	0.698	0.763	0.843	0.99	1.075	1.257	1.309	1.379
	EC	0.22	0.2	0.2	0.2	0.21	0.21	0.22	0.22	0.23
France	GDP	0.669	0.813	0.96	1.044	1.226	1.293	1.475	1.524	1.567
	EC	0.16	0.17	0.19	0.2	0.22	0.24	0.25	0.26	0.26
Italy	GDP	0.618	0.714	0.881	0.958	1.102	1.174	1.293	1.319	1.339
	EC	0.12	0.13	0.14	0.14	0.15	0.16	0.18	0.18	0.18
India	GDP	0.137	0.157	0.184	0.238	0.324	0.417	0.552	0.607	0.699
	EC	0.06	0.08	0.1	0.14	0.19	0.25	0.32	0.34	0.38
China	GDP	0.096	0.124	0.161	0.268	0.383	0.712	1.129	1.357	1.657
	EC	0.23	0.34	0.43	0.56	0.69	0.89	0.77	1.03	1.43
World	GDP	11.549	13.956	16.857	19.153	21.899	24.349	28.698	29.71	31.749
	EC	5.02	5.78	6.64	7.19	8.12	8.54	9.08	9.49	10.22

Sources: Calculated based on BP (2005), United Nations Statistic Database (UN, 2005), *Statistical Communiqué on the 2006 National Economic and Social Development* (in brief *Communiqué 2005*) (NBS, 2006a), *Bulletin on China's Historical GDP Data Revision* (in brief *GDP Revision*) (NBS, 2006b).

The change of energy intensity is influenced by industrial structure adjustment, as well as technical progress. As for developed countries, following post-industrialization, the service industry accounts for a large proportion of the economy, resulting in decreasing energy intensity. While for developing countries, due to their partial industrialization, it is necessary to develop energy intensive industries. As a result, their energy intensities are relatively high and often increasing.

1.1.2.2 Energy Consumption Structure

Energy consumption structures across countries differ greatly as a result of their resource endowments, stage of economic development, and differing energy strategies. As Table 1-3 shows, countries with lower oil resource reserves per capita, such as China and India, still regard coal as the main energy source. Countries such as Brazil and Canada have abundant water resources, therefore hydropower accounts for a large proportion of their energy consumption. The nuclear power industry in France is relatively developed, which accounts for 38.6% of the total energy consumption. Because of the non-renewable characteristic of fossil energy and environmental pollution it brought about, many countries are actively developing renewable energy and the sharp rise in world crude oil price over 2004–2005 accelerated this process.

1.1.3 World Energy Market

1.1.3.1 World Energy Production

The world crude oil production in 2004 was 3,870 million tons (an increase of 4.5% compared to 2003), among which the output of Organization of Petroleum Exporting Countries

(OPEC) was 1,590 million tons, accounting for 41.1% of the world total (BP, 2005). Saudi Arabia is the largest oil producer in the world, and its output set a new record in 2004, reaching 510 million tons, 13.1% of world output. Russia has already become one of the largest crude oil producers, exceeded only by OPEC, whose output was 460 million tons, some 11.8% of the world total. The oil production in Iraq and Venezuela also improved, up to 100 million tons and 150 million tons each, growing since 2003 by 50.8% and 13.8%, respectively. Nigeria, Kuwait, and Kazakhstan's oil output also increased by a significant amount.

Table 1-3 Energy consumption structure in the world and major countries (%)

Country	Oil	Natural gas	Coal	Nuclear energy	Hydropower
U.S.	40.2	25.0	24.2	8.1	2.6
Japan	46.9	12.6	23.5	12.6	4.4
Germany	37.4	23.4	25.9	11.4	1.8
Britain	35.6	38.9	16.8	8.0	0.7
France	35.8	15.3	4.8	38.6	5.6
Canada	32.4	26.2	9.9	6.7	24.8
Russia	19.2	54.1	15.8	4.8	6.0
India	31.7	7.7	54.5	1.0	5.1
Brazil	44.9	9.1	6.1	1.4	38.6
China	22.3	2.5	69.0	0.8	5.4
World Average	36.84	23.67	27.17	6.11	6.20

Source: BP (2005).

While the crude oil output increases, the output of world natural gas and coal also increase rapidly. The world natural gas production was 2,700 billion cubic meters in 2004 (an increase of 2.8% compared with that of 2003), among which that of Russia and United States were 580 billion and 540 billion cubic meters, respectively, corresponding to 21.9% and 20.2% of the world output. World coal production was up to 5,540 million tons in 2004, rising by 6.8% compared with that of 2003. And the coal production in China and United States were 1,960 million tons and 1,010 million tons, respectively, which accounted for 35.3% and 18.2% of the world total (BP, 2005).

1.1.3.2 World Energy Trade

(1) World Petroleum Trade.

To a large extent, the world energy trade is based on petroleum. The turnover of world crude oil trade was 1,860 million tons in 2004 and that of petroleum products was 530 million tons (BP, 2005). The United States is the largest oil consumer in the world, and also the largest net importer. Its net import of crude oil and petroleum products were 500 and 90 million tons, respectively, in 2004, accounting for 26.9% and 17.4% of the world total. Japan ranked second, whose net import of crude oil and petroleum products were 210 and 50 million tons, respectively, in 2004, accounting for 11.3% and 8.6% of the world total. China's were 117 million tons and 26 million tons, respectively, in 2004 (NBS, 2005a), which were 6.3% and 6.2% of the world total.

Due to the unbalanced oil resource reserves, geographic location, economic, and political relationship differences among countries, the import channels of oil importers differ. Approximately 45.5% of world crude oil imports came from the Middle East in 2004; the oil import of the United States mainly came from Canada, Middle and Southern America, the Middle East and West Africa; Europe's mainly came from Russia, the Middle East and North Africa; the majority of Japan's came from the Middle East area; and 63% of China's came from the Middle East, 27% from West Africa and 18% from Russia (BP, 2005).

The principal characteristic of world energy markets over 2004–2005 was the sharp increase in crude oil price. The crude oil future price in the New York Mercantile Exchange (NYMEX) on one occasion exceeded $70 per barrel. It has promoted the adjustment of world energy supply and demand. Moreover, in order to deal with the shortage in supply of crude oil, many International Energy Agency (IEA) members drew on their national strategic oil reserves. In 2004–2005, there were also oil supply shortages in some of China's local regions.

(2) World Natural Gas Trade.

Compared with the petroleum trade, natural gas is relatively small. Natural gas was 680 billion cubic meters in 2004, accounting for 25.3% of the world total output (BP, 2005). Russia is the largest natural gas exporter in the world, selling 148 billion cubic meters in 2004; and U.S. is the largest natural gas importer, receiving 126 billion cubic meters in 2004.

(3) World Coal Trade.

Data in the *World Coal Development Report 2004* (Huang, 2005) shows that the world coal trade was 718 million tons in 2003, accounting for 14.6% of the total output. Australia is the largest coal exporter in the world, exporting 208 million tons in 2003; China, ranked second in the world, exported 93 million tons, and the third place in the world was the export of Indonesia with 90 million tons. Japan is the biggest coal importer in the world, with imports of 162 million tons in 2003. With the influence of growth in domestic demand, China's net export was 68 million tons in 2004, declining by 18% when compared with that of 2003.

1.1.4 International Energy and Environmental Cooperation

1.1.4.1 International Energy Competition and Cooperation

The growth of energy demand and price intensified the international energy competition, and at the same time further promoted cooperation on energy. The world's main energy producers and consumers are constantly launching their energy diplomacy activities, and there is reorganizing or merging among large multinational energy enterprises.

The 2005 Global Upstream M&A Review (Harrison, John, 2005) showed that the global oil and gas upstream enterprises and asset mergers, reached 68 billion in 2004, increasing by more than 50% compared with that of the previous year. This represented a large rebound for the first time since 1998. World energy enterprise M&A activities were still intensifying in 2005, for example: ChevronTexaco Corporation, the second largest petroleum giant in the United States, annexed Unocal Corporation, and Valero Energy Corporation, the third largest oil refining enterprise in the United States annexed Premcor Refining Group Inc., becoming the largest oil refining enterprise in North America.

China participated actively in the multilateral energy cooperation under all kinds of international organizational frameworks, and has signed bilateral energy cooperative agreements with more than 30 countries. In June of 2005, the energy administrations of both China and the United States carried on an energy policy dialogue for the first time in Washington, discussing thoroughly cooperation on clean energy, oil, natural gas, nuclear power, energy conservation and energy efficiency, etc. China's energy enterprises have also participated actively in the M&A of overseas petroleum assets. In November 2005, China National Petroleum Corporation (CNPC) acquired PetroKazakstan (a Kazakhstan oil company) for US$4,180 million, setting a new record for China's petroleum enterprises acquiring an overseas oil company. In the complicated and changeable environment of international energy competition and cooperation, the Chinese government and enterprises are facing unprecedented opportunities and challenges.

1.1.4.2 *Kyoto Protocol* Comes into Effect

Energy and environmental issues are closely related, and need the joint effort and cooperation of all countries. The rapid growth of global energy consumption has resulted in a serious influence on greenhouse gas emissions. The *Kyoto Protocol*, which aims at controlling global warming, came into effect on February 16, 2005. It indicated that international environmental cooperation made great progress, and the international community entered a substantial stage of greenhouse gas emissions reductions. Until August 13, 2005, only 142 countries and regions in the world had signed the protocol. Except for the United States, all the industrialized countries assigned reduction quotas for greenhouse gas emission. The United States, which accounts for a quarter of the world carbon dioxide emission, refused to sanction the protocol.

Partly driven by the Kyoto Protocol, carbon emission reduction has already become one of the important goals for social and economic development of contracting parties. In order to realize the emission reduction goals and reduce the cost, the Kyoto Protocol offers three kinds of flexible mechanisms for contracting parties: Clean Development Mechanism (CDM), Joint Implementation (JI) and International Emissions Trading (IET). CDM aims at promoting cooperation between developed and developing countries; JI emphatically promotes cooperation between developed countries; IET allows the developed countries to transfer between each other some of their "emission allowances".

The developed countries which signed protocol are seeking more international cooperation to accomplish the predetermined obligation of reduction, with as low cost as possible. Developing countries are also participating actively in the international cooperation of the CDM project, expecting to improve the environment, optimize their energy consumption structure, increase employment and income, promote technical progress, and realize sustainable development by selling certified emission reductions (CERs).

1.2 Energy Production in China

1.2.1 Energy Reserves and Distribution in China

1.2.1.1 China's Coal Resources Are Mainly Distributed in the North, While Less in the South

China is rich in coal resources, according to the second coalfield forecast completed in 1981 (Wang, 2003). Coal total resource deposits, within 1,000 vertical meters below the surface, amounted to 2.6 trillion tons. Within 2,000 vertical meters below the surface, estimated prospective reserves are 5.06 trillion tons. At of the end of 2001, proven coal reserves totalled 1.0033 trillion tons, some 19.83% of total reserves. Studies indicate that at the end of 2002, the proven coal reserves that can be used reached 188.6 billion tons. According to accepted international criteria for classification, the remaining recoverable reserves of China's coal resource that can be developed economically reached 114.5 billion tons. This represents 11.6% of the world reserves (984.2 billion tons) (Institute of Industrial Economics of the Chinese Academy of Social Sciences, 2005). China's coal resources are mainly distributed in the north, with less in the south. Table 1-4 reflects the distribution of coal reserves.

1.2.1.2 China's Oil Resources Are Mainly in the Style of a Terrestrial Reservoir

China is one of the countries with abundant oil resources, but its average oil ownership per capita is very low. Research shows that (National Development and Reform Commission Goods Reserve, 2005), at the end of 2004, China's proven oil reserves reached 24.844 billion tons (including crude oil and condensate oil); total proven oil reserves that can be developed reached 6.791 billion tons; total produced volume reached 343 million tons; remaining recoverable reserves reached 2.491 billion tons.

Table 1-4 Proven reserves of coal resources in China

Region	Coal resource distribution		Proven rate/%	Resource guaranted/a
	Total reserves/%	Proven reserves/%		
North China	39.90	48.95	24.44	1486
Northwest	45.90	29.98	13.01	3734
Northeast	1.34	3.30	48.92	300
East China	4.44	5.90	26.47	341
Central South China	2.61	3.27	24.91	319
Southwest	5.80	8.61	29.56	1101

Source: *China Energy Development Report* (China Energy Development Report Editing Committee, 2001).

China's oil resources are distributed very widely, and mainly in the style of a terrestrial reservoir. A petroliferous basin can be divided into three basic types (State Power Information Network, 2005): east tension-type basin, central transition basin, and the western extrusion basin. The country can be divided into six zones: East, including the Northeast and North China; Central, including Shaanxi, Gansu, Ningxia and Sichuan; West, including Xinjiang, Gansu, Qinghai and the western region of Gansu; South, including Jiangsu, Zhejiang, Anhui, Fujian, Guangdong, Hunan, Jiangxi, Yunnan, Guizhou, Guangxi; Tibet region, including south of the Kunlun Mountains, west of the Hengduan Mountains region; Ocean zones, including the southeast coast of the South China Sea and the continental shelf waters.

1.2.1.3 Great Potential in Natural Gas Resources

The widely distributed sedimentary rock and continental basin in China has provided enhanced geological conditions for natural gas. A study shows that the prospect of natural gas reserve is about 47–54 trillion cubic meters (Li, 2004). Close to 2003, the proven recoverable gas reserves reached 2.5 trillion cubic meters, but only 18% of the total reserve can be exploited. This means that the gas industry of China is still in the early stages of exploration. There are 11.5 trillion cubic meters of natural gas reserves left to be proven.

With respect to China's natural gas geology and economic development, Zhou and Tang (2004) divided China into four gas-bearing zones: eastern, central, western zone and continental shelf. Natural gas resources concentrated in central and western areas, which are a distance away from the economically developed eastern areas. The recoverable natural gas resources of central and west areas are 4.1 trillion cubic meters and 3.52 trillion cubic meters, respectively, taking together to be about 66.15% of the national natural gas resources. In the ocean zones of China, the gas recoverable resources in the continental shelf offshore is about 2.98 trillion cubic meters, accounting for 25.71% of the country. Eastern gas resource is relatively small, only 8.14% of the total country.

1.2.1.4 Hydropower Resources Concentrated in Southwest China

With many rivers and lakes, China is one of the world's countries that are fortunate to have abundant water resources. According to the 2001 to 2004 China water resources census results, the theoretical potential reserves of China's hydropower resources is 6.89 million kilowatts, ranking first in the world, of which, able to be developed installed capacity, is 4.02 million kilowatts (He, 2005). However, China's hydropower resources distribution is very uneven, with 70% in the southwest region (Xinhua News Agency, 2005). Studies indicate that China's hydropower resources are concentrated mainly in the Yangtze River, the middle

1.2 Energy Production in China

reaches of the Yellow River, the middle and lower reaches of the Yarlung Zangbo River, the Pearl River, Lancang River, Nujiang River, and the upper reaches of Heilongjiang. Reasonable capacity of medium and large-sized hydropower developed from each of these seven rivers would be more than 1,000 kilowatts, which accounts for 90% of the national capacity. Table 1-5 shows the reserves and reasonable capacity of China's hydropower resources (He, 2002).

Table 1-5 The reserves and reasonable capacity of China's hydropower resources
(classified by river systems)

River system	Resource reserves		Reasonable capacity	
	10^4 kW	10^9 kWh/yr	Capacity/10^4 kW	Production/(10^9 kWh/yr)
Total	67 604.71	59 221.8	37 853.24	19 233.04
Yangtze River	26 801.77	23 478.4	19 724.33	10 274.98
Yellow River	4 054.80	3 552.0	2 800.39	1 169.91
Pearl River	3 348.37	2 933.2	2 485.02	1 124.78
Hailuan River	294.40	257.9	213.48	51.68
Huaihe River	144.60	127.0	66.01	18.94
Rivers in Northeast	1 530.60	1 340.8	1 370.75	439.42
Rivers in Southeast	2 066.78	1 810.5	1 389.68	547.41
Rivers in Southwest	9 690.15	8 488.6	3 768.41	2 098.68
Yarlung Zangbo River and other rivers in Tibet	15 974.33	13 993.5	5 038.23	2 968.58
Rivers in North and Xinjiang	3 698.55	3 239.9	996.94	538.66

Source: Website of Ministry of Water Resources (Ministry of Water Resources, 2005).

1.2.1.5 China's Wind Resources are Very Rich

China is in the east of Asia, facing the Pacific Ocean, where there are strong monsoons, many inland mountains, and with a complex terrain. Moreover, the towering Tibetan Plateau in western China changes the air pressure and atmospheric circulation, which increases the complexity of monsoons. Therefore, China's wind resources are very rich.

Information indicates that the speed of wind in existing wind farm sites of China reached an average speed of 6 m/s and above. The southeastern coast area between Nan'ao Island and the Yangtze River is the largest and most abundant wind resource area of China; the average wind speed there reaches 6 m/s and above. The areas with rich wind resources include: Shandong, Liaodong Peninsula and the Yellow Sea coast, west of Nan'ao Island in the South China Sea coast, the Hainan Island and the South China Sea islands, the areas from the north of Yinshan Mountains to the north of Greater Xing'an Mountains in the Inner Mongolia, Dabancheng of Xinjiang, Alashankou, the Hexi Corridor, the Songhua River downstream, northern areas in Zhangjiakou, and the mountain passes and the peaks all over the country (the former State Development Planning Commission, the Secretary for Basic Industries, 2000). Table 1-6 shows China's wind energy resources.

Table 1-6 Classification of China's wind resources

Index	Abundant	Rich	Available	Poor
Effective wind resource density/(W/m^2)	>200	200–150	<150–50	<50
⩾3m/s hours per year/h	>5000	5000–4000	<4000–2000	<2000
⩾6m/s hours per year/h	>2200	2200–1500	<1500–350	<350
Area percentage of entire country/%	8	18	50	24

Sources: The Secretary for Basic Industries (former SDPC, 2000).

1.2.1.6 Abundant Solar Energy Resources and Geothermal Energy Resources

China is located in the eastern part of northern Eurasia, mainly in a temperate and subtropical zone, with rich solar energy resources. According to the long-term observation data from more than 700 national meteorological stations (The Secretary for Basic Industries, the former State Development Planning Commission, 2000), the solar radiation throughout the year will be around 3.3–8.4bJ/m², an average of about 5.86bJ/m². Western Tibet, southeast Xinjiang, western Qinghai, and Gansu's western region have abundant solar energy resources, and the total radiation reaches 6.7–8.4 bJ/m² per year.

Geothermal resources are a source of clean energy, which use the earth's internal heat resources to be developed economically by mankind. The geothermal resources of China are very abundant and widely distributed. The energy resources within 2,000 vertical meters from the surface are equivalent to about 1.3711 trillion tce. If 1% of the energy can be developed, the adopted energy could reach 13.7 billion tons of coal equivalent (Zhu, 1999). China's geothermal resources can be divided into three types by their attributes: specifically, high temperature convection-type of geothermal resources, where the temperature is above 150°C, mainly located in the province of Taiwan, south Tibet, southern Yunnan and southern Sichuan; middle temperature convection-type of geothermal resources, where the temperature is from 90–150°C, mainly found in Fujian, Guangdong, Hunan, Hubei, Shandong, and Liaoning provinces; low-temperature conduction geothermal resources, located mainly in the north, Songliao, Sichuan, Ordos and other areas (Yu, 2005).

1.2.2 China Energy Production and Structure

After reform and the opening to the outside world, China has made significant progress with its energy industry, both in terms of quantitative and qualitative indices. It has become one of the world's powers in energy production and consumption.

Fig.1-3 shows the trend in development of energy production in China during 1949–2004. Energy production in China increased steadily on the whole. It started smoothly in the beginning, and increased soundly in the middle-phase. From 2001, with the rapid development of China's economy and speedy expansion of its high energy consumption industry, China has come into an accelerated developmental period. Although China, the second energy production power, took 2.06 billion tce of primary energy production in 2005 (NBS, 2006a), its per capita energy production still maintained a low level, owing to its large population.

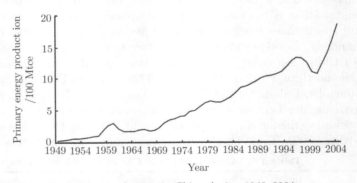

Fig. 1-3 Energy production in China during 1949–2004
Sources: China Energy Databook v6.0 (LBNL, 2004), *China Statistical Yearbook 2005* (NBS, 2005a)

Table 1-7 lists primary energy production and the composition of China from 1978 to the most recent 20 years. Table 1-8 lists the energy production and world ranking of China in 1949, 1980, 1990 and 1998–2003. After 50 years' development, the energy industry of China has established a complete energy production system, of which coal is the first priority, while other kinds of energy play an important role. The system supports China's economy to maintain a stable growth rate. Due to the expansion of a high energy consumption

1.2 Energy Production in China

industry in recent years, energy expenditure in China increased summarily, as did primary energy production.

Table 1-7 Primary energy production and composition of China

Year	Total energy production/Mtce	Proportion/%			
		Coal	Crude oil	Natural gas	Hydro power
1978	627.70	70.3	23.7	2.9	3.1
1980	637.35	69.4	23.8	3.0	3.8
1985	855.46	72.8	20.9	2.0	4.3
1989	1016.39	74.1	19.3	2.0	4.6
1990	1039.22	74.2	19.0	2.0	4.8
1991	1048.44	74.1	19.2	2.0	4.7
1992	1072.56	74.3	18.9	2.0	4.8
1993	1110.59	74.0	18.7	2.0	5.3
1994	1187.29	74.6	17.6	1.9	5.9
1995	1290.34	75.3	16.6	1.9	6.2
1996	1326.16	75.2	17.0	2.0	5.8
1997	1324.10	74.1	17.3	2.1	6.5
1998	1242.50	71.9	18.5	2.5	7.1
1999	1091.26	68.3	21.0	3.1	7.6
2000	1069.88	66.6	21.8	3.4	8.2
2001	1209.00	68.6	19.4	3.3	8.7
2002	1383.69	71.2	17.3	3.1	8.4
2003	1599.12	74.5	15.1	2.9	7.5
2004	1846.00	75.6	13.5	3.0	7.9

Source: *China Statistical Yearbook 2005*(NBS, 2005a).

Table 1-8 Energy production and world rank of China

		Year									
		1949	1980	1990	1998	1999	2000	2001	2002	2003	2004
Primary energy	Production/Mtce	23.7	637.4	1039.2	1242.5	1091.3	1069.9	1209.0	1383.6	1599.1	1846.0
	Rank	13	3	3	3	3	3	3	2	2	2
Coal	Production/10^6 ton	32	620	1080.0	1250.0	1045.0	998.0	1110.0	1380.0	1667.0	1956.4
	Rank	10	3	1	1	2	1	2	1	1	1
Crude oil	Production/10^6 ton	0.12	106.0	138.3	161	160	163	165	166.9	169.6	174.5
	Rank		6	5	5	5	5	5	5	5	6
Natural gas	Production/10^8 m^3	0.07	142.7	153.0	232.8	252	272	303.4	326.6	350.2	408
	Rank		12	20	18	17	19	19	17	17	16
Electricity	Production/10^8 kWh	4.3	300.6	621.2	1167.0	1239.3	1355.6	1478.0	1654.0	1905.2	2187.0
	Rank	25	6	4	2	2	2	2	2	2	2

Sources: *China Statistical Yearbook 2004* (NBS, 2004d), *China Statistical Yearbook 2005* (NBS, 2005a), BP (2005), *China Energy Databook V6.0* (LBNL, 2004), *Communiqué on the Main Data of the First National Economic Census (No.2)* (NBS, 2005b), *Statistical Communiqué 2005 of People's Republic of China on National Economy and Social Development* (NBS, 2006a).

In recent years, the main conflicts in China's energy production have been:

(1) Coal production augmented greatly against the grim situation of coal mine security.

As one of the major coal producers, China always maintained its coal production at about 70 per cent of its primary energy production. With the primary energy production increasing, coal production remained stable, and raw coal production reached 2.19 billion tons (NBS, 2006a). Because the domestic demand on energy increased and international oil price soared swiftly, the substitution effect of coal to oil has promoted coal production. Simultaneously, frequent accidents and a dismal security situation in coal mines threatened the stable growth of coal production.

(2) Proportion of oil production decreased, the proportion needs to be improved.

In recent years, oil production kept relatively stable with slow growth. In contrast to the high growth of other primary energy production, the proportion of oil production decreased recently. In 2005, China's crude oil production reached 0.181 billion tons (NBS, 2006a), and the dependency of crude oil on import exceeded 40%. Taking account the special role of oil in modern industry and national energy security, China should take a international cooperation strategy on the basis of keeping and strengthening domestic oil production.

(3) The asymmetry between natural gas production and regional economies is unbalanced, the exploitation potential of natural gas is large.

For some time, natural gas production has been a low amount of China's primary energy production. However, since 1999, it has remained stable and has achieved above 3 % annual growth. Since natural gas production broke through 20 billion cubic meters in 1998, it has kept increasing annually by 2–3 billion cubic meters, and reached 50 billion cubic meters in 2005 (NBS, 2006a). China's natural gas exploitation only accounts for 10 % of her proven reserves, therefore, it possesses great potential. Natural gas reserves are distributed in central and western China. There is huge gap between the natural gas supply and demand in developed eastern China. Consequently, such initiatives as the West-East natural gas transmission project will boost China's natural gas production.

(4) Electricity power production increased swiftly, and therefore it is necessary to increase renewable energy power.

Electricity power production increased rapidly in China. National electricity installed capacity broke through 500 million kW by the end of 2005 (NDRC, 2005), and power generation reached 2470 billion kWh, including hydro power generation of 400 billion kWh, equating to an increase of 12.3 %(NBS, 2006a). Power generation in China still depends mainly on traditional energy sources. Thermal power generation accounts for 81.5 % of the total power generation. Excessive emission of such gases as carbon dioxide (CO_2), sulfur dioxide (SO_2) and so on, caused by vast combustion of fossil fuel, has brought about huge environmental anxiety.

1.2.3 Production and Development of Renewable Energy in China

China possesses abundant renewable energy resources. Although the proportion of renewable energy production to primary energy production is small, the Chinese government has begun to support energy renewable production in policy and its legal system. We can expect that renewable energy exploitation and employment, especially renewable energy power generation, will develop more, and renewable energy resources will play an important role in China's energy production.

Law of the People's Republic of China on Renewable Energy took effect on January 1st, 2006. Renewable energy exploitation and employment were singled out as a priority in energy development. Focusing on renewable energy exploitation, and taking other effective measures, will promote the establishment and development of a renewable energy market.

China's New Energy and Renewable Energy Development Program (1996–2010) (former State Planning Commission et al., 1995) suggested that new energy technologies should

Table 1-9 Main goals in
China New Energy and Renewable Energy Development Program (1996–2010)

Index/Unit	Wind power	Biomass power	Solar power	Geothermal power	Small hydropower	
	Installed capacity /million kW		Applied capacity /Mtce		Installed capacity /million kW	Power generation /kWh
Goal (by 2010)	1.0–1.1	0.3	467	151	27.88	117

Source: former State Planning Commission (1995).

eventually reach the level of mass-production. Table 1-9 lists main goals by 2010.

According to the outline of China's National Development and Reform Commission, the proportion of renewable energy to primary energy consumption should rise from 7 % to 15 % by 2020. In future, installed capacity of hydro power, wind power and solar power should reach 300 million kW, 30 million kW and 100 million kW, respectively, and production of biomass briquette fuel should reach 50 million tons.

In recent years, renewable energy exploitation in China developed at the rate of over 25 % annually. Installed capacity of hydro power, grid-connected wind power, and solar PV has reached 110 million kW, 760 thousand kW and 60 thousand kW, respectively, while solar water heating has accounted for over 40 per cent of global usage. Civil biogas pools have exceeded 110 million (Liu, 2005). To accelerate the development of new energy and renewable energy, headed by wind power, the National Development and Reform Commission, (formerly State Planning Commission, and formerly National Development and Planning Commission) has successively carried out such new energy and renewable energy projects as Ride Wind Program, Brightness Project, National Demonstration Project of Straw Gasification, and so on.

The persistent and speedy development of the economy means that China needs the support and driving force from the ample energy. Owing to the restriction of technology and the economy, China has not fully taken advantage of the abundant renewable energy resources. The soaring price of oil created an opportunity massive usage of renewable energy, and made it more economically feasible to take advantage of renewable energy. There are many advantages to becoming the catalyst in the transformation from the traditional energy to the renewable energy. For example, advantages as the promotion of environmental protection by government and the public, opportunities, and duty and responsibility of the government on environmental protection. Therefore, renewable energy has excellent potential and a bright future in China.

1.3 Energy Consumption in China

1.3.1 Energy Consumption Quantity and Structure in China

1.3.1.1 Quick Growth in Energy Consumption

China is the largest developing country in the world. It has maintained extensive growth in both its economy and energy consumption since the foundation of the new country. China's GDP has grown from RMB 161.5 billion yuan in 1953 to RMB 658.4 billion yuan in 1978, and RMB 7867.8 billion yuan in 2005; which can be seen from Table 1-10. The average rate of increase in these two phases is 5.8% and 9.7%, respectively. Energy consumption grew from 540 million tce in 1953 to 571 million tce in 1978, and 2225 million tce in 2005, representing an average increase rate of 9.9% and 5.3%, respectively.

The energy consumption per capita is also growing fast, from 0.09 tce in 1953 to 0.59 tce in 1978 and 1.70 tce in 2005. The average electricity consumption per capita in rural and urban households was 173.7 kWh in 2003, while only 10.7 kWh in 1980 (NBS, 1999a; 2005a). China's energy consumption level per capita is still low compared to developed countries.

As China became a petroleum net importer since 1993, and a crude oil net importer since 1996, it did not affect the Chinese economy too much in three world oil crises. During the period 1958–1960, energy consumption has grown rapidly, because of the "Big Leap Forward" economy development strategy. However, energy consumption decreased, but the economy still grew stably in the period 1997–1999, which caused great conjecture around the world. China argued it was the sum effect of several reasons, such as the requirement market was weak, energy product consumption decreased; lots of "five small" companies closed due to their large energy consumption, air pollution; change in industrial structure; technical progress; data underestimate, and so on (Shi, 2005).

Table 1-10 Economy and energy consumption in China, 1953–2005
[unit: Energy consumption: Mtce; GDP: RMB 100 million Yuan (1990 constant RMB); Energy intensity: tce/RMB 10 thousand Yuan GDP]

Year	Energy consumption	GDP	Energy intensity	Year	Energy consumption	GDP	Energy intensity
1953	54	1615	3.34	1980	603	7638	7.89
1954	62	1683	3.68	1981	594	8038	7.39
1955	70	1798	3.89	1982	621	8766	7.08
1956	88	2069	4.25	1983	660	9719	6.79
1957	96	2173	4.42	1984	709	11192	6.33
1958	176	2634	6.68	1985	767	12700	6.04
1959	239	2868	8.33	1986	809	13824	5.85
1960	302	2858	10.57	1987	866	15425	5.61
1961	204	2077	9.82	1988	930	17165	5.42
1962	165	1961	8.41	1989	969	17862	5.42
1963	156	2161	7.22	1990	987	18548	5.32
1964	166	2555	6.5	1991	1038	20251	5.13
1965	189	2991	6.32	1992	1092	23135	4.72
1966	203	3321	6.13	1993	1160	26361	4.4
1967	183	3123	5.86	1994	1227	29811	4.12
1968	184	2995	6.14	1995	1312	33070	3.97
1969	227	3500	6.49	1996	1389	36381	3.82
1970	293	4181	7.01	1997	1378	39759	3.47
1971	345	4475	7.71	1998	1322	42873	3.08
1972	373	4643	8.03	1999	1301	46145	2.82
1973	391	5007	7.81	2000	1303	50029	2.6
1974	401	5123	7.83	2001	1432	54183	2.64
1975	454	5569	8.15	2002	1518	59108	2.57
1976	478	5478	8.73	2003	1750	65033	2.69
1977	524	5896	8.89	2004	2032	71591	2.84
1978	571	6584	8.67	2005	2225	78678	2.83
1979	586	7083	8.27				

Sources: AMR (2005), NBS (1999a), NBS (2006a), NBS (2006b).

Since 2002, China's economy has entered a new spiraling increase. Fixed assets investment grows rapidly; proportion of heavy industry is in a increasing trend; heavy energy consumption industry, such as steel, architectural material and electrolytic aluminum has expanded rapidly, greater than the economy's growth. Energy demand elasticity coefficient has been continuously larger than in 2002–2004.

1.3.1.2 Energy Consumption Structure Gives Priority to Coal

There are abundant coal resources in China, and coal is the largest fuel in the primary energy consumption. As Table 1-11 shows, coal consumption accounts for more than 90 % in total primary energy consumption in the early days of China's foundation. The proportion of coal in the total energy consumption reduces alongside the development of petroleum and natural gas industry, and improved hydroelectricity source. In 2005, total coal consumption was 2.14 billion tons, with an increase of 10.6%; crude oil consumption was 0.3 billion tons, with an increase of 2.1%; natural gas consumption was 50 billion cubic meters, an increase of 20.6%; hydroelectricity consumption was 401 billion kWh, an increase of 13.4%; nuclear electricity consumption was 52.3 billion kWh, an increase of 3.7%, which compares with the consumption in 2004 (NBS, 2006a). Since 1978, the proportion of hydroelectricity usage of the total primary energy consumption increased incrementally, increasing from 3.4% in 1978 to 7.0% in 2004. China has already formed an energy production and consumption structure of "based on coal, developing multiple energy resources", but coal still takes a greater part in total energy consumption, and brings serious environmental and pollution problems.

1.3 Energy Consumption in China

Table 1-11 Energy consumption structure in China (%)

Year	Coal	Petroleum	Natural gas	Renewable energy (mainly hydro-electricity)	Year	Coal	Petroleum	Natural gas	Renewable energy (mainly hydro-electricity)
1953	94.33	3.81	0.02	1.84	1993	74.7	18.2	1.9	5.2
1955	92.94	4.91	0.03	2.12	1994	75.0	17.4	1.9	5.7
1960	93.90	4.11	0.45	1.54	1995	74.6	17.5	1.8	6.1
1965	86.45	10.27	0.63	2.65	1996	74.7	18.0	1.8	5.5
1970	80.89	14.67	0.92	3.52	1997	71.7	20.4	1.7	6.2
1975	71.85	21.07	2.51	4.57	1998	69.6	21.5	2.2	6.7
1978	70.7	22.7	3.2	3.4	1999	68.0	23.2	2.2	6.6
1980	72.2	20.7	3.1	4	2000	66.1	24.6	2.5	6.8
1985	75.8	17.1	2.2	4.9	2001	65.3	24.3	2.7	7.7
1990	76.2	16.6	2.1	5.1	2002	65.6	24.0	2.6	7.8
1991	76.1	17.1	2.0	4.8	2003	67.6	22.7	2.7	7.0
1992	75.7	17.5	1.9	4.9	2004	67.7	22.7	2.6	7.0

Sources: AMR (2005), NBS (2005a).

1.3.1.3 Increasing of China's Energy Import Dependence

China is a major player in both energy production and energy consumption. From the "total amount" aspect, it is *balanceable* in energy production and consumption. According to the data in NBS (2006a), it is calculated that the primary energy consumption at a self-supplying level calculates to 92.8%, and import dependence rate is only 7.2%. However, when the impact of resource distribution characteristics is considered, the structure of basic primary energy supply and demand is not balanced. As we can see from Table 1-12, since 1993, China has become an oil products net importer and a crude oil net importer since 1996. Its net imports increase constantly. In 2005, the net imports of crude oil and oil production were 118.75 million tons and 17.42 million tons, respectively, and petroleum import dependency was beyond 40%. China's energy safety situation becomes more and more austere because of the rapid increase of petroleum import dependency. There is abundant coal in China, the net export of coal in 2005 was 45.51 million tons, and export of coke was 12.76 million tons. Impacted by the domestic market demand, the coal and coke net export amount reduced during 2004–2005.

Table 1-12 Energy import and export in China (unit: 10^4 ton)

Year	Net import of crude oil	Net import of oil production	Net export of coal	Net export of coke
1993	−376	1345	1838	261
1994	−614	969	2309	404
1995	−176	1024	2701	886
1996	222	1166	2580	769
1997	1564	1820	2872	1058
1998	1172	1739	3071	1146
1999	2944	1437	3574	997
2000	5996	978	5293	1520
2001	5271	1226	8763	1385
2002	6220	966	7303	1357
2003	8289	1442	8312	1472
2004	11723	2642	6805	1501
2005	11875	1742	4551	1276

Sources: Huatongren Data Center (2005), NBS(1998).

1.3.2 Energy Consumption in Industrial Sectors

Development of the civil economy in various sectors and improvement of people's living standard can not grow without energy. Table 1-13 reflects proportions of different industries' energy consumption in terms of the total country's energy consumption during 1994–2003.

China has already been in the process of industrial development. Industrializing has always consumed lots of energy, and the energy consumption takes about 70% of total energy. Although in 1999 industrial energy consumption dropped, it returned to the level of 70% in 2003, and the energy consumption was about 1.196 billion tons of coal equivalents (State Statistic, 2005a). Within the industry sector, manufacturing consumes most of the energy, accounting for more than 50% of the whole country's energy consumption, in which Processing of Petroleum, Coking; Manufacture of Raw Chemical Materials and Chemical Products (chemical industry); Smelting and Pressing of Ferrous Metals (steel industry) and Manufacture of Non-metallic Mineral Products (construction material industry) consume the greater part of energy, which accounted for 5.3%, 10.0%, 14.1% and 7.4%, respectively, of the total energy consumption. Their energy consumption amounts were 0.9 billion tons, 1.711 billion tons, 2.407 billion tons, 1.266 billion tons of coal equivalents, respectively. Electric Power, and Gas and Water Production and Supply also account for a large part, which was 8.4% of total energy consumption in 2003.

Beside the industry sector, energy consumption in the transport sector also shows increases. The proportion increased from around 4.6% in 1994 to 7.5% in 2003. Along with the "development of a well-off society" happening, the number of family owned cars increased quickly, and therefore energy consumption in the transport sector will increase.

Table 1-13 Proportion of different industries' energy consumption in China (%)

Industry	Year									
	1994	1995	1996	1997	1998	1999	2000	2001	2002	2003
Farming, Forestry, Animal Husbandry, Fishing and Water Conservancy	4.2	4.2	4.1	4.3	4.4	4.5	4.4	4.6	4.4	3.9
Industry	71.6	73.3	72.2	72.4	71.4	69.8	68.8	68.5	68.9	70.0
Mining	7.8	7.6	7.1	8.0	7.9	7.1	7.1	7.1	7.0	7.1
Manufacture	58.4	59.7	58.4	56.4	55.7	54.4	53.4	53.3	53.7	54.5
In which:										
(1) Processing of Petroleum, Coking	2.9	4.2	2.6	5.3	5.2	5.5	5.7	5.8	5.7	5.3
(2) Manufacture of Raw Chemical Materials and Chemical Products	13.2	12.1	14.5	11.3	10.6	9.9	9.8	9.6	9.8	10.0
(3) Smelting and Pressing of Ferrous Metals	12.5	14.1	13.1	13.2	12.9	13.0	12.9	12.7	13.0	14.1
(4) Manufacture of Non-metallic Mineral Products	10.2	10.0	9.9	8.9	8.8	8.4	7.8	7.4	7.2	7.4
Electric Power, Gas and Water Production and Supply	5.4	6.0	6.7	8.0	7.9	8.3	8.3	8.0	8.3	8.4
Construction	1.1	1.0	1.0	0.9	1.2	1.1	1.1	1.1	1.1	1.0
Transport, Storage and Post	4.6	4.5	4.3	5.5	6.2	7.1	7.6	7.6	7.5	7.5
Wholesale, Retail Trade, Hotels and Catering Services	1.5	1.5	1.6	1.7	1.9	2.2	2.2	2.4	2.3	2.4
Others	4.5	3.5	4.0	3.4	3.9	4.2	4.4	4.5	4.3	4.0
Non-Production Consumption	12.6	12.0	12.8	11.9	10.9	11.2	11.4	11.4	11.5	11.3

Sources: AMR (2005), NBS(2005a).

1.3.3 Energy Efficiency and Conservation

Energy efficiency can be separated into energy economic efficiency and energy technological efficiency. The first can be expressed as energy intensity (energy consumption per unit of GDP), and the second can be expressed as energy consumption per production. Energy economic efficiency can be affected by energy technological efficiency, and also by economic structure. Energy economic efficiency has advanced since 1978, because of the improvement of technology and regulation of the economy's structure. But compared with developed countries, there are still great disparities, and energy consumption per RMB 10^4 yuan GDP was 2.83 tce in 2005 (1990 constant RMB, the same below).

In terms of energy technological efficiency, its increase is also rapid. With strong management of energy, combined with the introduction of overseas advanced technology and domestic exploitation, the unit production of energy consumption of the main production reduces year by year. Moreover, the difference between China and overseas developed countries is shrinking. Table 1-14 gives unit production energy consumption in the main high energy intensive industries since 1996. Even though, when compared to the developed countries, there is still a large shortfall in improving energy technological efficiency in China.

Table 1-14 Unit production energy consumption of main high energy consuming production

Index	Unit	Year					2003/1996 Reduced rate annual average/%
		1996	1998	2000	2002	2003	
Integrated energy consumption per ton steel	tce/t	1.392	1.29	1.18	1.06	1.08	3.6
Electricity supply energy consumption	gce/kWh	410	404	392	383	380	1.1
Integrated energy consumption for aluminum	tce/t	10.88	10.11	9.56	9.30	9.17	2.4
Integrated energy consumption for cuprum	tce/t	5.572	5.039	4.613	4.441	4.373	3.4
Integrated energy consumption for cement	kgce/t	175	168	162	159	150	2.2
Integrated energy consumption for glass	kgce/heavy box	29.4	25.6	25.0	24.0	23.2	3.3
Synthetic ammonia (large sized)	kgce/t	1320	1294	1273	1250	1222	1.1
Synthetic ammonia (medium-sized)	kgce/t	1977	1919	1892	1881	1946	0.2
Synthetic ammonia (mini-sized)	kgce/t	2050	1868	1801	1790	1782	2.0
Alkali (dissepiment)	kgce/t	1696	1661	1563	1538	1463	2.1
Alkali (ionic membrane)	kgce/t	1163	1110	1090	1074	1071	1.2
Energy consumption for oil refining	kgoe/t factor	14.44	14.46	14.10	12.97	12.64	1.9
Energy consumption for ethane	kgoe/t	871.1	809.5	787.4	724.8	711.7	2.8
Coal consumption for electricity generating	gce/kWh			363	356	355	

Source: Constructing Forehanded Society International Workshop Background Report Group (2005).

The major reasons for China's low energy utilization efficiency are the extensive types of economic growth, industrial structure, lagging technological level, and low energy management level, etc. The National Development and Reformation Committee (2004) enacted *Middle-long Term Special Layout for Energy Conservation*, which chiefly identifies the aim and developing stress point for 2010, and brings forward the aim to 2020. It states that energy consumption per 10 thousand GDP should be reduced to 2.25 tons of coal equivalent, the average energy conservation rate during 2003–2010 be 2.2%, and the ability of energy conservation be 0.4 billion tons of coal equivalent. In 2020, energy consumption per 10 thousand GDP should be reduced to 1.54 tons of coal equivalent, the average energy conservation rate during 2003–2020 to be 3%, and the ability of energy conservation be 1.4 billion tons of coal equivalent. Saving energy, improving energy efficiency and developing new industrial methods are significant for developing a conservation-minded society, protecting the environment and creating sustainable development.

1.4 Energy Consumption, Resources and Economy in Regions

The irregular economic development among regions is one of the fundamental realities of China. This reflects not only the unevenness of social and economic development of different regions, but also the energy consumption. Along with development of the economy and population increase, the problems between energy consumption and economic development, and resource distribution are causing increasing tensions. Therefore, research on regional energy consumption, resource distribution, economic development, and population in China is deemed necessary.

Research on energy consumption has mainly focused on the relationship between consumption and economic growth (Kraft J, Kraft A, 1978; Erol, Yu, 1987; Glasure, Lee, 1997; Yu, Choi, 1985; Asafu-Adjaye, 2002), as well as the relationship between energy consumption and the environment (Wang, 2001; Xu, Chen, Qi, 2000). Research has been principally based on time series data at the national level (Jiang, Zhang, 2005; Han, Wei, Jiao, 2004). Some other studies have reported the relationship between energy consumption and resources at the regional level. Research on energy structure is absent.

Based on this research we try to investigate the spatial distribution characteristic of the regional energy total consumption and the structure, as well as the relationship between regional energy consumption and economic development. Such work would support our understanding on relationships between regional energy consumption and other key factors. We hope that our research can provide information for policy implications at the level of regional energy development.

1.4.1 Data and Processing

To analyze the relationship between regional energy expense, resources, and economy, data was selected for the total quantity of regional energy expenses, the structure of nonrenewable energy costs, and the energy resource reserves and other important economic factors.

1.4.1.1 Energy Expenses Data

The data on energy expenses include the total quantity of regional energy expenses, the proportion of regional energy expenses for total energy expenditure (total quantity of coal, petroleum, natural gas, electricity power), and the structure of regional energy expenditure (coal, petroleum, natural gas, electric power). The data for the total quantity of energy expenses for coal, petroleum, natural gas, water, and electricity originate from *The Chinese Energy Statistics Yearbook* (NBS, 2004b). The unit of measure for the total energy expenses is ten thousand tce, the unit of coal and petroleum is ten thousand tons; the unit of natural

1.4 Energy Consumption, Resources and Economy in Regions

gas is hundred million cubic meters; and the unit of water and electricity is hundred million kilowatt-hours.

1.4.1.2 Economic Data

The data on economic development includes the per capita GDP, industry structure, and employment structure, industrialization rate, urbanization rate, and energy intensity. The industrialization rate is identified by the composition of value-added secondary industry. The urbanization is defined as the ratio of urban resident population in the total population. The industrial structure is defined as the first, second and tertiary industries GDP ratio. The employment structure for the industry engaged in the first, second and the tertiary industries accounted for the conclusion of the ratio of number of industry.

The data for GDP of the first, second and tertiary industry originates from *The Chinese Statistics Yearbook* (NBS, 2003c). The data for employment originates from *The Chinese Statistics Yearbook* (NBS, 2003c). The data for the population originates from *The Chinese Population Statistics Yearbook* (NBS, 2004a). The unit of measure for the GDP is hundred million RMB; the unit of employment person is ten thousand; and the unit of population is one person.

1.4.1.3 Energy Resources Data

The description of our secondary data sources is as follows: In Hainan, the total quantity of energy expense in 2001 was used; the data from Ningxia in 2002 was amended according to the data change rule of Qinghai from 1998–2002. A conversion factor was adopted for each type of non-renewable energy, using a conversion coefficient. This coefficient was used to measure the structure of each region's energy expenditure, and to obtain the share of each type of non-renewable energy.

The data of energy resources includes the proven and prospective reserves of regional coal, petroleum, natural gas, water, and electricity. The reserves data for regional coal resources were collected from the forecast data of each province at coalworld.net (SCCI, 2005). Due to the data's commensurability, the forecast data of coal from the same organization were selected. The reserves data of petroleum were collected from the proven reserves bulletin statistics of petroleum and natural gas in 2000 (Zhou, Zhang, Tang, 2003). The natural gas data were sourced from the national third gas resources appraisal (Zhai, 2002); the third oil gas resources appraisal started from 1998 and was carried on by various oil gas companies.

1.4.2 Regional Energy Consumption

Along with the sharp economic growth in China and the steadily improving living standards of the Chinese people, energy consumption has increased dramatically. In 2003, the total energy consumption reached 1.678 billion tce, an increase of 10.1% on the previous year. Two-thirds of the provinces appeared to be "switching out for limiting power consumption". In 2004, it reached 1.97 billion tce, and became the world's second largest energy consumer after the U.S. In 2004, twenty-six provinces and autonomous regions switched out to limit power consumption.

The energy structure in China is dominated by coal; with some 66% of total energy consumption. In 2004, in terms of the world's energy consumption, coal, crude oil and natural gas each accounted for approximately 30% each (coal 27.2%, crude oil 36.8%, natural gas 23.7%, water and electricity 6.2%, and nuclear 6.1%). However, for China, coal usage was 67.7%, petroleum 22.7%, whilst water, electricity, nuclear power, wind power, and solar energy were 7%, and natural gas only 2.6%. Clearly, coal and petroleum are major energy sources in China. Coal occupies more than 90% of total energy reserves; so coal will be the most important energy in China for quite a long time. The structure of energy consumption has been mainly affected by the condition of natural resources, the ability of participating international resources assignment, the economic structure of the state, as well as people's consumption level. The regional energy consumption and structure have many differences, which are caused by the resources disposition, level of development, economic structure, and inhabitants' life expectancy.

1.4.2.1 Difference Energy Consumptions between the Provinces and Cities

Fig. 1-4 shows the energy consumption in provinces. The energy consumption in many provinces exceeded 100 million tce. For example, in Hebei, Guangdong, Shandong, Liaoning, Shanxi, and Henan et al., the energy consumption of these provinces was around 45% of the total in China. At the same time, in the western area (Guangxi, Inner Mongolia, Chongqing, Sichuan, Yunnan, Guizhou, Tibet, Shaanxi, Gansu, Qinghai, Ningxia, Xinjiang), the number is 23%. Whilst Hebei, Guangdong, Shandong, Liaoning, Shanxi are nearly 20 times larger than Hainan, and 10 times larger than Qinghai or Ningxia.

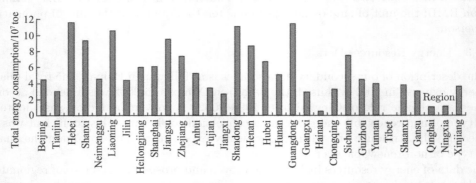

Fig. 1-4 Regional energy consumption

The regional energy consumption scale is shown in Fig.1-5, from which we can see that the primary region of energy consumption is the eastern coastal areas, such as Hebei, Guangdong, Shandong, Liaoning, Jiangsu and so on. The quantity was 5% of the total consumption. The middle-located provinces (such as Shanxi, Henan), was above 5%. In the west (Qinghai, Ningxia, Guangxi, Gansu, Chongqing, Xinjiang, Shaanxi, Yunnan) and middle-eastern areas (Tianjin, Fujian, Jiangxi, Hainan), the proportion was under 2.5%.

The areas in which the consumption is under 1% are Hainan, Qinghai, Ningxia. The number of Hainan is only 0.31%; Qinghai and Ningxia are 0.61% and 0.65%. However, the quantity for Hebei, Guangdong, Shandong and Liaoning are 6.97%, 6.83%, 6.64%, 6.37%, respectively.

If there are obvious differences and characteristics between regional energy consumption, one might ask: What about the regional energy structure?

1.4.2.2 Regional Distributions of Primary Energy

Coal is mostly distributed in the northeast, eastern and middle areas, oil in the east-west areas, such as Liaoning, Guangdong, Shandong, Heilongjiang, natural gas mainly in the mid-west area, and electric power primarily in the eastern area.

The analysis of energy structure shows that energy consumption in China is dependent principally on coal and petroleum, among which coal occupies 66% and above. In recent years, along with economic development and energy structure adjustment, the proportion of coal has also been dropping year by year. Energy structure varies greatly from region to region because of the differences in the energy resources condition and level of development.

1.4 Energy Consumption, Resources and Economy in Regions

Therefore, it is possible to identify the regional difference from the proportion of regions by energy consumption.

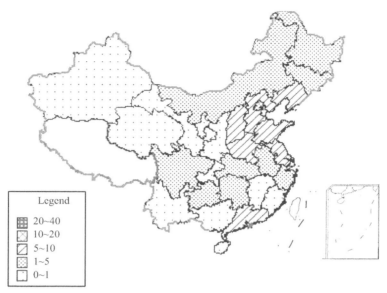

Fig. 1-5 The degree of regional energy consumption scale (Unit: %)
(Note: Without Tibet data)

The regional distribution of coal, petroleum, natural gas and electric power is shown in Fig. 1-6. It can be seen that there are significant differences among regions, and the province that consumed more coal may be not the larger consumer of petroleum.

The spatial distribution rules of each kind of energy are as follows:

(1) The provinces which consume most coal, are mainly located in the northeast, eastern and middle areas, while the proportions of west, southwest and part of middle area are small. The proportion of Shanxi Province is more than 10%; Hebei and Shandong Province are 8.23% and 7.75%; Henan, Jiangsu and Liaoning are 6.19%, 5.79% and 5.61%; but Hainan, Qinghai, Ningxia, Guangxi, Beijing, Jiangxi, Fujian, Gansu, Xinjiang, Tianjin, Chongqing, Yunnan, Shaanxi are less than 2.5%, Hainan, Qinghai, Ningxia are only 0.13%, 0.37% and 0.62%, respectively.

(2) The locations of provinces which expended their petroleum use are different. The proportion of Liaoning is 19.03%, and Guangdong province occupied 8.85%; Shandong, Heilongjiang, Shanghai, Jiangsu, Zhejiang, Xinjiang account for 7.34%, 7.15%, 6.43%, 6.35%, 5.60%, and 5.09%; while Guizhou, Shanxi, Chongqing, Hainan, Yunnan, Sichuan, Qinghai, Guangxi, Ningxia, Inner Mongolia are less than 1%.

(3) The provinces which expended more natural gas are mostly located in the mid-west area. Sichuan province is 25% of total quantity, and Xinjiang occupies 12.5%.

(4) The regularity of electric power's spatial distribution is evident. The provinces are chiefly located in the eastern area, for example, the proportion of Guangdong province is 10% of total quantity; Jiangsu, Zhejiang, Shandong, Hebei, Henan, Liaoning are 7.68%, 7.59%, 6.27%, 5.95%, 5.72% and 5.30%, respectively; the quantity of these seven provinces and cities is nearly half of total consumption in China.

From the analysis of regional primary energy, we find that Shanxi is the province that consumed most coal. The energy consumption of Guangdong province is primarily petroleum and electric power. High consumption in Liaoning is coal, petroleum and electric power.

Consumption in Hebei is coal and power; Jiangsu, Henan is coal, petroleum and electric power; Zhejiang is petroleum and electric power. Thus, it can be seen that the pattern of energy consumption in various provinces differs greatly.

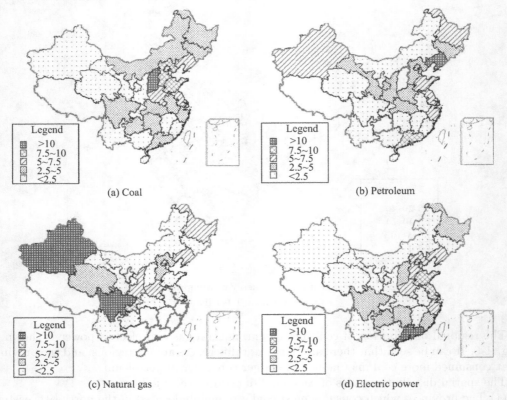

Fig. 1-6 The proportion of energy consumption by region (Unit:%)
(Note: Without Tibet data)

China's energy source is dominated by coal, but the pattern in each province is different. There are three kinds of patterns.

Coal is the most important primary energy in China. Its proportion is more than 67%; while petroleum is the most prominent energy source, followed by coal and natural gas in other countries. Through the analysis of regional energy consumption patterns (shown in Fig. 1-7), it can be seen that the principal energy consumption in China is coal, and this is why there is over-dependence on coal. In a number of provinces, the dependence has reached more than 80%, whilst in some areas it has reached more than 90% (Shanxi, Inner Mongolia, Guizhou, Guangxi, Anhui, and Yunnan). The dependence level of Shanxi Province has reached as high as 99%. This kind of pattern will bring environmental problems in future. At the same time, coal's usage in Xinjiang, Liaoning, Qinghai and Beijing is approximately 50%. The proportion of Heilongjiang, Gansu, Shanghai, Guangdong, Shaanxi, Tianjin and Sichuan is some 60%.

From Fig.1-7, we can also see that the proportional relativity of regional natural gas and electric power cost is high. The natural gas and electric power's proportions of region energy expense in Xinjiang, Qinghai, Sichuan and Chongqing are 10% or above. So three kinds of regional energy expense pattern can be identified:

(1) The first type is coal dominated, namely, coal accounts for the absolute proportion (in regional energy consumption, the proportion of coal accounts for more than 90%), such

1.4 Energy Consumption, Resources and Economy in Regions

as Shanxi, Inner Mongolia, Guizhou, Guangxi, Anhui, Yunnan provinces.

(2) The second type is where coal and petroleum are equal. Namely, petroleum's proportion is more than 30%; coal's proportion is less than 65%. This is seen in Liaoning, Gansu, Guangdong, Shanghai, Xinjiang, Heilongjiang, Beijing, Tianjin and others.

(3) The third is an effective and high quality energy structure. The proportion of coal is small, and the proportion of petroleum and natural gas is high. This can be seen in Qinghai, Xinjiang and Beijing. The proportions of coal, petroleum, natural gas and power are: Qinghai 51.13:13.3:17.31:21.27; Xinjiang 43.98:34.24:9.77:12.01; Beijing 51.67:30.54:7.98:9.81.

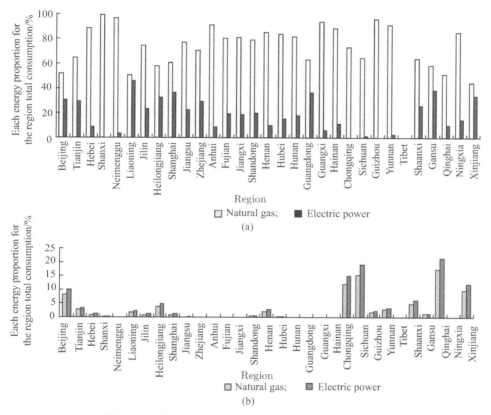

Fig. 1-7 The structure of regional energy consumption

1.4.3 Regional Differences of Energy Resources and Consumption

Generally speaking, coal and hydropower resources are rich in China, whilst petroleum and natural gas are relatively deficient. In regional distribution, the energy resources are largely concentrated in the west and north areas in which the economic development was laggard. The eastern area, in which the economy developed quickly, consumed more than other areas, but the energy resource is scarce.

The coal resource is mainly distributed in the west and mid-north of China; petroleum is mainly distributed in the eastern, northen, and western areas, the middle area's petroleum resource is, however, limited. The natural gas resources are mainly distributed in the eastern Songliao Basin, the Persian Gulf Basin, the middle Sichuan and Ordos Basin, the western Tarim Basin, the Dzungar Basin and the Tsaidam Basin. The water resource is mainly distributed in the southwest area.

Coal is mainly distributed in the western and the mid-north areas. Oil and gas are mainly distributed in the eastern and west Basins as well as middle Erdos and the Sichuan Basin. The hydropower resources are mainly concentrated in the southwest area.

From Fig. 1-8, we can see that the coal resources are largely distributed in the west and mid-north area. Xinjiang and the Inner Mongolia's coal resources occupy 66% of the total quantity. The next is Shanxi Province, where the coal reserves account for 9% of total quantity. There are some other provinces in which the coal resource reserves are relatively high, such as Shaanxi, Guizhou, Qinghai, Gansu, Henan, Anhui, Hebei, where the proportions are 4.4%, 4.17%, 3.78%, 3.14%, 2.02%, 1.34% and 1.32%, respectively. The proportion of Yunnan, Shandong, and Ningxia also approaches 1%. In the south area, coal resource is relatively few; only the coal resources of Guizhou and Yunnan province are comparatively rich.

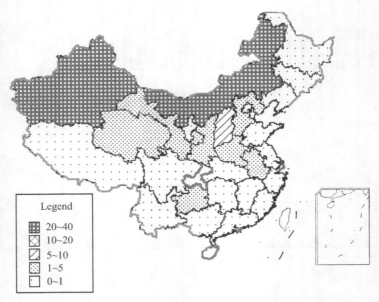

Fig. 1-8 The graduation chart of regional coal resources (Unit: %)

According to the result of the Third National Oil Gas Resources Assessment (Table 1-15), the quantity of petroleum in China reaches 100 billion tons, where 77.5% was distributed approximately on land, and 22.5% in the sea. The distribution of oil resource on land is 39.292 billion tons; and 39.9% were in the eastern area, mainly concentrating in the Songliao and Bohai Gulf Basin. Petroleum's quantity in the midwest area is 38.21 billion tons, which occupies 38.2% of the total. The oil resource in the midwest area is mainly in the western Tarim Basin, the Dzungar Basin, Tsaidam, Kazak, the middle Ordos Basin and the Sichuan Basin. The petroleum resource in the south area is limited. The marine oil resource is mainly in the Bohai, Zhujiangkou and the east China Sea.

By the end of 2000, the oil proven reserves that have been verified are 21.29 billion tons, which account for 21% of the total oil resource reserves. Along with the technology enhancement of exploration, the proven reserves are also increasing. The spatial distribution of proven reserves of oil resource was different from the total quantity of petroleum (Table 1-16). In the proven reserves, 93.4% are distributed on land; the reserves in the sea only occupy 6.6%, and are much lower than the total quantity of petroleum resource in the sea. In the proven reserves, 74.2% of oils distribute in the eastern area, and are mainly distributed

1.4 Energy Consumption, Resources and Economy in Regions

in the Songliao and Bohai Gulf Basin; while 5.77% distribute in the middle area; 13.3% distribute in the western area.

According to the Third Appraisal of Natural Gas Resources (Table 1-15), the spatial distribution of natural gas resources in China are: In the eastern area the natural gas resources are 491 billion cubic meters and 8.9% of total quantity located in the eastern area, which is mainly concentrated in the Songliao and Bohai Gulf Basin; 18.06 billion cubic meters and 32.74% located in the middle area, and mainly concentrated in the Sichuan and Erdos Basin; 13.2 billion cubic meters and 23.93% located in the west, being mainly concentrated in the Tarim, Dzungar and Qaidam Basin; 3.2 billion cubic meters and 5.8% located in the south. Sea area resources are 15.79 billion cubic meters, representing 28.63% located in the sea, and mainly distribute in Zhujiangkou, Yingge Sea and the east-south of Qiong Basin.

Table 1-15 The third oil and gas resources appraisal result (total amount of resources)

Region	East	Songliao	Bohai gulf	Mid-west	Tarim	Dzungar	Erdos
Oil/10^{12} ton	392.92	128.88	198.76	382.10	107.6	85.87	66.38
Gas/10^{12}m^3	4.91	0.88	2.67	31.26	8.39	2.09	10.70
Region	Tsaidam	Turfanhami	Sichuan	South	Total on land	Total in sea	Total
Oil/10^{12}ton	20.05	15.75	11.35	25.00	775.02	225	1000
Gas/10^{12}m^3	1.92	0.37	7.36	3.20	39.37	15.79	55.16

Data source: according to 2000 year National Petroleum Gas Reserves Bulletin statistics (Zhai, 2002).

Hydroelectric resources in China are abundant, and the total quantity is the highest in the world. However, the hydroelectric resources are uneven in the regional distribution. Generally speaking, the hydroelectric resources in the west area are plentiful, and in the east area are poor. The hydroelectric resources are concentrated in the southwest area, but the quantity in the eastern area is limited, where economy is developed and the demand for energy is large. According to the Hydro electric Resources Re-examination Achievement (NDRC, 2005), the technology potential for hydro-electric resources is rich in Sichuan, Tibet and Yunnan, where the proportion of total quantity is 22%, 20% and 19%, respectively. The quantity of rivers and streams include: the Yangtze valley, the Yarlung Zangbo River, and the Huanghe River that account for 47%, 13% and 7% of the total quantity in China.

Table 1-16 The petroleum proven reserves regional distribution

Region	East	Songliao	Bohai Gulf	Erlian	Middule	Erdos	Sichuan
Proven reserves /10^6 ton	1 579 904	650 091	851 451	19 719	122 860	115 934	6926
Accounts for the national proportion/%	74.21	30.54	39.99	0.93	5.77	5.45	0.33
Region	West	Tarim	Dzungar	Turfanhami	Tsaidam	Total in sea	Total
Proven reserves /10^6 ton	283 101	44 811	176 210	23 885	25 377	141 086	2 128 956
Accounts for the national proportion/%	13.3	2.1	8.28	1.12	1.19	6.63	100

Data source: according to 2000 year National Petroleum Gas Reserves Bulletin statistics (Zhou, Zhang, Tang, 2003).

From the analysis of regional energy consumption and distribution, the relationship between the regional energy consumption and resources is identified. First, from a quantity perspective, the energy consumption and resource distribution are uneven in the spatial distribution. In other words, China's energy distribution is highly unbalanced. Next, there is strong relativity between the energy consumption pattern and distribution structure. This is to say, the energy consumption pattern is mainly decided by the structure of resource distribution.

From the total quantity of energy, it can be seen that the energy consumption is primarily distributed in the eastern area, while the energy resources are mainly distributed in the western area.

From the regional distribution of energy resources and consumption, it is evident that the area which consumed most energy mainly distributes on the southeast coast and middle partial provinces and cities. Consumption in the western area is small; yet the energy resources are mainly distributed in the western area. Energy resources of the eastern area are wanting. While the eastern area consumed most energy, its own resources are limited, and the production of resources cannot satisfy the demand. At the same time, in the western areas, in which the energy resources are relatively abundant, its own demand of energy is actually little.

For example, the energy cost for Guangdong is as much as 6.83% of total energy expenses in China. However, Guangdong's own energy resources are poor; and the fiscal status does not match its energy resources: specifically, the petroleum, natural gas, water and electricity resources are seriously deficient; its coal reserves also only account for 0.02% of total reserves. Nevertheless, in Qinghai, Ningxia, Gansu, Xinjiang and Inner Mongolia and other western areas, the coal, petroleum and natural gas resources are relatively rich, but energy consumption in proportion of total quantity, is actually low.

The relativity between the regional consumption pattern and the resources characteristics structure is strong. In the case of Shanxi, Inner Mongolia, Guizhou, Guangxi, Anhui and Yunnan province, coal is more than 90% in the energy consumption pattern. The energy resource is mainly coal. Because the Songliao Basin has extensive oil reserves, the proportion of oil energy consumption is high in Liaoning and Heilongjiang.

If there are greater choices of energy type and quantity, people always select a relatively effective and good quality consumption pattern. For example, the coal, petroleum and gas reserves are rich in Xinjiang; the coal reserves occupy 40% of the total recoverable resources, and the oil and gas resources in Talimu, Dzungar, Tsaidam and Tuha Basin are rich, so when there is choice of energy type and quantity, the consumption pattern in Xinjiang favours petroleum, natural gas and electric power.

1.4.4 Regional Differences of Energy Consumption, Resources Distribution and Economic Development

In recent years, there has been considerable research on irregular economic development between regions. Overall, east, middle and west, are the three major areas of China's economic development level of regional differences. For example, economically developed and the more developed provinces and municipalities are all located in the eastern region, and economic underdevelopment and economically backward provinces are mainly in central and western regions. China's economic development level is regionally significantly different.

The regional economic development indexes scale is shown in Fig. 1-9.

Fig.1-9(a) indicates that the differences of per capita GDP is located between the eastern areas and the middle-west provinces, as well as southeast coast and middle-partial provinces and cities; whilst the difference between the middle-partial provinces and west areas is small. The irregular economic development is not only among the eastern areas, middle-partial and

1.4 Energy Consumption, Resources and Economy in Regions

west areas, but also in inner areas. There is a large difference among eastern provinces in per capital GDP, while the difference between the middle-partial provinces and west areas is more balanced development compared to the eastern areas.

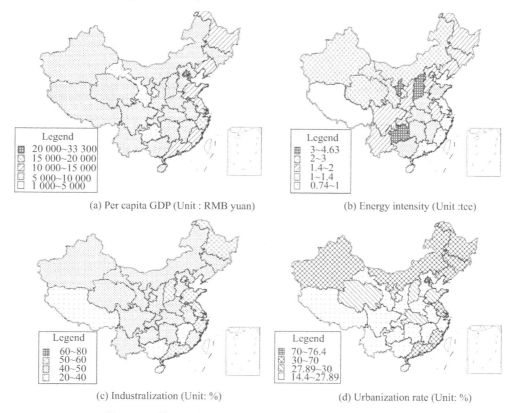

Fig. 1-9 The graduation chart of regional economic indexes

Compared to the regional analyses of per capital GDP and energy consumption, we can see that per capita GDP in the eastern area is higher than the west and the middle-partial regions, and the spatial rule of energy consumption is similar to GDP. However, we cannot say that in every province, the spatial distribution of GDP and energy consumption is correlated. Because there are some provinces in which per capita energy consumption and GDP are both low or high, but there are some provinces in which per capital GDP is lower and energy consumption is higher (for example Shanxi, Henan); and some provinces in which per capital GDP is higher and energy consumption is lower (for example, Hainan).

The regional distribution of energy intensity is shown in Fig. 1-9(b). It can be seen that there are significant differences between regions, and also the regional spatial rule is very clear. The regional distribution of energy intensity is following a basic law. From the eastern area, the middle-partial regions, the west and the north area, the energy intensity increased from low to high, in which Shanxi's energy intensity is the highest, this is correlative with Shanxi's energy consumption structure.

General industrial standards for industrialization are: a rate of 20%–40% is early industrialization; 40%–60% for the semi-industrialized countries; more than 60% for the industrialized countries.

The regional distribution of classification of industrialization is shown in Fig. 1-9(c). It can be seen that the majority of provinces and cities in China are in the semi-industrialized regions, and only a few provinces and cities in the industrialized regions or early industrialization. Therefore, on the whole, China's overall stage is semi-industrialized. From the industrialization of the regional classification map, we can clearly see that the regional distribution and the level of industrialization and economic development are similar, and the regional distribution of energy consumption trends are also alike.

Generally, the development of cities marks off three stages. In the initial stage of the city, the urbanization rate is under 30%; in the city of intermediate stages, the urbanization rate is between 30% and 70%; in the city of an advanced stage, the urbanization rate is above 70%. Now, the urbanization rate is 27.89% in China, this means that China is in the initial stage. At this stage, because of the enormous differences among regions, China is at the initial stage, but Beijing and Shanghai have entered the advanced stage, whose urbanization has reached more than 70%.

From the Fig. 1-9(d), we can see that China has 20 provinces and cities in the initial phase, 9 provinces in the intermediate stage, and 2 provinces in the advanced stage. Cities in the initial stage are mainly distributed in the west and middle-partial region, and the cities at intermediate stage, mainly distributed in the northern and eastern coastal areas, including Tianjin, Heilongjiang, Jilin, Liaoning, Guangdong, Jiangsu, Fujian, Inner Mongolia, Xinjiang, etc.

According to the analysis of the relationship between the regional energy consumption and the urbanization, we observe that there is no obvious correlation.

From Fig. 1-10, we can see that most regional industrial structures are at the second and tertiary industry level, but there are individual provinces (Hainan) whose industry structure is nearly fully built by the first industrial and tertiary industries. In varying provinces, the regional industrial structure is very different.

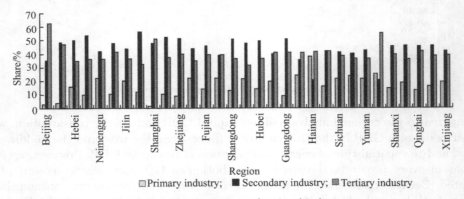

Fig. 1-10 The share chart of regional industry

According to the analyses of the relationship between the regional energy consumption and the industrial structure, it can be found that the regions in which the second industry occupies an absolute proportion of the region, its energy consumption is relatively high. Therefore, we can say that there is certain relevance between regional energy consumption and the industrial structure.

Comparing the regional analysis of economic development and energy consumption, we can see that economic development in the eastern region is higher than the west and the middle-partial regions, and the spatial rule of energy consumption is similar to economic development. Nevertheless, we cannot say that in every province the spatial distribution of economic development and energy consumption is correlated. At present, some research shows that there is cointegration and causal relationship between energy consumption and GDP in China. Clearly, there is need for further study about the relationship among energy

1.5 Summary

Energy is one of the basic factors driving economic growth. As the largest developing country in the world and the second energy producer and consumer, China is undertaking a massive process of building an affluent society in a systemic way. Its energy strategy and policy have already become the focus of attention from different fields, both at home and abroad. As the basic study background for the remainder of this book, this chapter introduced the background of world energy development and China's energy production and consumption.

(1) World energy resource reserves are distributed in imbalance.

World proven petroleum reserves were about 1 188.6 billion barrels by the end of 2004, in which that of the Middle East area accounted for 61.7%; in China it were 17.1 billion barrels, accounting for 1.4% of the world total. World proven natural gas reserves were about 179,530 billion cubic meters, mainly distributed in Russia and the Middle East; China's were about 2,230 billion cubic meters. World proven coal reserves were about 909.1 billion tons, 67% concentrated in U.S., Russia, China, India; and that in China were about 114.5 billion tons, accounting for 12.6% of the world total. China's total energy resource reserves were abundant, whereas the per capita reserves were deficient.

(2) World energy consumption increases continuously.

Since 1970, the world economy and energy consumption both maintained persistent growth, and the global gross domestic product (GDP) and energy consumption increased annually by 3.0% and 2.1%, respectively. World energy consumption was chiefly petroleum, coal, and natural gas, which represented 37%, 24% and 27%, respectively, of the total energy consumption in 2004. The U.S. is the largest energy-consuming country in the world, which consumed 23% of the world's total.

(3) World energy supply and demand situation is being adjusted.

The world energy market had changed greatly over 2004–2005. World crude oil price has run high, and the crude oil future price of the New York Mercantile Exchange (NYMEX) once exceeded US$70 per barrel. The world energy supply and demand pattern is being adjusted. Principal energy producers and consumers were launching energy diplomatic activity constantly in 2005, and mergers and acquisitions among large multinational energy enterprises frequently took place. The Kyoto Protocol came into effect on February 16, 2005, which indicated great progress in international environmental cooperation. In the complicated and changeable environment of international energy competition and cooperation, the Chinese government and enterprise are facing enormous challenges and opportunities.

(4) China has realized its fast economic growth with relatively low growth in energy consumption.

Since its reform and opening-up in 1978, China has witnessed fast economic growth, whose GDP and energy consumption increased rapidly, the annual rate being 9.6% and 5.2%, respectively. The annual energy consumption was 2,220 Mtces in 2005, increasing 9.5% compared with that of the previous year, and energy consumption per ten thousand GDP was 1.43 tce (2000 constant RMB), essentially the same as that in the previous year. Coal consumption was 2,140 million tons, increasing by 10.6%; crude oil consumption was 300 million tons, an increase of 2.1%; and natural gas consumption was 50 billion cubic meters, a growth of 20.6%; hydro and electricity consumption was 401 billion kWh, rising by 13.4%; nuclear power consumption was 52,300 million kWh increasing by 3.7%.

(5) China's energy production and demand are basically in balance, while structural differences are distinct.

China's primary energy production was 2,060 Mtce in 2005, increasing by 9.5% compared with that of the previous year; raw coal production was 2,190 million tons, rising by 9.9%; and 181 million tons of crude oil, increasing by 2.8%. The annual power generation in 2005 was 2,470 billion kWh, rising by 12.3%. By the end of 2005, the installed capacity for power generation in China set a new record, reaching 500 million kilowatts. Renewable energy was being paid more and more attention in China. On February 28, 2005, the 14th Session of Standing Committee of the Tenth National People's Congress passed the Renewable Energy Law.

China's energy production and consumption are basically in balance; nearly 92.8% can be self-supported. China is a large oil importer, and large coal exporter too, whose net import of crude oil and refined oil were 119 million tons and 17 million tons respectively, in 2005, and net export of coal and coke were 46 million tons and 13 million tons, respectively. China had entered a new period of economic growth since 2002, energy demand increasing sharply, and there were also some energy supply shortage in some of China's local regions. The contradiction between energy supply and demand was beginning to alleviate in 2005.

Industrial sectors are energy intensive consumers in China. In 2003, energy consumption of the iron and steel industry, chemical industry and building materials took 14.1%, 10.0% and 7.4%, respectively, of the national total. The energy use per unit of China's high energy-consuming products declined fast and the disparity with the developed countries is shrinking progressively, whereas there was still great opportunity for energy conservation. Improving energy efficiency of industrial sectors and taking the road to new industrialization are of significant importance for reducing energy, building the saving-type society, protecting the ecological environment and realizing sustainable development.

China's energy consumption differed greatly from one area to another. The analysis of regional differences in energy consumption, and that between energy consumption, resource distributions and economic development showed us:

① China's energy consumption mainly relies on coal and petroleum, and the distribution of energy resources in different regions differed greatly. So the realization of pluralism of energy consumption structure has great potential. We should invoke different energy policies that suit the local conditions in different areas. ② The imbalance of energy consumption and resource distribution between western and eastern areas has determined the necessity of regional energy cooperation, which could turn resource advantages of the western area into economic advantage.

CHAPTER 2

Structural Relationship between Chinese Energy and Economic Growth

Since China started its reform and opening-up policy, its economy has kept up a continual and rapid growth, while China's energy consumption maintained less growth. In particular, 1997–1999 saw China's total energy consumption decrease, while its economy still kept growing. This is different from the experience of most other countries, and therefore attracted broad attention. Now the relationship between China's economy and energy consumption has become a focus of international studies.

- In the 21st century, can China maintain a slowing down of its energy consumption, while its economy keeps growing rapidly?
- What effects will the Chinese economy growth target have on energy demand?
- What kind of causality is there between the Chinese economy and energy consumption?
- Will the control of energy consumption impact on China's economic growth?
- What influence has the change of China's economic structure exerted on its energy intensity?
- Will the decreasing trend of China's energy intensity be continued?
- How much can the marginal energy input support China's economic growth?
- What is the difference between the effects of different energy resources in supporting economic growth?
- How can China improve its economy structure and decrease energy consumption while maintaining economic growth?

Above are some basic questions in the study of China's energy economy. These topics are also factors that the Chinese government has to give consideration to when developing its energy strategy and policy. This chapter will explore empirical studies focusing on these questions, and put forward our conclusions, so as to provide some suggestions and evidence for supporting Chinese energy strategy and policy.

2.1 The Cointegration and Causality between Chinese GDP and Energy Consumption

Since China started its reform and opening-up policy in 1978, the rapid growth of the Chinese economy has attracted extensive attention. However, different from the experience of most other countries, China's energy consumption maintained a low momentum of growth, while its economy continually grew. The period 1997–1999 saw the total energy consumption decrease by 6.35%, while the economy increased 25.7% in the same period. The coexistence of the economy's rapid growth and energy consumption's slow increase, attracted extensive attention, and thus makes the reality and sustainability of the relationship between China's economy and energy consumption a focus of international studies.

In the 21st century, many fundamental questions will need to be asked. Such questions as: whether the reduction in China's energy intensity will continue, how the annual growth target of 7% will affect China's energy demand, and whether the energy consumption controls will impact on China's economic growth target. These questions are all critical for China's Government when it decides energy strategy and policies. Therefore, this chapter discusses the cointegration and causality between China's economy and energy consumption. This chapter will provide results of an empirical study for Chinese considering energy strategy and policies.

2.1.1 Method

Since being proposed by Granger (1969), cointegration and causality analysis have been applied broadly and become important tools in international studies of the relationship between energy consumption and economic growth. In 1978, Kraft revealed the causality running from American GDP to energy consumption. After this, empirical studies of the causality between economy and energy have been conducted in many countries (Erol, et al., 1987; Ghali, et al., 2004), such as the United Kingdom, Germany, Italy, Canada, France and Japan, as well as some developing countries.

In the last two decades, causality study has been gradually applied to Asian countries. Using the method of the Granger Causality Test, Yu and Choi (1985) revealed the causality running from GDP to energy consumption in South Korea, while Glasure and Lee (1997) opened out an inverse causal relation, from energy to GDP, in Singapore. In Asafu-Adjaye's (2000) research based on the Cointegration Test and Error Correction Model, unidirectional causality from energy consumption to GDP was found in India and Indonesia, while bi-directional causality was used between energy and the economy in the Philippines and Thailand. However, research results of the same area are not always in agreement; some of the studies drew completely different, even opposite, conclusions. For example, among the studies of causality between energy and economy in Taiwan, Hwang and Gum (1992) concluded that there was bi-directional causality between GDP and energy consumption, while Chen and Lai (1997) argued that there was only unidirectional causality from GDP to energy consumption. However, Yang (2000) reached the same conclusion as Hwang and Gum, and ascribed the difference between his result and that of Chen and Lai to different sample periods and price indexes adopted in their studies.

Although the methods of cointegration and causality have been adopted extensively, the study on China's energy economy in this area is still in its infancy.

2.1.1.1 Stationarity and Integration

Before the 1970s, the models of econometrics were basically built on the assumption of stationary time series; however, this assumption is obviously too simplistic. According to Nelson and Plosser (1982), most macro-economic time series are non-stationary. When an economic time series is non-stationary, the coefficients may have different distribution, and therefore the regression may not be reliable.

According to Stock and Watson's research (1989), the causality test is sensitive to the stationarity of time series. Thus, the first step in our study is to test the stationarity of the time series of China's GDP and energy consumption.

Stationarity requires the Mean, Variance and Auto-covariance of a series to be stationary. A series x_t is said to be stationary, if it has a constant mean $E(x_t)$, and its variance $\text{Var}(x_t)$ does not appear systematically to change over time. In this case, it will tend to fluctuate around the mean $E(x_t)$ steadily. Whereas, a series x_t is said to be non-stationary if it has a non-constant mean $E(x_t)$, and its variance $\text{Var}(x_t)$ appears to systematically change over time. If the difference of a nonstationary series is stationary, the series is said to be integrated, i.e., $I(1)$. If a nonstationary series has to be differenced d times to become stationary, then it is said to be integrated of d order: i.e., $I(d)$ (Li, 2000). Only when two series are integrated of the same order, can it be proceeded to test for the presence of cointegration.

Early in 1976, Dickey and Fuller developed the DF method to test the stationarity of time series. In 1979–1980, they improved the DF method to ADF method (Dickey, Fuller, 1979;

2.1 The Cointegration and Causality between Chinese GDP and Energy Consumption

Said, Dickey, 1984). Because actual series are usually not 1 order autoregression series, the augmented Dickey-Fuller (ADF) test is broadly applied to examine unit root and stationarity of series here.

Firstly, set up the regression equation:

$$\Delta x_t = (\rho - 1)x_{t-1} + \sum_{j=1}^{p} \lambda_j \Delta x_{t-j} + \varepsilon_t \tag{2-1}$$

Where: ε_t is the residual (the same as follows).

Then test the null hypothesis H_0: $\rho = 1$, that x_t is nonstationary, against H_1: $\rho < 1$, that x_t is stationary.

2.1.1.2 Cointegration

Statistically, the long-term equilibrium between nonstationary series is called cointegration. That indicates, even though two series have their own fluctuating features, there can be a long-term equilibrium between them, as long as they are cointegrated. Only in this case, it is reliable to conduct the regression on them.

According to Engle and Grange (1978), if two series are both nonstationary, but integrated of the same order, and there is a linear combination of them which is stationary, then the two series are cointegrated, and the relationship between them is defined as cointegration. Only when two series are integrated of the same order, can it be proceed to test for the presence of cointegration.

According to the two-step method developed by Engle and Granger in 1987, if two series, x_t and y_t, have been tested to be nonstationary, but both of them are integrated of the same order, the regression equation can be set up as:

$$x_t = \alpha + \beta y_t + \varepsilon_t \tag{2-2}$$

and the cointegration between x_t and y_t can thereby be tested by examining the stationarity of the residual ε_t. If x_t and y_t are not cointegrated, any one of their linear combinations will be nonstationary, and therefore the residual ε_t will also be nonstationary. Contrariwise, if the residual ε_t is tested to be stationary, then the cointegration between x_t and y_t can be justified.

2.1.1.3 Causality

In our study, the Granger Causality test is adopted to examine the causality between two series. When the past information is collected to forecast variable y_t, we can use only the past information of y_t, or use past information of both x_t and y_t. According to the Granger Causality test, there is causality from x_t to y_t if the past information of x_t can help us to forecast y_t more precisely.

When applying the Granger Causality test, we firstly set up the bi-variable autoregression model:

$$y_t = \alpha_0 + \sum_{i=1}^{m} \alpha_i y_{t-i} + \sum_{i=1}^{m} \beta_i x_{t-i} + \varepsilon_t \tag{2-3}$$

$$x_t = \alpha_0 + \sum_{j=1}^{m} \alpha_j x_{t-j} + \sum_{j=1}^{m} \beta_j y_{t-j} + \varepsilon_t \tag{2-4}$$

Then F Test is carried out to test the null hypothesis H_0:$\beta_i(i = 1, 2, \cdots, m) = 0$, which is equal to the hypothesis that "x_t has no Granger Causality to y_t". If H_0:$\beta_i(i = 1, 2, \cdots, m) = 0$ is rejected, then we can also reject the hypothesis "x_t has no Granger Causality to y_t", and thereby conclude that x_t has Granger Causality to y_t. Similarly, the hypothesis H_0:$\beta_j(j = 1, 2, \cdots, m) = 0$ can be tested to verify whether there is Granger Causality from y_t to x_t.

2.1.2 Cointegration and Causality in Chinese Energy Economy

2.1.2.1 Data

In our empirical study on cointegration and causality between China's energy and economy, the time series of real GDP and energy consumption from 1978 to 2003 were adopted. We did not use the data before 1978, because the Chinese economy before this point is a "self-supply economy" and often impacted by political events. Moreover, we studied the period of 1978–2000 and 1978–2003 separately, so that we can make a comparison and find more useful conclusions.

The data of energy consumption came from *China's Energy Statistical Yearbook* (1986; 1989; 1991; 1996; 2002) and *China's Statistical Yearbook* (2004), and is expressed in terms of million-ton coal equivalent (Mtce). Real GDP data were obtained from *China's Statistical Yearbook* (2004), recalculated by the price index, and expressed in terms of RMB billion Yuan. Fig. 2-1 shows the trend of China's energy consumption and real GDP in 1978–2003. All the regressions and statistical tests were processed by econometrics software Eviews 3.1.

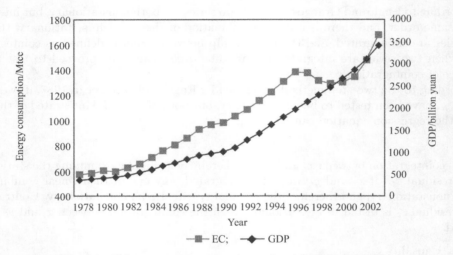

Fig. 2-1 Chinese energy consumption and real GDP in 1978–2003

2.1.2.2 Stationarity of China's Energy Consumption and Real GDP

The unit root test is initially carried out on the two series. Here we test three different forms of test equation, i.e., without intercept or trend, with intercept but no trend, with intercept and trend. We find different equation forms have no effect on the result. Lag lengths are selected according to Akaike Information Criteria (AIC) (Akaike, 1974); due to the limited sample size, maximum lags length is chosen as 4.

Define, $\text{AIC}(k) = \ln \sigma_k^2 + 2k/T \quad (k = 0, \cdots, m)$

Then, $\text{AIC}(\hat{k}) = \min \{\text{AIC}(k) \,|\, k = 0, \cdots, m\}$

Based on the AIC, optimal lag lengths are chosen as 1 for energy consumption, and 2 for real GDP.

Thereby, two test equations are set up,

$$\Delta \text{GDP}_t = (\rho - 1)\text{GDP}_{t-1} + \lambda_1 \Delta \text{GDP}_{t-1} + \lambda_2 \Delta \text{GDP}_{t-2} + \varepsilon_t$$

$$\Delta \text{EC}_t = (\rho - 1)\text{EC}_{t-1} + \lambda_1 \text{EC}_{t-1} + \varepsilon_t$$

2.1 The Cointegration and Causality between Chinese GDP and Energy Consumption

Then unit root tests were carried out respectively on the series of the levels, 1st differentials and 2nd differentials. Table 2-1 shows the test result.

Table 2-1 ADF unit root test of real GDP and energy consumption in 1978—2000

Series	ADF value	5% critical values	1% critical values	Conclusion
GDP	3.1178	−1.9592	−2.6889	non-stationary
GDP d (1)	1.1189	−1.9602	−2.6968	non-stationary
GDP d (2)	−2.1966	−1.9614	−2.7057	stationary
EC	0.6866	−1.9583	−2.6819	non-stationary
EC d (1)	−1.6341	−1.9592	−2.6899	non-stationary
EC d (2)	−3.9167	−1.9602	−2.6968	stationary

Note: MacKinnon critical values of 1% and 5% reported here for unit root test (MacKinnon, 1991) (the same as follows).

Similarly, for the period of 1978–2003, optimal lag lengths are chosen as 1, for both energy consumption and real GDP using the Akaike Information Criteria. Then unit root tests were also carried out respectively on the series of the levels, 1st differentials and 2nd differentials. Table 2-2 shows the test result.

Table 2-2 ADF unit root test of real GDP and energy consumption in 1978—2003

Series	ADF value	5% critical values	1% critical values	Conclusion
GDP	2.1514	−1.9559	−2.6649	non-stationary
GDP d (1)	1.1576	−1.9566	−2.6700	non-stationary
GDP d (2)	−3.0414	−1.9574	−2.6756	stationary
EC	1.2086	−1.9559	−2.6649	non-stationary
EC d (1)	−0.8856	−1.9566	−2.6700	non-stationary
EC d (2)	−2.2428	−1.9574	−2.6756	stationary

According to the test results in Tables 2-1 and 2-2, both the series of China's real GDP and energy consumption in the two periods are found non-stationary, while the series of their 2nd differentials are both stationary. That means the series of China's real GDP and energy consumption are both integrated of order 2.

2.1.2.3 Cointegration between China's Energy Consumption and Real GDP

Because the series of China's real GDP and energy consumption are both integrated of order 2, it can proceed to test for the presence of cointegration. The regression equation is set up as:[①]

$$\text{GDP} = 0.236\,148\,3409\text{EC} - 109\,04.274\,27 + \varepsilon_t \qquad (2\text{-}5)$$

$$(t = -0.45^{**}) \quad (t = 1.31^{**})$$

$$R^2 = 0.9325, \quad F = 71.08^{**}$$

The regression result and residuals are shown as Fig. 2-2.

Then the unit roots of the series of residuals are tested. Based on Akaike Information Criteria (AIC), the optimal lag length is chosen as 1. Then regression equations are set up in 3 different forms, i.e. with intercept and trend, with intercept but no trend, and without intercept or trend. The test results of the 3 different forms are shown as Table 2-3.

Table 2-3 Unit root test result of the residuals (1978—2000)

Regression equation	ADF value	5% critical values	1% critical values	Conclusion
Without intercept or trend	−1.7257	−1.9583	−2.6819	Non-stationary
With intercept but no trend	−1.4628	−3.0114	−3.7856	Non-stationary
With intercept and trend	−0.2678	−3.6454	−4.4691	Non-stationary

① * means at 5% significance; ** means at 1% significance.

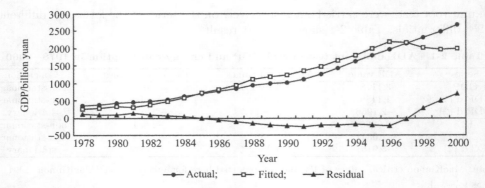

Fig. 2-2 Regression result and residuals of Eq. (2-5)

According to the unit root test result, the time series of residuals is non-stationary in all the three regression equations; thereby, it can be concluded that, for 1978–2000, there was no cointegration between the series of China's real GDP and energy consumption.

With the same procedure, the cointegration between China's real GDP and energy consumption in 1978–2003 can also be tested. The regression equation is set up as:

$$\text{GDP} = -13\ 652.663\ 46 + 0.270\ 519\ 699\ 5\text{EC} + \varepsilon_t \qquad (2\text{-}6)$$

$$(t = -1.32^{**}) \quad (t = 0.69^{**})$$

$$R^2 = 0.9461, \quad F = 271.08^{**}$$

Then conduct the ADF unit root test. Based on Akaike Information Criteria (AIC), the optimal lag length is chosen as 1. The test results are shown as Table 2-4.

Table 2-4 Unit root test result of the residuals (1978—2003)

Regression equation	ADF value	5% critical values	1% critical values	Conclusion
Without intercept or trend	−3.8316	−1.9559	−2.6649	Non-stationary
With intercept but no trend	−3.8836	−2.9907	−2.6348	Non-stationary
With intercept and trend	−3.8913	−3.6118	−4.3942	Non-stationary

According to the unit root test result, the time series of residuals is stationary in all the three regression equations; thereby, it can be concluded that, for 1978–2003, there was cointegration between the series of China's real GDP and energy consumption. This conclusion is totally different from the conclusion of 1978–2000, the reason of which will be discussed in the following sections.

2.1.2.4 Causality Between China's Energy Consumption and Real GDP

According to the method of Granger causality test, a bi-variable autoregression model is set up as:

$$\text{GDP}_t = \alpha_0 + \sum_{i=1}^{m} \alpha_i \text{GDP}_{t-i} + \sum_{i=1}^{m} \beta_i \text{EC}_{t-i} + \varepsilon_t \qquad (2\text{-}7)$$

$$\text{EC}_t = \alpha_0 + \sum_{j=1}^{m} \alpha_j \text{EC}_{t-j} + \sum_{j=1}^{m} \beta_j \text{GDP}_{t-j} + \varepsilon_t \qquad (2\text{-}8)$$

And the null hypotheses H_0: $\beta_i(i = 1, 2, \cdots, m) = 0$ and $\beta_j(j = 1, 2, \cdots, m) = 0$ were tested. Based on the Akaike Information Criteria (AIC), the optimal lag length is chosen as 1. Table 2-5 shows the result of the Granger causality test.

2.1 The Cointegration and Causality between Chinese GDP and Energy Consumption

Table 2-5　Result of the Granger causality test (1978—2000)

Pairwise Granger Causality Tests Sample: 1978–2000 Lags: 1			
Null hypothesis:	Obs	F-Statistic	Probability
EC does not Granger Cause GDP	22	3.135 71	0.092 64
GDP does not Granger Cause EC		7.507 78	0.013 01

According to the test result of Table 2-5, we can reject, at 10% significance level, the null hypothesis of "EC does not Granger Cause GDP" and "GDP does not Granger Cause EC", and therefore accept the conclusion that there is bi-directional Granger causality between the Chinese energy consumption and GDP in 1978–2000.

Similarly, the Grange causality between the Chinese energy consumption and GDP in 1978–2003 can also be tested. The result is shown as Table 2-6.

Table 2-6　Result of the Granger causality test (1978—2003)

Pairwise Granger causality tests Sample: 1978–2003 Lags: 2			
Null hypothesis:	Obs	F-Statistic	Probability
EC does not Granger Cause GDP	24	0.743 52	0.488 77
GDP does not Granger Cause EC		7.495 48	0.003 98

According to the test result of Table 2-6, we can reject, at 1% significance level, the null hypothesis of "GDP does not Granger Cause EC", and therefore accept the conclusion that there is Granger causality from GDP to the Chinese energy consumption in 1978–2000; while we cannot reject the hypothesis of "EC does not Granger Cause GDP", and therefore cannot accept the conclusion that there is Granger causality from the Chinese energy consumption to GDP in 1978–2000.

2.1.3　Cointegration and Causality in Three Industries' Energy Economy

From the study on the causality between the Chinese energy consumption and GDP, we draw different conclusions from the data in two different periods. This implies that we need more studies and evidence to verify our conclusion. Thus we will extend our causality test into the level of three industries in this part, so that we can get a more reliable conclusion regarding the causality in the Chinese energy economy.

Three industries are divided according to the statistics, i.e., primary industry; secondary industry, including industry and construction; and tertiary industry, including transport, storage, post, wholesale and retail trades.

Because of the non-availability of the data before 1980, the study on causality of the energy economy at the level of three industries, is limited to the period 1980–2002.

2.1.3.1　Cointegration and Causality in the Energy Economy of the Primary Industry

In 1980–2002, the trends of GDP (GDP1) and energy consumption (EC1) of the primary industry are displayed as Fig. 2-3.

Firstly, test the stationarity of GDP1, EC1, the result is listed as the Table 2-7.

Table 2-7　ADF unit root test results of GDP1, EC1

Series	Lag	ADF value	5% critical values	1% critical values	Conclusion
GDP1	3	−2.77094	−3.6746	−4.5348	Non-stationary
GDP d (1)	3	−4.1514	−3.6902	−4.5743	Stationary
EC1	3	−1.8683	−3.6746	−4.5348	Non-stationary
EC d (1)	3	−4.2961	−3.6902	−4.5743	Stationary

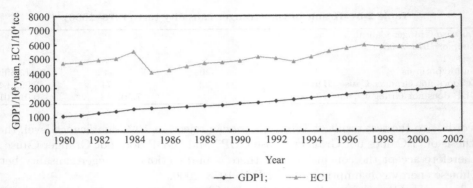

Fig. 2-3 GDP (GDP1) and energy consumption (EC1) of the primary industry

According to the test results, both the series of GDP1 and EC1 are found to be integrated of order (1). Therefore the cointegration test can be conducted. The regression equation is set up as:

$$\text{GDP1} = -1775.409\,073 + 0.736\,587\,580\,3\text{EC1} + \varepsilon_t \tag{2-9}$$

$$(t = -2.8603^{**}) \quad (t = 6.1947^{**})$$

$$R^2 = 0.6463, \quad F = 38.37^{**}$$

Then test the unit roots of the series of residuals. The result is shown as Table 2-8.

Table 2-8 ADF unit root test results of the residual (the primary industry)

Series	Lag	ADF value	5% critical values	1% critical values	Conclusion
Residual 1	1	−2.446194	−1.9583	−2.6819	Stationary

The conclusion is that, in 1980–2002, there was cointegration relationship between the GDP and energy consumption of the primary industry.

Furthermore, test the Granger causality between the GDP and energy consumption of the primary industry. The test result is shown as the Table 2-9.

Table 2-9 Granger causality test result of the GDP1, EC1

Pairwise Granger Causality Tests Sample: 1980–2002 Lags: 1			
Null hypothesis:	Obs	F-Statistic	Probability
EC1 does not Granger cause GDP1	22	1.06818	0.31433
GDP1 does not Granger cause EC1		4.29062	0.05218

The conclusion is that, for the primary industry in 1980–2002, there was Granger causality from the GDP to the energy consumption.

2.1.3.2 Cointegration and Causality in the Energy Economy of the Secondary Industry

In 1980–2002, the trends of GDP (GDP2) and energy consumption (EC2) of the secondary industry are displayed as Fig. 2-4.

Test the stationarity of GDP2, EC2, the result is listed as Table 2-10.

Table 2-10 ADF unit root test results of GDP2, EC2

Series	Lag	ADF value	5% critical values	1% critical values	Conclusion
GDP2	1	1.5726	−1.9583	−2.6819	Non-stationary
GDP2 d (1)	1	0.7024	−1.9592	−2.6889	Non-stationary
GDP2 d (2)	1	−2.6737	−1.9602	−2.6968	Stationary
EC2	3	2.4511	−1.9602	−2.6968	Non-stationary
EC2 d (1)	3	−1.3480	−1.9614	−2.7057	Non-stationary
EC2 d (2)	3	−3.1169	−1.9627	−2.7158	Stationary

2.1 The Cointegration and Causality between Chinese GDP and Energy Consumption

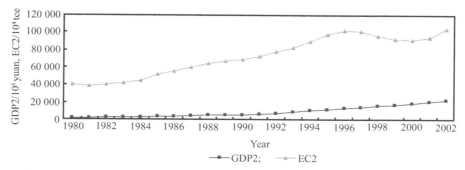

Fig. 2-4 GDP (GDP2) and energy consumption (EC2) of the secondary industry

According to the test results, both the series of GDP2 and EC2 are found to be integrated of order (2). Therefore the cointegration test can be conducted. The regression equation is set up as:

$$\text{GDP2} = -9659.434\,727 + 0.255\,966\,308\,9\text{EC2} + \varepsilon_t \tag{2-10}$$

$$(t = -4.4241^{**}) \quad (t = 8.9218^{**})$$

$$R^2 = 0.7912, \quad F = 79.60^{**}$$

Then test the unit roots of the series of residuals. The result is as Table 2-11.

Table 2-11 ADF unit root test results of the residual (the secondary industry)

Series	Lag	ADF value	5% critical values	1% critical values	Conclusion
Residual 2	1	-3.4505	-1.9583	-2.6819	stationary

The conclusion is that, in 1980–2002, there was cointegration relationship between the GDP and energy consumption of the secondary industry.

Furthermore, test the Granger causality between the GDP and energy consumption of the secondary industry. The test results are seen in Tables 2-12 and 2-13.

Table 2-12 Granger causality test 1 result of the GDP2, EC2

Pairwise Granger Causality Tests Sample: 1980–2002 Lags: 1			
Null hypothesis:	Obs	F-Statistic	Probability
EC2 does not Granger Cause GDP2	22	5.310 62	0.032 65
GDP2 does not Granger Cause EC2		0.028 39	0.867 97

Table 2-13 Granger causality test 2 result of the GDP2, EC2

Pairwise Granger Causality Tests Sample: 1980–2002 Lags: 2			
Null hypothesis:	Obs	F-Statistic	Probability
EC2 does not Granger Cause GDP2	21	0.012 83	0.987 26
GDP2 does not Granger Cause EC2		3.819 04	0.044 06

According to the test results, it can be concluded that, for the secondary industry, when the lag is 1, there is Granger causality from energy consumption to economy, but no causality from the economy to energy consumption; when the lag is 2, the Granger causality from energy consumption to the economy disappeared, while the causality from the economy to energy consumption can be found.

2.1.3.3 Cointegration and Causality in the Energy Economy of the Tertiary Industry

The trends of GDP (GDP3) and energy consumption (EC3) of the tertiary industry in 1980–2002 are displayed as Fig. 2-5.

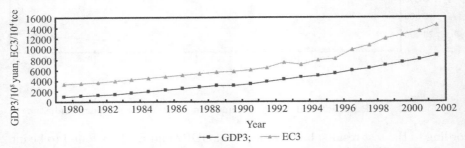

Fig. 2-5 GDP (GDP3) and energy consumption (EC3) of the tertiary industry

Test the stationarity of GDP3, EC3, the result is listed as Table 2-14.

Table 2-14 ADF unit root test results of GDP3, EC3

Series	Lag	ADF value	5% critical values	1% critical values	Conclusion
GDP3	1	2.1785	−1.9583	−2.6819	Non-stationary
GDP3 d (1)	1	0.7250	−1.9592	−2.6889	Non-stationary
GDP3 d (2)	1	−4.4844	−1.9602	−2.6968	Stationary
EC3	3	4.7887	−1.9583	−2.6819	Non-stationary
EC3 d (1)	3	−0.2905	−1.9592	−2.6889	Non-stationary
EC3 d (2)	3	−3.9084	−1.9602	−2.6968	Stationary

According to the test results, both the series of GDP3 and EC3 are found to be integrated of order 2. Therefore the cointegration test can be conducted. The regression equation is set up as:

$$GDP3 = -955.693\,867\,4 + 0.684\,729\,371\,7^{*}EC3 + \varepsilon_t \quad (2\text{-}11)$$

$$(t = -5.4161^{**}) \quad (t = 30.9960^{**})$$

$$R^2 = 0.9786, \quad F = 960.72^{**}$$

Then test the unit roots of the series of residuals. The result is as Table 2-15.

Table 2-15 ADF unit root test results of the residual (the tertiary industry)

Series	Lag	ADF value	5% critical values	1% critical values	Conclusion
Residual 3	2	−2.2498	−1.9592	−2.6889	Stationary

The conclusion is that, in 1980–2002, there was cointegration relationship between the GDP and energy consumption of the tertiary industry.

Furthermore, test the Granger causality between the GDP and energy consumption of the tertiary industry. The test result is shown as Table 2-16.

Table 2-16 Granger causality test result of the GDP3, EC3

Pairwise Granger Causality Tests Sample: 1980–2002 Lags: 1			
Null hypothesis:	Obs	F-Statistic	Probability
EC3 does not Granger Cause GDP3	22	0.282 88	0.600 98
GDP3 does not Granger Cause EC3		4.086 47	0.057 54

The conclusion is that, for the tertiary industry in 1980–2002, there was Granger causality from the GDP to energy consumption.

2.1.4 Conclusions and Policy Implications

2.1.4.1 Discussion of the Conclusions

(1) The uncertainty of the long-term trend of the Chinese energy economy.

According to the cointegration test results of the Chinese energy economy, data of different periods lead to completely different conclusions. Using the data of 1978–2000, no cointegration relationship can be found between the economic growth and energy consumption. However, when the data were extended to 1978–2003, the cointegration relationship appeared in the Chinese energy economy. From Fig. 2-1, it can be found that, there was an obvious and sharp decrease of the energy consumption just before 2000, which changed the trend of energy consumption and therefore removed the cointegration relationship of the Chinese energy economy. After 2000, the energy consumption, with a rapid increase, returned to its original trend, and thus recovered the cointegration relationship with economic growth.

The contradiction of the test results implies that the available data, with only 26 years, are not enough to get a consolidated conclusion. More data are to be accumulated to reach a more reliable conclusion.

(2) The diversity of the relationship between energy consumption and economic growth of three industries.

The different results of the cointegration tests also imply particularity of the energy economic data of 2000–2003. The change of the energy economy in three years bears some special meaning for China, and is worth further studying. The analysis of energy structure and energy efficiency in section 2.3 will give more discussion on the rapid increase of energy consumption in this period.

In order to gain greater evidence about the Chinese energy economy, we extend our research into the level of three industries. Because of the availability of data, we test the cointegration of the energy economy in three industries for 1980–2002. The test results show that there is significant cointegration relationship in all the three industries.

Based on all the results above, the final conclusion is that there is cointegration relationship between the energy consumption and economic growth over the last 25 years. However, allowing for the significant influence of the data for 2000–2003, it is necessary to carry out further study to verify the sustainability of the cointegration relationship.

(3) The causality between the energy consumption and economic growth.

According to the results of the Granger causality test, the same conclusion can be drawn from the two periods, i.e., there is Granger causality from economy to energy consumption, but not from energy consumption to economy. The additional studies of the three industries had more interesting findings. For the primary and tertiary industries, the conclusion is, as above. While for the secondary industry, the Granger causality appeared to be bi-directional, and the time lags of different directions are different. Energy consumption influenced the economic growth one year after, while the economic growth's influence on energy consumption appeared after two years.

In sum, it can be concluded that economic growth has Granger causality to energy consumption, while the energy consumption's Granger causality to energy only exists in the secondary industry. The different Granger causalities in three industries imply the different features of the three industries' energy consumption.

2.1.4.2 Policy Implications

(1) The pressure of energy supply needs more emphasis.

During 1978–2000, China's economy kept growing rapidly, while its energy consumption has kept increasing at a lower rate, and even decreased over 1997–2000. However, according to our study, this relationship appeared between the economy and energy consumption in 1978–2000 is proven to be not a long-term relationship. Contrarily, after the sharp increase of energy consumption in 2001–2003, the long run relationship between the economy and energy consumption appeared to be cointegrated. This implies that the coexistence of rapid growth of economy and slow increase of energy consumption in 1978–2000 will most probably be unsustainable in the future. In the long-term, the relationship between the economy and energy consumption will be a common rapid growth.

Therefore, we cannot be too optimistic on energy consumption in the future. China has set its target of continuing economic growth, but it is hard to keep the energy intensity decreasing as previously done, which will inevitably accelerate the increase of energy consumption. Therefore, the pressure of energy supply may be much heavier, and also come earlier, than estimated.

How to enlarge the energy supply channel, and ensure the energy security of China has become a very pressing strategic problem. China has to improve its economic structure and energy structure simultaneously, to push the energy intensity decreasing, so as to sustain its economy growth continually with as low energy increase as possible.

(2) New policies and technologies should be promoted to increase energy efficiency.

The uncertainty of the cointegration in China's energy economy also implies that, the rapid increase of energy efficiency in 1978–2000 was probably produced by some temporary factors, e.g. the openness of the Chinese economy, the establishment of a market economic system, the introduction and promotion of (foreign-sourced) advanced technology and management, and the innovation of the management system at both macro and micro level. With the improvement of China's market economic system, and the narrowing of the gap between local and foreign technology and management, the potential of further improvement of technology and management will gradually decrease and finally disappear, which will consequently lead to a lower and lower pace in the increase of energy efficiency. Therefore, China has to put more effort to implement and promote new policies and technologies to push the energy efficiency, so as to counteract the pressure of energy consumption's rapid growth.

(3) Implement distinct policies in different industries.

Our study reveals the one-way causality from economy to energy consumption, which means that the growth of the Chinese economy led to the increase of energy consumption, while the drop of energy consumption did not lead to the fluctuation of economy. This may be due to the control ability of Chinese government, e.g., the sharp decrease of energy consumption after 1997 did not lead to obvious fluctuation of Chinese economic growth.

However, bi-directional causality appeared in the secondary industry's energy economy, which means the growth of the secondary industry has a strong dependence on energy consumption, and the fluctuation of energy supply will impact on the economic growth of the secondary industry. Allowing for the one-way directional causality in the primary and tertiary industries, the government should pay greater attention to the differences among the three industries, and implement distinct policies for different industries, so as to avoid possible negative impact on the economic growth.

2.2 Energy Intensity and Economic Structure of China

Energy intensity means the total energy consumption input to produce a unit economic output. In the 20 years before 2000, China's energy intensity had kept decreasing steadily and continually, which attracted many scholars' attention, and someone even criticized the reliability of the statistics of the Chinese energy and economy (Wang, Meng, 2001). Thomas G. Rawski (2001) analyzed the relationship between China's economy and energy after 1990, compared with some other Asian countries. He concluded that it was impossible for China to achieve the economic growth when energy consumption decreased, and attributed the

2.2 Energy Intensity and Economic Structure of China

mistake to the unreliability of the statistics of Chinese economic growth. On the other hand, Jonathan E. Sinton (2001) put forward his criticism from the aspect of energy statistics. He argued that the statistics of small scale collieries' output was probably unreliable, and the real outputs may be larger than the statistics, so the energy consumption of China was most probably underestimated. However, he also admitted, if the statistics were correct, it will be valuable to make further study on the continual decreasing of Chinese energy intensity.

Many Chinese scholars have also made this the focus of their studies. Wei and Xie (1991) undertook econometric analysis on the growth rates of Chinese economy and energy. Chen and Geng (1996) studied the role of energy consumption and energy saving in the economic growth from some aspects, such as economic structure, industrial structure, product structure, international trade, and so on. Zhao and Wei studied the models of energy and economic growth. Shi (2002) analyzed the energy efficiency increase from the perspectives of open economy, structural change, and market reformation, and concluded that the improvement of energy efficiency was the main reason for the slow increase of China's energy consumption after the 1980s.

However, because there is no empirical study on the change of economic structure and energy efficiency, the viewpoint that China's economic structure change has pushed the energy consumption to decrease still appears in much literature about China's energy economy. At the same time, some other scholars insisted that China's structural change helped to decrease the energy intensity, but the role is becoming smaller and smaller (Shi, 1999; Wang, 2001). In order to provide a reliable and empirical conclusion, we conduct the study regarding the impact of economic structural change on energy intensity. Firstly, the trend of energy intensity and economic structure were reviewed; then the change of energy intensity was decomposed to show share of economic structure change and efficiency increase; furthermore, the decomposition was extended into the level of sub-industry structure within the secondary industry.

2.2.1 Decomposition of Energy Intensity Based on SDA

2.2.1.1 Structural Decomposition Analysis

Structural Decomposition Analysis (SDA) was applied broadly from the 1980s. In the energy economy area, SDA has been mostly applied to analyze energy consumption, energy intensity, and carbon emission. Statistics show, in 2000 and 2001, there were more than 40 papers using the SDA published in the literature on the energy economy (Choi, Ang, 2003).

The basic idea of SDA is to decompose the change of a variable into a combination of the changes of a few factors, and thereby distinguish the different shares of the factors. Once the main factor is found, the decomposition can be conducted at the next level, and finally, the impact of each factor can be estimated separately.

There are different methods for undertaking structural decomposition. One is the Laspeyres method, in which a factor is assumed to change from the time (0) to time (n), while the other factors are keep unchanged at time (0). Another is the Paasche method, in which a factor is assumed to change from the time (n) to time (0), while the other factors are kept unchanged at time (n). However, there is a common problem of these two methods, i.e., the change of variable can not be completely decomposed and there is always a residual in both methods. Therefore both of them are approximate decompositions. Our study adopted another decomposition method introduced by Sun (1998), which can decompose a variable completely, avoid the residual, and therefore improve the decomposition results.

2.2.1.2 Decomposition of Energy Intensity: Structure and Efficiency Share

According to the definition given above, energy intensity e is the total energy consumption input to produce a unit of economic output, i.e.,

$$e = \frac{E}{Y} \tag{2-12}$$

Where: E means energy consumption; Y means GDP.

E and Y can be decomposed to three industries., i.e.,

$$E = \sum_i E_i, \quad Y = \sum_i Y_i \quad (i = 1, 2, 3)$$

Thereby e can be decomposed as:

$$e = \frac{\sum_i E_i}{\sum_i Y_i} = \frac{\sum_i e_i \cdot Y_i}{\sum_i Y_i} = \sum_i e_i \cdot y_i \quad (i = 1, 2, 3) \tag{2-13}$$

Where: e_i means the energy intensity of the industry (i); y_i means the share of the outputs of industry (i) in GDP.

From $e = \sum_i e_i y_i$, we can see that, the aggregate energy intensity is affected by two factors, one is the energy intensity of each industrial energy intensity, which reflects the efficiency of each industry in using energy; the other is economic structure, which reflects the shares of each industry in the total GDP. If we conduct correlation analysis, we can find the correlation between the decrease of energy intensity and many factors, such as market system, international trade, and capital assets, etc. However, these factors actually impact on the economic structure or industrial energy intensity, and then indirectly influence the aggregated energy intensity. Therefore, energy intensity can be firstly decomposed into two factors, structure share and efficiency share.

Let e^0 represent the energy intensity of time (0), and $e^n (n=1, 2, \cdots, N)$ the energy intensity of time (n). Then e^n can be decomposed as:

$$e^n = \sum_i e_i^n \cdot y_i^n = \sum_i e_i^0 \cdot y_i^0 + \sum_i e_i^0 \cdot (y_i^n - y_i^0) + \sum_i (e_i^n - e_i^0) \cdot y_i^n$$

$$(i = 1, 2, 3; n = 1, 2, \cdots, N) \tag{2-14}$$

Thereby, the energy intensity can be decomposed as:

$$\Delta e = e^n - e^0 = \sum_i e_i^n \cdot y_i^n - \sum_i e_i^0 \cdot y_i^0 = \sum_i e_i^0 \cdot (y_i^n - y_i^0) + \sum_i (e_i^n - e_i^0) \cdot y_i^n$$

$$(i = 1, 2, 3; n = 1, 2, \cdots, N) \tag{2-15}$$

Where: $e_i^0 \cdot (y_i^n - y_i^0)$ means the change of energy intensity due to the change of the share of industry (i) in total GDP; and $\sum_i e_i^0 \cdot (y_i^n - y_i^0)$ means impact of the change of the whole economic structure. Thus the share of structure can be formulated as:

$$\frac{\sum_i e_i^0 \cdot (y_i^n - y_i^0)}{\sum_i e_i^n \cdot y_i^n - \sum_i e_i^0 \cdot y_i^0} \quad (i = 1, 2, 3; n = 1, 2, \cdots, N) \tag{2-16}$$

2.2 Energy Intensity and Economic Structure of China

Similarly, $(e_i^n - e_i^0) \cdot y_i^n$ means the change of energy intensity due to the change of the share of efficiency in industry (i); and $\sum_i (e_i^n - e_i^0) \cdot y_i^n$ means impact of the change of the efficiency of all industries economic structure. Thus the share of efficiency can be formulated as:

$$\frac{\sum_i (e_i^n - e_i^0) \cdot y_i^n}{\sum_i e_i^n \cdot y_i^n - \sum_i e_i^0 \cdot y_i^0} \quad (i = 1, 2, 3; n = 1, 2, \cdots, N) \quad (2\text{-}17)$$

The structure share of time (n) can be formulated as:

$$\frac{\sum_i e_i^{n-1} \cdot (y_i^n - y_i^{n-1})}{\sum_i e_i^n \cdot y_i^n - \sum_i e_i^{n-1} \cdot y_i^{n-1}} \quad (i = 1, 2, 3; n = 1, 2, \cdots, N) \quad (2\text{-}18)$$

and the efficiency share of time (n) can be formulated as:

$$\frac{\sum_i (e_i^n - e_i^{n-1}) \cdot y_i^n}{\sum_i e_i^n \cdot y_i^n - \sum_i e_i^{n-1} \cdot y_i^{n-1}} \quad (i = 1, 2, 3; n = 1, 2, \cdots, N) \quad (2\text{-}19)$$

Respectively, Eqs. (2-16) and (2-17) reflect the contribution of the change of structure and efficiency to the change of energy intensity from time(0) to time(n); while Eqs. (2-18) and (2-19) reflect the contribution of the change of structure and efficiency to the change of energy intensity from time($n-1$) to time(n). If the value of share is positive, it means the change of structure or efficiency has facilitated the energy intensity change; if the value is negative, it means the change of structure or efficiency has counteracted the energy intensity change.

2.2.2 Impact of Chinese Economic Structure Change on Energy Intensity

2.2.2.1 The Trends in the Chinese Economic Structure and Energy Consumption

(1) Chinese economic structure.

Over 1978–2000, the Chinese economy kept continually growing. The GDP in the price of 1978 has increased 6.3 times. Especially after China quickened its establishment of the Socialist market system in 1993, the annual growth of GDP exceeded 10%. The secondary industry increased annually greater than 20% in some years. Although the economic growth became slow after the Asian financial crisis in 1997, it still kept a growth rate above 7% annually, and the secondary industry more than 10%. Fig. 2-6 displays the trends of Chinese and structural GDP in 1980–2000.

As for the economic structure, just as shown as Fig. 2-7, the share of primary industry has been continually decreasing except for the beginning of 1980s; the share of secondary industry, after it decreased from 48.5% in 1980s to 41.6% in 1990, had been increasing continually in the first half of 1990s, and then keeping around 50% after 1996; the share of tertiary industry had kept increasing before 1992, but thereafter started wavering because of the increase of secondary industry.

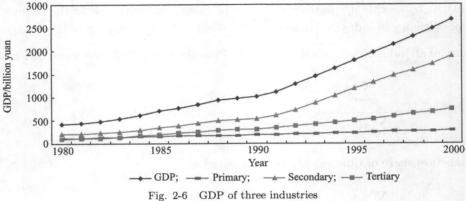

Fig. 2-6 GDP of three industries

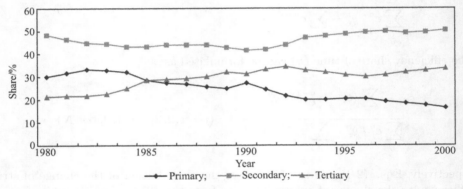

Fig. 2-7 Shares of three industries

(2) Energy consumption of industries.

While its economy kept growing, China's energy consumption steadily increased before 1996, and the annual increase rate was 5.36%. After reaching its peak of 1389.48 million tons coal equivalent in 1996, the energy consumption dropped successively in 1997–1999, at an annual rate of −2.16%. At the level of industries, shown as Fig. 2-8, the drop of energy consumption was mainly due to the secondary industry consumption and non-production consumption, while the consumption of the other industries basically kept steady.

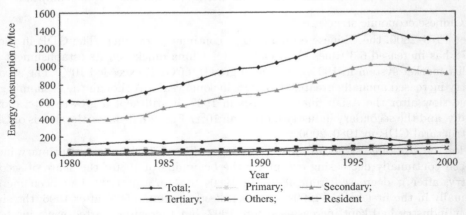

Fig. 2-8 Energy consumption of different sectors

(3) Change of energy intensity.

The trends and fluctuation of energy intensity are affected by the economic outputs and energy consumption. According to Fig. 2-9, the decrease of Chinese energy intensity became slower after 1988, which was mainly due to the slowdown of economic growth, especially the secondary industry's growth. In 1993 and 1996, the decrease of energy intensity was accelerated. In 1993, it was mainly due to the acceleration of economic growth. While in 1996, the main reason was the drop of energy consumption caused by the closing of small-scale collieries. Although the economic growth also decreased, the drop of energy consumption dominated the change of energy intensity. After 1999, the trend in decline of energy consumption, consequently the energy intensity, gradually disappeared. In this period, the energy of primary and tertiary industries was only slightly affected by the fluctuation of the economy and energy consumption, and kept slowly decreasing.

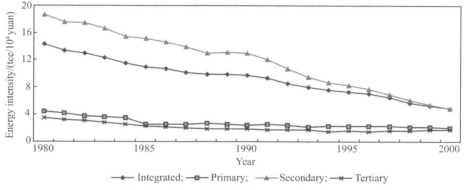

Fig. 2-9 Trend of China's energy intensity

In our opinion, the change in nature of energy intensity is the result of the improving of techniques, management and system arrangement, although it may be affected statistically by economic outputs and energy consumption. Therefore energy intensity should be deemed as an exogenous variable to economic output and energy consumption. In other words, it is the economic outputs and energy intensity that jointly decide the energy consumption, rather than the economic outputs and energy consumption that decide the energy intensity; energy intensity decreases intrinsically with the improvement of techniques and management. However, in the short term, the change of energy price due to the fluctuation of energy supply and demand can also result in savings in energy consumption or substitution between energy resources, which will decrease energy intensity.

From 1980 to 2000, China's energy intensity decreased from 14.34 to 4.87, i.e., annually 0.52. In 1997–1999, it decreased by 26.64%. The rapid decrease of energy intensity has attracted much discussion. However, we can see from Fig. 2-10, the differentials of energy intensity have been steadily fluctuating between −0.2 to −0.8 except 1990, and appeared periodically. Therefore, it can not be concluded that the Chinese economy or energy statistics were wrong, simply because the energy consumption decreased sharply while the economy continued growing over 1997–1999. The rapid decrease of energy intensity after 1997 was reasonable allowing for the closing of small scale collieries, and therefore encouraged energy saving and substitution.

However, the continual decrease of energy intensity has not been found in other countries, so it is meaningful to study the factors behind the phenomena, not only for China's future energy economy but also other developing countries' economy development.

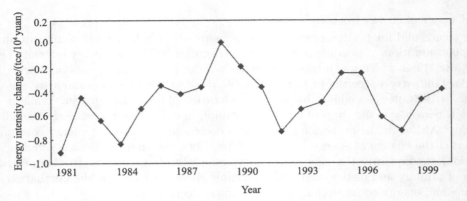

Fig. 2-10 Change of the integrated energy intensity

2.2.2.2 Decomposition of China's Energy Intensity

(1) Structure and efficiency share.

Using the method above and historical statistics, we studied the efficiency and structure share in the changes of China's energy intensity. Table 2-17 and Fig. 2-11 display the decomposition results.

Table 2-17 Structure share and efficiency share in change of China's energy intensity

Year	1981	1982	1983	1984	1985	1986	1987	1988	1989	1990
Structure share	0.32	0.47	0.11	0.23	0.12	−0.78	0.04	−0.04	2.02	0.56
Efficiency share	0.68	0.53	0.89	0.77	0.88	1.78	0.96	1.04	−1.02	0.44
Year	1991	1992	1993	1994	1995	1996	1997	1998	1999	2000
Structure share	−0.11	−0.31	−1.03	−0.10	−0.70	−0.18	−0.07	0.08	−0.01	−0.09
Efficiency share	1.11	1.31	2.03	1.10	1.70	1.18	1.07	0.92	1.01	1.09

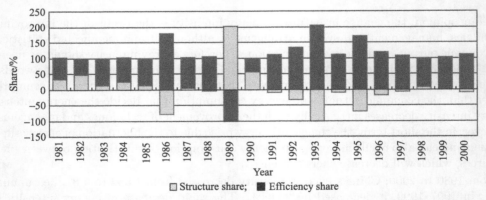

Fig. 2-11 Structure share and efficiency share in change of China's energy intensity

Drawing on the results, the efficiency improvement has always been the chief reason of intensity decrease, except for the economic wave around 1989. After 1993, the values of structure share were mostly negative, i.e., the change of economic structure actually pushed an intensity increase. In 1980–2000, the sum of structure share is −1.76%, while the sum of efficiency share is 101.7%, which means that the decrease of energy intensity in 1980–2000 was totally produced by the efficiency improvement of industries.

(2) The efficiency shares of three industries.

According to the result above, the decrease of China's energy intensity for 1980–2000 was completely produced by the efficiency improvement of three industries. Therefore, it is

2.2 Energy Intensity and Economic Structure of China

necessary to make an analysis of the efficiency share of different industries.

Based on the formula of efficiency share Eq. (2-17),

$$\frac{(e_i^n - e_i^{n-1}) \cdot y_i^n}{\sum_i e_i^n \cdot y_i^n - \sum_i e_i^{n-1} \cdot y_i^{n-1}} \quad (i = 1, 2, 3; n = 1, 2, \cdots, N)$$

Where: Y_i^n represents the efficiency share of the industry (i) in time (n). Thereby the efficiency share of three industries can be calculated.

Fig. 2-12 displays the results. We can see that, after 1985 (except for the fluctuation around 1989), the efficiency share of the secondary industry was mostly above 100%. Therefore, the reason of intensity decrease can be further attributed to the efficiency improvement of the secondary industry. For 1991–2000, the average efficiency share of the secondary industry was 124%, which counteracted the negative impacts of other industries, and pushed the continual and rapid decrease of aggregate energy intensity.

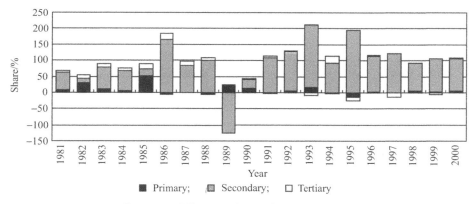

Fig. 2-12 Efficiency shares of three industries

2.2.3 Impact of the Secondary Industry Structure on Energy Intensity

The study above has proved that the main drive of the decrease of China's energy intensity came from improvement in industrial efficiency, within which the efficiency share of the secondary industry played a dominant role. Therefore, further study should focus on the reasons of the efficiency improvement of the secondary industry. Which factor has pushed the improvement of secondary industry efficiency, the change of subindustries structure or the efficiency improvement of subindustries? This is the research question of this section.

After 1998, China adjusted the statistics of the outputs of secondary industry, so the data before and after 1998 are not comparable. Therefore, the study on the subindustries' structure and efficiency only adopted the data between 1998 and 2002.

According to the *Chinese Statistical Yearbook*, the secondary industry is divided into 36 subindustries. The outputs of subindustries are calculated in nominal price, so the value of energy intensity in this section is different from that in previous sections. However, because the study only compares the value between subindustries, the price has little significant effect on the results.

2.2.3.1 The Secondary Industrial Structure and Energy Intensity

(1) The secondary industry's energy intensity.

From Fig. 2-13, we can see that most subindustries' energy consumption increased over 1998–2002, among which the subindustry of "Mining and Processing of Ferrous Metal Ores" increased the most, from 170.12 million tce in 1998 to 193.27 million tce in 2002. While nine subindustries' energy consumption decreased, among which the subindustries of "Mining and Washing of Coal", "Manufacture of Foods", "Manufacture of Beverages", "Mining and Processing of Nonmetal Ores" decreased the most, 13.03, 2.84, 1.72, and 10.10 million tons coal equivalent respectively.

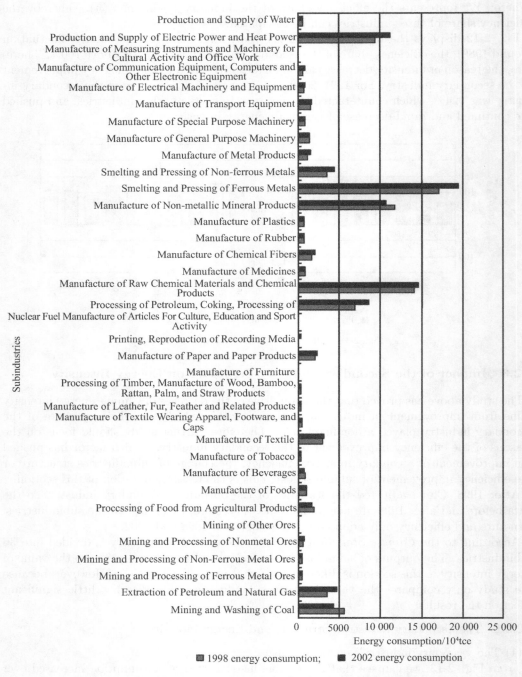

Fig. 2-13 Energy consumption of subindustries in 1998 and 2002

2.2 Energy Intensity and Economic Structure of China

While the total energy consumption increased from 1998 to 2002, all the subindustries' energy intensities decreased (Fig. 2-14). The subindustry of "Mining and Processing of Ferrous Metal Ores" decreased the most, from 17.31tce/RMB10^4 yuan in 1998 to 10.74 tce/RMB10^4 yuan in 2002. Besides this, the subindustries of "Mining and Washing of Coal", "Mining and Processing of Nonmetal Ores", "Manufacture of Raw Chemical Materials and Chemical Products", and "Processing of Petroleum, Coking, Processing of Nuclear Fuel" all decreased more than 4tce/RMB10^4 yuan, and played important role in pushing the decrease of secondary industry's energy intensity.

(2) The secondary industry's structure.

During 1998–2002, the outputs of almost all the subindustries, except "Processing of Timber, Manufacture of Wood, Bamboo, Rattan, Palm and Straw Products" had increased (Fig. 2-15). Some of them, e.g., "Manufacture of Transport Equipment" and "Manufacture of Communication Equipment, Computers and Other Electronic Equipment" increased more than 100%, and "Processing of Petroleum, Coking, Processing of Nuclear Fuel", "Extraction of Petroleum and Natural Gas", "Manufacture of Raw Chemical Materials and Chemical Products", "Mining and Processing of Ferrous Metal Ores", "Manufacture of Electrical Machinery and Equipment" increased 80%. Increasing of these subindustries formed the main drive for the growing of the secondary industry.

When the outputs increased, the structure of the secondary industry did not change significantly (Fig. 2-16). "Manufacture of Transport Equipment" and "Manufacture of Communication Equipment, Computers and Other Electronic Equipment" changed the most, but still only 1.04% and 1.88%. While the other subindustries all changed no more than 1%.

2.2.3.2 Decomposition of Energy Intensity of the Secondary Industry

(1) Structure share and efficiency share.

Using the statistics and decomposition method proposed above, we can calculate the efficiency share and structure share in the change of Chinese secondary industry's energy intensity over 1998–2002, i.e. 1.87% and 98.13%, respectively (Fig. 2-17). This is the same as the decomposition result in 2.2.2, i.e., the decrease of energy intensity, both of the macro economy and the secondary industry, can be attributed almost entirely to the efficiency share, and the structure change had little influence on the decease of energy intensity.

Since the decrease of the secondary industry's energy intensity for 1980–2000 was mainly produced by the efficiency improvement of the subindustries, it is necessary to make an analysis of the efficiency share of the subindustries.

(2) The industrial efficiency and structural transformation, and the comparisons between their impact on energy intensity.

According to decomposition method,

$e_i^0 \cdot (y_i^n - y_i^0)$ is the change of energy intensity due to structure share of industry(i),

$(e_i^n - e_i^0) \cdot y_i^n$ is the change of energy intensity due to efficiency share of industry(i).

Thereby we can calculate the change of energy intensity due to structural share and efficiency share of each industry.

According to the results displayed as Fig. 2-18, the structure share of most subindustries is small. Structural change of some subindustries, such as "Manufacture of Non-metallic Mineral Products", "Manufacture of Chemical Fibers", "Manufacture of Textile", "Manufacture of Beverages", and "Mining and Washing of Coal", had pushed the decrease of energy intensity, but counteracted by some others, such as "Manufacture of Communication Equipment, Computers and Other Electronic Equipment", "Manufacture of Transport Equipment", "Mining and Processing of Ferrous Metal Ores", "Mining and Processing of

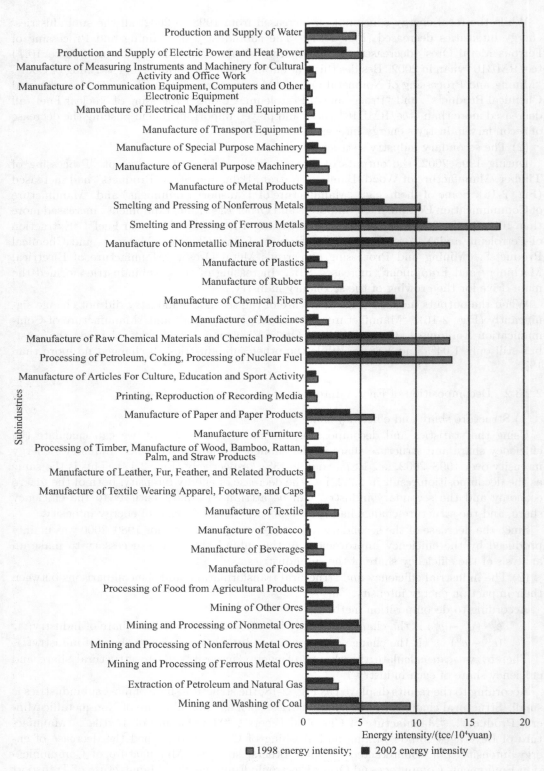

Fig. 2-14 Energy intensity of subindustries in 1998 and 2002

2.2 Energy Intensity and Economic Structure of China

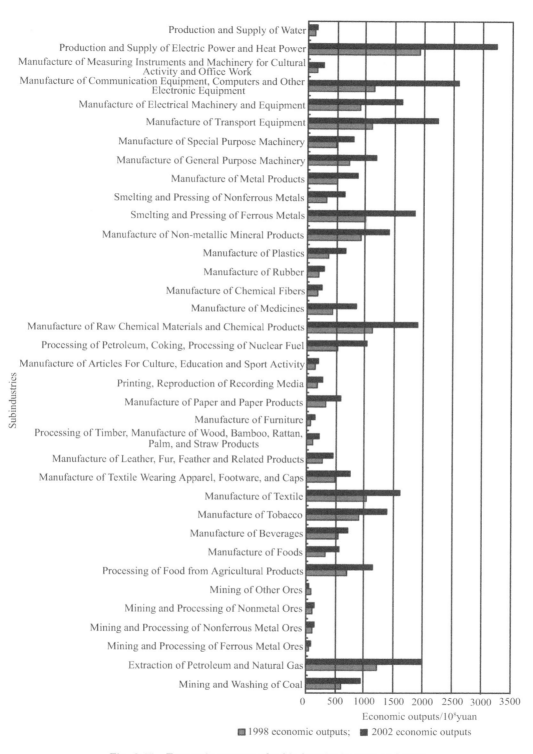

Fig. 2-15 Economic outputs of subindustries in 1998 and 2002

Chapter 2 Structural Relationship between Chinese Energy and Economic Growth

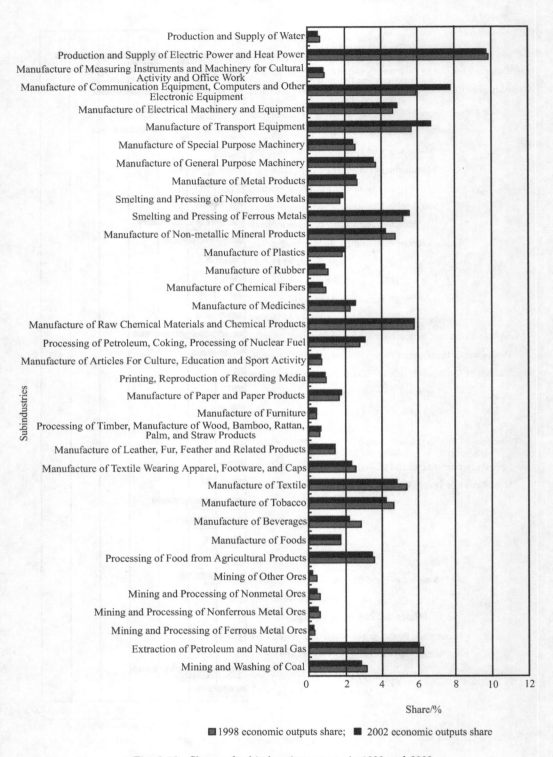

Fig. 2-16 Shares of subindustries outputs in 1998 and 2002

2.2 Energy Intensity and Economic Structure of China

Nonferrous Metal Ores", and "Processing of Petroleum, Coking, Processing of Nuclear Fuel". In total, the structural share in the change of the secondary industry's energy intensity is only 1.87%.

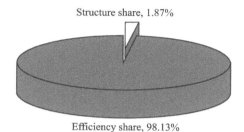

Fig. 2-17 Structure and efficiency share in the change of energy intensity for the secondary industry

On the other hand, the efficiency of all subindustries increased remarkably, and six of them formed the main driver in pushing the decease of energy intensity, including "Mining and Processing of Ferrous Metal Ores", "Mining and Processing of Nonmetal Ores", "Manufacture of Raw Chemical Materials and Chemical Products", "Processing of Petroleum, Coking, Processing of Nuclear Fuel", "Mining and Washing of Coal", and "Production and Supply of Electric Power and Heat Power". The efficiency share of these six subindustries contributed 71.85%, while the other 30 contributed only 26.38%. From the results, we can find that, to keep the energy intensity decreasing, the key is to push the efficiency of these six subindustries to continually improve.

2.2.4 Conclusions

2.2.4.1 Discussion of the Results

According to the study of the structural decomposition analysis in this section, the structural shares of industries and subindustries are both very small. The structural change over 1980–2000 had pushed the aggregate energy intensity to an increase of 1.76%; while the substructural change, although it did provide positive impact in the decrease of energy intensity of the secondary industry, had a share of only 1.87%. Therefore, the structural change, in both industry and subindustry level, did not become a main driver for decreasing energy intensity.

The results make clear that the drive of Chinese energy intensity's decrease for the most part came from the improving of efficiency. The efficiency share was 101.76% in the aggregate energy intensity's change and 98.13% in the secondary industry.

The efficiency improvement of the secondary industry played the most important role in pushing the decreasing of energy intensity. The efficiency share of the secondary industry had been beyond 100% continuously until after 1985, and an average 124% after 1990. This clearly indicates that, since the structural share tended to lead the intensity to increase and the efficiency share of the other industries had just been fluctuating, the rapid and continual decrease of energy intensity was entirely due to the efficiency improvement of the secondary industry.

Within the secondary industry, six subindustries formed the main driver in pushing the decease of energy intensity. The efficiency share of them is a total of 71.85%, almost three times the efficiency share of the other 30 subindustries.

Fig. 2-18 Structure and efficiency share of subindustries in change of energy intensity of the secondary industry

2.2.4.2 Policy Implication

(1) Potential of structure improvement is to be explored to decrease energy intensity. The study in this section proved that the structure efficiency does little to decrease energy

intensity, which disproves the viewpoint that the change of economic structure helps decrease the energy intensity after the Opening-up and Reform in the early 1980s. However, it also implies that there is much potential unexplored in improving the structure, and so decrease the energy intensity. Therefore, for Chinese energy strategies and policies, more attention should be paid to improving structure. In the future, with any decrease of Chinese energy intensity, the improved structure should play an as important role, if not more, as the efficiency increasing.

(2) China should change its economy development direction from heavy industrialization to high tech industrialization.

After 1990, secondary industry increased steadily in the structure of the Chinese economy, while the tertiary industry did not grow so much. This indicates that the change of Chinese economic structure was still in the direction of heavy-industrialization, which resulted from the foreign techniques transfer, and the Chinese economic development strategy at present. The energy intensity of secondary industry is much higher than in the other industries, so the rapid growth of the secondary industry had inevitably absorbed the cost of much more energy consumption. With the increasingly hot competition in energy supply, such an economic development route is harder and harder to sustain. In order to secure the economic development target of double GDP in 2020 than in 2000, the Chinese economic strategy is required to be adjusted. China should make more effort to develop its tertiary industry, especially the high-tech industry based on new technologies, so that the economic development can be continually supported by lower growth in energy intensity and energy consumption.

(3) At present, management creation and technical learning are still the main driver of decreasing energy intensity.

Increase of energy efficiency is derived from technology and management's improvement. As the analysis of 2.1 indicates, the management creation and technical learning have kept the Chinese economy growing rapidly with slowly increasing energy consumption. At present, efficiency improvement based on management creation and technical learning is still the main driver of decreasing energy intensity, although the potential of it is gradually declining. China should make more effort to encourage the management creation and technical learning to thrust its energy intensity to decrease continually, and thereby control the fast growth of energy consumption.

(4) The efficiency improvements in six subindustries are the key factors of the decease of energy intensity, on which China should focus more attention.

From the decomposition of the energy intensity change of secondary industry, we can see that the efficiency improvement in six subindustries, including "Mining and Processing of Ferrous Metal Ores", "Mining and Processing of Nonmetal Ores", "Manufacture of Raw Chemical Materials and Chemical Products", "Processing of Petroleum, Coking, Processing of Nuclear Fuel", "Mining and Washing of Coal", and "Production and Supply of Electric Power and Heat Power", are the key factors for the decease of energy intensity. However, there is still a wide gap between the efficiency of subindustries in China and that of developed countries. So China should focus more attention on narrowing the gap, which will be the most potential and effective way in decreasing China's energy intensity.

2.3 Energy Structure and Energy Efficiency in China

The relationship between energy input and economic output can be studied from two directions: energy intensity or energy efficiency. This reflects the identical measure of the relationship between energy consumption and economic growth from different perspectives. Energy intensity, from the perspective of energy demand, represents the extent to which economic outputs consume energy resources (energy input/economic output), while energy efficiency, from the perspective of factor supply, represents the extent to which energy resources support economic output (economic output/energy input). In our opinion, it is more suitable to apply energy structure analysis from the perspective of energy efficiency.

Energy efficiency, defined as the economic output produced by a unit of energy input, represents the extent to which energy resources support economic output. Intrinsically, the increase of energy efficiency is driven by the technology improvement. However, because of the different ability of different energy resources in supporting economic output, the aggregate energy efficiency can also be affected by the change of energy structure.

In the area of Chinese energy economy, relevant empirical studies were mainly conducted from the perspective of economic structure and energy intensity, while there are few empirical studies on the relationship between energy structure and energy efficiency. Moreover, relevant studies have primarily focused on the aggregate energy efficiency, but rarely touched on the marginal efficiency of energy resources and their comparative substitution rate.

There are a number of studies focusing on energy productivity, such as the analysis of improved energy efficiency resulting from an energy structural shifting (Edenhofer, 1998), the relationship between energy efficiency and economic structure in Holland (Farla, et al., 1998), the interaction between British industry structure and energy efficiency (Jenne, 1983), to name but a few. Such studies have largely related energy efficiency to economic structure, rather than energy structure. In China, there are also a few studies on energy efficiency. For example: Zhu (2004) studied the influence of electricity development on improving energy efficiency; Wang (2003) made a comparison between Chinese energy efficiency and selected countries; and Wen (2001) discussed the market incapability and government's role in improving energy efficiency. However, these studies rarely gave quantitative analysis to support their conclusions.

Thus it can be seen that there are still few studies focusing on the energy structure and energy efficiency, although it is an important part of the study on energy economy. In the period of 1978–2003, China's energy efficiency constantly increased at a rapid ratio, which has attracted broad attention. Undoubtedly, the study on the Chinese energy economy from the perspective of energy structure and energy efficiency is significant for both China and other developing countries. This section reports our empirical study on this topic.

Our study firstly made a brief review of the economic development and energy structure of China from 1978 to 2003, then analyzed the impact of China's energy structure on its energy efficiency, and finally made a primary estimation of the marginal efficiency (ME) of coal, petroleum and electricity in China, as well as the marginal rate of substitution (MRS) between them.

2.3.1 China's Energy Economy and Energy Efficiency

2.3.1.1 Economic Output and Its Growth in China

From 1978 to 2003, China maintained a continuous, stable economic growth. Its real GDP at the price of 1978 has increased 840% in total, with average annual growth rate of 9.38%. In 1984, 1987 and 1993, the Chinese economy had three waves of fast growing, and the annual growth rate of real GDP reached 15%, 11% and 14%, respectively. After 1997, the Chinese government introduced a series of policies to control its economic growth; consequently the growth rate was kept between 7% and 9%. However, in 2003, the Chinese economy showed a new flourish of rapid growth, and the annual growth rate once again exceeded 9% (Fig. 2-19).

2.3.1.2 Energy Consumption and Its Growth in China

When China's economy maintained a rapid growth, its energy consumption increased at

2.3 Energy Structure and Energy Efficiency in China

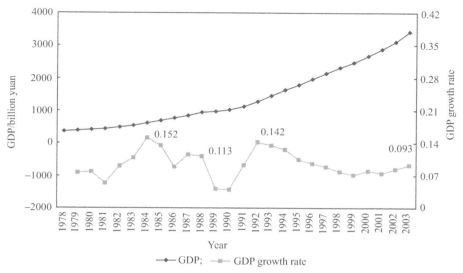

Fig. 2-19 Real GDP and growth rate of China in 1978–2003

a much lower rate. Before 1996, the annual growth rate of China's energy use was generally between 5% and 6%. In 1996, China's total energy use reached its peak at 1389.48Mtce, and from then kept decreasing over the next three years at an average annual rate of −2.16%. However, after 2000 its energy consumption started to increase at an accelerated rate. From 2000 to 2003, the annual growth rate was 0.14%, 3.54%, 9.86% and 13.21%, respectively (Fig. 2-20).

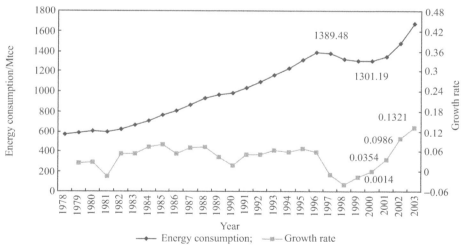

Fig. 2-20 Chinese energy consumption and its growth rate

2.3.1.3 Trend of Chinese Energy Efficiency

After 2001, the continual improvement of China's energy efficiency ended after a 23-year period. Between 1978 and 2001, China's aggregate energy efficiency increased from 0.0634 yuan/tce to RMB 0.2133×10^4 yuan/tce. However, in 2002 and 2003, it decreased from its peak to RMB 0.2103yuan/tce and RMB0.2030×10^4yuan/tce successively (Fig. 2-21).

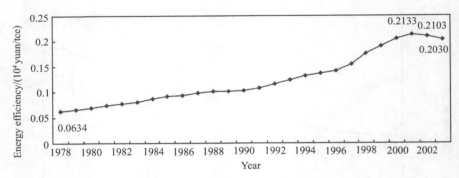

Fig. 2-21 Chinese energy efficiency

The continual increasing of China's energy efficiency over 23 years had resulted in many criticisms about its stability and reliability. With the drop of the energy efficiency in recent years, the criticisms gradually declined. However, the reasons behind such a drop need to be seriously studied.

Energy efficiency represents the economic output produced by a unit energy input, which estimates the extent to which energy resources support economic output. Intrinsically, it reflects the improvement of management and technology, so it should be an exogenous variable driven by the level of management and technology of an economic system. If there was no notable decline of the management and technology productivity of China in such a short period, then what factors had pushed the decrease of China's energy efficiency after 2000? Undoubtedly, the analysis on this topic is meaningful for China to increase its energy efficiency in the future.

2.3.2 Empirical Study on the Impact of China's Energy Structure on Energy Efficiency

2.3.2.1 Model

In statistics, energy efficiency is calculated as the ratio of economic outputs and energy resources inputs, i.e.

$$ef = \frac{Y}{E} \tag{2-20}$$

Where: Y means the economic outputs (GDP); E means the energy consumption.

According to this formula, although energy efficiency is theoretically decided by management and technology, and thus is an exogenous variable, in realistic statistics it is calculated as the ratio of economic outputs and energy resources inputs, and becomes an endogenous variable. Therefore, some other factors, for example, energy structure or economic structure, could indirectly influence energy efficiency.

In Chinese resource statistics, the total primary energy use includes four resources: coal, petroleum, natural gas, and hydro and nuclear power. Dissimilar energy types are converted into coal equivalent and then summed together as total energy inputs. This method of quantifying total energy use has an implicit presumption that different energy resources of the same coal equivalent would produce the same economic outputs. However, this presumption is not always the case in a realistic economic system.

A practical economic system comprises different industries, which are supported by diverse energy resources. For example, if high-tech industries are mainly supported by one energy resource, low-tech industries are most probably supported by a different one. Generally speaking, high-tech is associated with high value added and low-tech is usually linked with

2.3 Energy Structure and Energy Efficiency in China

low value added. Therefore, the energy resource supporting high-tech industries will produce more economic output compared to the energy resource supporting low-tech industries (even if they are the same coal equivalent).

Let E_i ($i = 1, 2, 3, 4$) represent the input of different energy resources, i.e., coal, petroleum, natural gas, and hydro and nuclear power respectively; and Y_i ($i = 1, 2, 3, 4$) represent the economic output supported by different energy resources. Then the energy efficiency of different energy resources can be formulated as:

$$ef_i = \frac{Y_i}{E_i} \quad (i = 1, 2, 3, 4)$$

Therefore, the total economic output is:

$$Y = \sum_i ef_i \cdot E_i \quad (i = 1, 2, 3, 4) \tag{2-21}$$

Finally, the aggregate energy efficiency ef_i can be formulated as:

$$ef = \frac{Y}{E} = \frac{\sum_i ef_i \cdot E_i}{E} = \sum_i ef_i \cdot S_i \quad (i = 1, 2, 3, 4) \tag{2-22}$$

Where: S_i ($i = 1, 2, 3, 4$) represents the share of different energy resources in total energy inputs.

Because the data of Y_i ($i = 1, 2, 3, 4$) is not available from practical statistics, the variable ef_i in Eq. (2-22) cannot be calculated with the method of SDA proposed in 2.2. Therefore, a multi-variables regression model is used to analyze the impact of energy structure on aggregate energy efficiency.

To allow for the increase of energy efficiency due to improvement of technology and management, the time variable t is introduced into the model. It is assumed that the efficiency increase due to improvement of technology and management is linear with the time variable t.

Let ef be the dependent variable, and let t and s_i ($i = 1, 2, 3, 4$) be independent variables, the regression model is set up as:

$$ef = b \cdot t + \sum_i a_i \cdot s_i + \varepsilon \quad (i = 1, 2, 3, 4) \tag{2-23}$$

Where: b and a_i ($i = 1, 2, 3, 4$) are the coefficients of the regression equation.

2.3.2.2 Impact of Energy Structure on Energy Efficiency

Using the data of China's energy structure (Fig. 2-22) and real GDP from 1978 to 2003, we get a regression equation:

$$ef = 0.004\,27t - 0.000\,634S_c + 0.003\,632S_o - 0.001\,306S_g + 0.008\,199S_e + \varepsilon$$

$$(t = 10.73^{**})(t = -8.14^{**})(t = 4.78^{**})(t = -0.27)(t = 3.94^{**})$$

$$R^2 = 0.9933, \quad F = 773.01^{**} \tag{2-24}$$

Where: S_c, S_o, S_g, S_e represent the share of coal, oil, gas, and hydro and nuclear power in total energy inputs respectively.

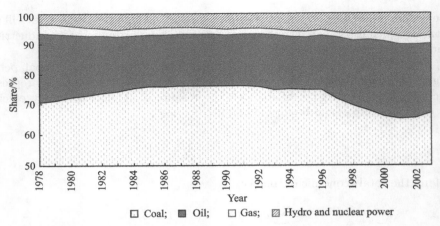

Fig. 2-22 Change of Chinese energy structure

In Eq. (2-24), although the coefficients of t, S_c, S_o, S_e and the equation all passed the significance test at 1% significance, the coefficients of S_g failed to pass the test. This infers that there is no significant correlation between the aggregate energy efficiency and S_g, the share of natural gas in total energy inputs. Therefore, exclude S_g from the equation and set up the model as:

$$ef = b \cdot t + \sum_i a_i \cdot s_i + \varepsilon \qquad (i = 1, 2, 4) \tag{2-25}$$

Still using the data above, we get a new regression equation.

$$ef = 0.00427t - 0.000632S_c + 0.003451S_o + 0.008087S_e + \varepsilon \tag{2-26}$$

$$(t = 15.00^{**})(t = -8.33^{**})(t = 9.90^{**})(t = 4.05^{**})$$

$$R^2 = 0.9932, \quad F = 1076.03^{**}$$

The actual, fitted, and residual values are displayed as Fig. 2-23.

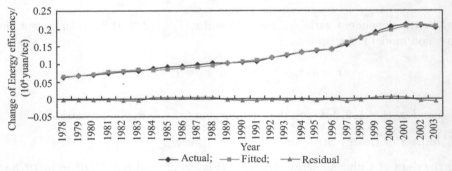

Fig. 2-23 Actual, fitted, and residual of Eq. (2-26)

Based on Eq.(2-26), the change of energy efficiency can be formulated as:

$$\Delta ef = 0.004\,341\Delta t - 0.000\,632\Delta S_c + 0.003\,451\Delta S_o + 0.008\,087\Delta S_e + \Delta\varepsilon \tag{2-27}$$

Where: t represents energy efficiency's change due to technology and management improvement, and ε represents the impact of random factors. Therefore, the influence of energy structure on aggregate energy efficiency, Δef_s, can be formulated as:

$$\Delta ef_s = 0.000\,632\Delta S_c + 0.003\,451\Delta S_o + 0.008\,087\Delta S_e \tag{2-28}$$

2.3 Energy Structure and Energy Efficiency in China

According to the Eq.(2-28), the impact of China's energy structure on energy efficiency can be calculated for each year during the period 1978–2003, which is displayed in Fig. 2-24.

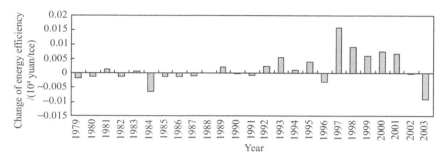

Fig. 2-24 Energy efficiency's change due to energy structure in 1978—2003

In the Eq. (2-26), the coefficients of S_c, S_o, S_e reflect the extent in which the change of energy structure influenced China's energy efficiency.

$$ef = 0.004\ 27t - 0.000\ 632S_c + 0.003\ 451S_o + 0.008\ 087S_e + \varepsilon$$

We can see from Eq. (2-26), the increase of share in coal will decrease the energy efficiency, while the increase of share of petroleum and hydro and nuclear power will make it increase. This implies that a shift of energy structure from coal to petroleum and hydro and nuclear would improve the aggregate energy efficiency in China.

2.3.3 ME and MRS in Chinese Energy Economy

It is important to estimate the efficiency of energy resources, not only for research, but also for policy development. However, limited by the available statistics, most relevant studies focus on the aggregate energy efficiency rather than marginal efficiencies of a certain energy resource. Similarly, the MRS between different energy resources has hardly been studied. This section will make an initial estimation of the ME and MRS of coal, petroleum and electricity in China, using the data from 1978 to 2003.

2.3.3.1 Marginal Efficiency

(1) ME of coal.

There are numerous factors involved in producing economic outputs, of which energy is just one. Thus, if the model is developed with only economic output and energy input, the role of energy will inevitably be exaggerated. If economic output and other factors, such as capital, labor, energy, amongst others can be included in the model, it would be more plausible. Because the inputs of capital, labor, and other factors are basically decided by the economic outputs in the previous year, the economic outputs in the previous year representing the inputs of all factors (except energy) have been used.

Based on the analysis above, the model is:

$$dY_t = a_1 + a_2 Y_{t-1} + a_3 dC_t + \varepsilon \quad (2\text{-}29)$$

Where: dY_t represents the increment of the economic outputs (GDP) between the year (t) and $(t-1)$; Y_{t-1} represents the economic outputs (GDP) in the year $(t-1)$; dC_t represents the increment of the input of coal between the year (t) and $(t-1)$ and a_1, a_2, a_3 are coefficients.

Using the data of China from 1978 to 2003, the following regression equation is derived,

$$dY_t = 80.2017 + 0.0786Y_{t-1} + 0.0307dC_t + \varepsilon \quad (2\text{-}30)$$

$$(t = 0.7448)\ (t = 11.8903^{**})\ (t = 2.4053^{**})$$

$$R^2 = 0.8751, \quad F = 77.04^{**}$$

In Eq. (2-30), the constant a_1 cannot pass the significance test. Therefore, it was excluded from the equation and the model became:

$$dY_t = a_1 Y_{t-1} + a_2 dC_t + \varepsilon \tag{2-31}$$

Using the same data, the following regression equation is derived:

$$dY_t = 0.0825 Y_{t-1} + 0.0332 dC_t + \varepsilon \tag{2-32}$$

$$(t = 20.3922^{**})\quad (t = 2.7239^{*})$$

$$R^2 = 0.8719, \quad F = 156.56^{**}$$

According to Eq. (2-32), all the coefficients and the equation as a whole pass the significance test. The actual, fitted and residual data are displayed in Fig. 2-25.

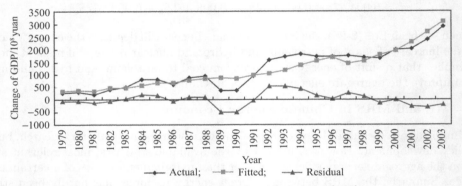

Fig. 2-25 Actual, fitted, and residual of Eq. (2-32)

In Eq. (2-32), the coefficient of dC_t can be regarded as an estimation of the ME of coal input, i.e., 1 tce marginal input of coal can produce economic output of RMB 0.0332×10^4 yuan. Energy efficiency is changing with time, so this estimation is an average of the changing energy efficiency over 1978–2003. If it is presumed that energy efficiency is linear with time, this estimation would be equal to the actual efficiency at the end of 1990, that is to say the middle point of the period 1978–2003.

(2) ME of petroleum.

Based on the same method, the model to estimate the ME of petroleum is:

$$dY_t = a_1 Y_{t-1} + a_2 dO_t + \varepsilon \tag{2-33}$$

Where: dO_t represents the increment of the input of petroleum between the year (t) and $(t-1)$.

Using the same data, we got the following regression equation:

$$dY_t = 0.0743 Y_{t-1} + 0.1782 dO_t + \varepsilon \tag{2-34}$$

$$(t = 10.0837^{**})\quad (t = 2.1669^{*})$$

$$R^2 = 0.8593, \quad F = 140.48^{**}$$

Fig. 2-26 shows the actual, fitted, and residual data.

2.3 Energy Structure and Energy Efficiency in China

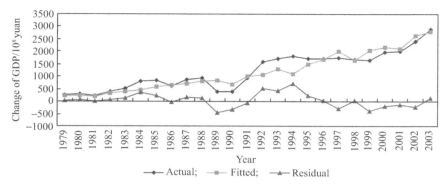

Fig. 2-26 Actual, fitted, and residual of Eq.(2-34)

Similarly, the coefficient of dO_t in Eq. (2-34) can be thought of as an estimation of the ME of petroleum input, i.e., 1 tce input of petroleum resource can produce economic output of RMB0.1782×10^4yuan.

(3) ME of electricity.

In the estimation of ME of electricity, we used the total electricity consumption, which came from the *Chinese Statistical Yearbook* and *Energy Statistical Yearbook*, and was converted to coal equivalent using the converting index of 0.1229 kgce/kWh.

The trend of electricity consumption is displayed as Fig. 2-27.

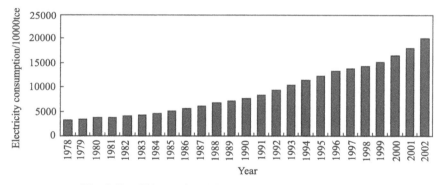

Fig. 2-27 Chinese electricity consumption in 1978–2002

Still using the method above, the model to estimate the ME of electricity is

$$dY_t = a_1 Y_{t-1} + a_2 dE_t + \varepsilon \tag{2-35}$$

Where: dE_t represents the increment of the input of electricity between the year (t) and ($t-1$).

The following regression equation is derived.

$$dY_t = 0.0563 Y_{t-1} + 0.5733 dE_t + \varepsilon \tag{2-36}$$

$$(t = 4.5584^{**}) \quad (t = 2.6377^*)$$

$$R^2 = 0.8375, \quad F = 113.40^{**}$$

Fig. 2-28 shows the actual, fitted, and residual data.

Similarly, the coefficient of dE_t in Eq. (2-36) can be thought of as an estimation of the ME of electricity input, i.e., 1 tce input of electricity can produce economic output of RMB0.5733×10^4yuan.

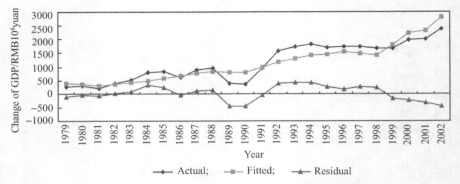

Fig. 2-28 Actual, fitted, and residual of Eq.(2-36)

2.3.3.2 MRS Between Coal, Petroleum, and Electricity

Based on the estimation above, the ME of coal, petroleum, and electricity in China in 1978–2003 are 0.0332×10^4/tce, 0.1782×10^4/tce and RMB 0.5733×10^4yuan/tce, respectively. Then the MRS between them can be calculated.

According to the formula of MRS:

$$\text{MRS} = \text{ME1}/\text{ME2} \tag{2-37}$$

The MRS between coal, petroleum, and electricity can be calculated as:

$$0.5733 : 0.1782 : 0.0332 = 17.27 : 5.38 : 1$$

i.e.,

MRS between electricity and coal is $0.5733/0.0332 = 17.27$
MRS between petroleum and coal is $0.1782/0.0332 = 5.38$
MRS between electricity and petroleum is $0.5733/0.1782 = 3.22$

Its economic implication is that the marginal input of petroleum and electricity of 1 tce can substitute the marginal input of coal of 5.38 tce and 17.27 tce, and the marginal input of electricity of 1 tce can substitute the marginal input of coal of 3.22 tce.

2.3.4 Conclusions

2.3.4.1 Discussion of Results

(1) Impact of energy structure on energy efficiency.

According to Fig. 2-24, the change of China's energy structure imposed a negative influence on its energy efficiency improvement during 1978–1991. In this period, the growth of the Chinese economy was an extensive growth, and the economic structure did not improve with the economic expansion. Without the push of the economy structure, the energy structure likewise did not improve. Therefore, the share of coal in total energy consumption increased in most years, and consequently counteracted the increase of energy efficiency.

In 1991–2000, China's energy structure started to change towards a more efficient direction, and thus played a positive role in improving its energy efficiency. Especially in 1997, the control on small-scale colliery decreased the energy supply, and therefore pushed the structure to change from coal to other energy resources. However, this effect declined in the following years.

(2) Degeneracy of energy structure behind the economic highly growing.

In 1984–1987, 1996 and 2003, the change of energy structure imposed considerable negative impact on energy efficiency. In most of these years, the Chinese economy appeared fast growing, and the annual growth rate of real GDP reached 10%. In 2003, the annual growth rate once again exceeded 9%, and the Chinese economy showed a new wave of rapid growth. A common feature of these years is that the share of coal notably increased, while the petroleum and electricity decreased, which led to the degeneracy of energy structure and consequently decrease of energy efficiency. This indicates the problem behind the high growth of the Chinese economy, i.e., lots of low efficiency energy were input into the economy, and the high growth was at the cost of efficiency.

(3) ME and MRS.

According to our study, the marginal efficiency of coal, petroleum and electricity in China during 1978–2003 are RMB 0.0332×10^4 yuan/tce, RMB 0.1782×10^4 yuan/tce and RMB 0.5733×10^4 yuan/tce respectively. The value is even lower than the present price of these energy resources. This can be due to two reasons, one is that the real GDP is in the price of 1978, another is that the value is an average estimation. The result would be equal to the actual efficiency at the middle point of the period 1978–2003, if it can be presumed that energy efficiency is linear with time.

This line of thought should be extended with further empirical study, as the methods and models adopted here are exploratory and experimental. On this theme, more effort is required to devise additional comprehensive methods and models to gain a more reliable estimation.

2.3.4.2 Policy Implication

(1) Increase production of petroleum and electricity.

From the study, it can be found that petroleum and electricity are more efficient energies than coal. Higher share of petroleum and electricity can increase the efficiency, while higher share of coal will decrease the efficiency. Therefore, China should build its strategy and policy to push the production and consumption of petroleum and electricity. Based on the energy resources of China, the direction should be focused on the electricity, especially hydro and nuclear electricity. China should make more efforts to increase the share of petroleum and electricity, push improvement to its energy structure, and help to increase its energy efficiency.

(2) Properly control the economic growth.

Generally, once China's annual economic growth rate exceeded 9%, its energy structure tended to become much less efficient. In 2003, China's annual economic growth rate exceeded 9% once more, and its energy efficiency decreased sharply. This implied that China's economy was most likely to start another wave of high growth but low efficient growth. Therefore, it is imperative for the Chinese central government to construct new policies to control and reform the economy. China should curb its economic growth rate to below 9%, so as to avoid reducing its energy efficiency.

(3) Encourage the transformation from coal to petroleum and electricity.

The marginal efficiency of coal, petroleum and electricity in China further indicates that petroleum and electricity are more efficient energies than coal. When driving the production of petroleum and electricity, China should also encourage the transformation from coal to petroleum and electricity. Such transformation is not only profitable for the electricity industry, but also beneficial for the improvement of energy efficiency.

2.4 Summary

Focusing on some fundamental questions in the Chinese energy economy, this chapter conducted empirical studies from three perspectives, and put forward conclusions.

The first section introduced the cointegration and causality into the study of China's energy economy and discussed the cointegration and causality between the economy and energy consumption over 1978–2003.

According to the test results, data of different periods lead to completely different conclusions. Using the data of 1978–2000, no cointegration relationship could be found between the economic growth and energy consumption. However, when the data were extended to 1978–2003, the conitegration relationship appeared in the Chinese energy economy.

In order to get more evidence about the Chinese energy economy, we extended our research into the level of three industries. The test results show that, there is significant cointegration relationship in all the three industries, and for the primary and tertiary industries, the Granger causality appeared to be from economy to energy consumption, while the secondary industry gave bi-direction.

The result above bears significant policy implications:

The different results of cointegration test imply the particularity of the energy economic data of 2000–2003. The change of the energy economy in the three years bears some special meaning for China, and is worth further study. The coexistence of rapid growth of the economy and slow increase of energy consumption over 1978–2000 will be most probably unsustainable in the future, and the long-term relationship between the economy and energy consumption will be of common rapid growth.

Study reveals the one-way causality from economy to energy consumption, which also means that the growth of the Chinese economy led to the increase of energy consumption. The drop of energy consumption did not necessarily lead to the fluctuation of the economy. This may be due to the control ability of the Chinese government. However, it appeared to be bi-directional causality in the secondary industry's energy economy, which means the growth of secondary industry has strong dependence on energy consumption. Hence, the Chinese government should pay more attention to the differences among the three industries, and implement distinct policies for different industries, so as to avoid the possible negative impact on the economic growth of secondary industry.

The second section conducts study on the impact of the economic structure's change on energy intensity.

Energy intensity, in nature, is driven by techniques, management, and system arrangement. From the 1980s, the reformation of system and management, as well as the introduction of new technology, has influenced the decrease of energy intensity. The decrease of energy intensity reflects the combination of China with the international economy. In the context of energy intensity, the trend and speed of its decrease has been basically consistent within the period. Therefore, it can not be concluded that the Chinese economy or energy statistics were wrong, just because the energy consumption decreased sharply while economy continued growing over 1997–1999.

Based on the SDA method, we decomposed the change of energy intensity into efficiency share and structure share, proposed formulas, and made empirical study using the data over 1980–2000. The results make clear that the drive of Chinese energy intensity decrease mainly came from improving efficiency, while within the secondary industry, six subindustries formed the main drive in pushing the decease of energy intensity. China should focus more attention on the efficiency and improvement of the six subindustries. This will give the most potential and be an effective way of decreasing the energy intensity. Meanwhile, China should make more effort to develop its tertiary industry, especially the high-tech industry based on new technologies, so that the economy development can be continually supported by lower energy intensity and less growth in energy consumption.

The third section established a model to analyze the impact of energy structure on energy efficiency, and conducted an empirical study using the data over 1978–2003.

2.4 Summary

According to our study, the change of China's energy structure imposed a negative influence on its energy efficiency during 1978–1991. While over 1991–2000, China's energy structure started to change towards a more efficient direction, and thus played a positive role in improving its energy efficiency. The marginal efficiency of coal, petroleum and electricity in China during 1978–2003 are estimated to be RMB 0.0332×10^4yuan/tce, RMB 0.1782×10^4yuan/tce and RMB 0.5733×10^4yuan/tce, respectively, and the MRS between them is 17.27 : 5.38 : 1. Based on our study, we suggest China designs its strategy and policy to push the production and consumption of petroleum and electricity, and also encourage the transformation from coal to petroleum or electricity.

We discussed the economic growth from the perspective of energy efficiency and energy structure. In 2003, China's annual economic growth rate exceeded 9% once more, and its energy efficiency decreased sharply, which implied that China's economy was most likely to start another wave of low efficient growth. Chinese central government should construct new policies to control and reform its economy and to curb its economic growth rate to below 9%, so as to avoid reducing its energy efficiency.

It must be acknowledged that the discussion raised in the third section should be extended with further empirical study, as the methods and models adopted here are exploratory and experimental. On this theme, more effort is required to devise additional comprehensive methods and models to gain more reliable estimations.

CHAPTER 3

Analysis and Forecasting of Energy Supply and Demands in China

Entering the 21st century, China's socialist-market economic system has taken shape, while the energy industry in China still remains in the process of transforming from a planned economy to a market economy. Such a transformation induces fundamental changes in how the energy strategy and policy is established. Demand and supply of energy can no longer be controlled by state planning. In this case, it is necessary to perform scientific and accurate analysis and forecasting on energy supply and demand.

Therefore, this chapter will forecast energy demands for all regions for China's rapidly developing society in the year 2020. Being aware of the situation that coal is, and will be, taking a major position in China's energy consumption, this chapter will shed special light on the coal supply situation in 2020. The analysis of this chapter focuses on:
- How much energy will be demanded in China in 2020?
- In what aspects can China control the rapid growth of energy demand?
- How to facilitate the effectiveness of inter-regional energy transfer?
- What will be the coal supply capacity during the period when China's building an affluent society in a holistic way?
- What has caused the overproduction of township coal mines in 2001–2010?
- How to reduce the production of township coal mines? How to solve the overload operation of state-owned coal mines?

3.1 Forecasting of Energy Requirements in 2020

The economy of China is under intensive development. The coming years, until 2020, will be the most important period for China to build a "well-off society" in an "allaround way". The government set the objectives that "On the basis of optimized structure and better economic returns, efforts will be made to quadruple the GDP of the year 2000 by 2020" and "in the main achieve industrialization by 2020". During this period, with the improvement of people's living standards and the accelerated advancement of urbanization, the consumption of household electric appliances will increase rapidly, and more and more cars will come into families; all these will consequently bring the increase of household energy consumption. The object "in the main achieve industrialization by 2020" implies that the manufacturing scale will still go on expanding. At the same time the accelerated advancement of urbanization will greatly speed up the development of services and transportation. As a result, production energy consumption in this period will also mount up significantly.

On the other hand, with a vast territory, regions in China are different in geography, economy, population, industry structure, etc. In particular, there are prominent differences between the most developed coastal regions and the central and western part of China, which further lead to obvious discrepancies between energy reliance and energy requirement structure among these regions. In order to facilitate harmonious development of energy and economy structure in all regions, and to avoid the biases caused by only taking scenarios at the national level into account, it is of great importance to further study the future energy demands at the regional-level as well as at the national level (Wei, et al., 2006). Such a study is of great help to support regional energy development programs according to local conditions, and to harmonize resource advantages among regions.

Based on an input-output method, this study aims to perform a scenario analysis of China's future energy requirements at both national and regional level for year 2010 and 2020, under different growth paths of major driving forces of energy consumption.

3.1.1 Present Research on Energy Requirement

Up to now there have been many studies focusing on future energy requirements (Clinch, et al., 2001; Gielen, et al., 2002; Hsu, et al., 2004; Roca, et al., 2001; Silberglitt, et al., 2003). Similar studies on China include presenting forecasts of energy consumption and related emissions (Lu, Ma, 2004; Chen, 2005; Crompton, Wu, 2005; Gielen, Chen, 2001; Han, 2005); analyzing strategies for developing a sustainable energy system (Ni, Thomas, 2004; Qu, 1992; Wu, Li, 1995; Xu, et al., 2002); assessing impact of driving forces on historical emissions. (Wu, et al., 2005; Zhang, 2000); exploring various types of energy technology (Eric, et al., 2003; Feng, et al., 2004; Mu, et al., 2004; Solveig, Wei, 2005; Wu, et al., 1994; Yan, Kong 1997); and energy efficiency standards (Lang, 2004; Yao, et al., 2005). Most of these studies focused either on a small region or on China as a whole. As an exception, Han (2005) divided China into six regions and forecast up to year 2015 for petroleum supply and demand for each region. However, his study primarily focused on petroleum and did not shed light on coal, which accounts for the largest share of total primary energy consumption in China. Therefore, in this study all the three fossil fuels are taken into account, including coal, crude oil and natural gas.

3.1.2 Scenario Analysis and Input–Output Analysis

3.1.2.1 Scenario Analysis

The energy system is a complex giant system, with many uncertainties due to the driving forces of energy requirements, such as the economy, population, technology, etc. The traditional trend-extrapolation approach works only when the changes of driving forces insistently follow the established paths, but can shed little light on the case of driving forces moving in a brand-new orbit, e.g., certain risks or challenges baffle economic development.

The current popular scenario analysis operates in a different way in that "it does not try to predict the future but rather to envision what kind of future is possible"(Silberglitt, et al., 2003). Through the description of various possible future scenarios representing different growth paths, driving forces' uncertainties can be taken into account.

3.1.2.2 Input-output Method

In this study, a multi-regional input-output model is developed to assess China's energy requirements under each scenario.

The IO model (Miller, Blair, 1985) is an analytical framework developed by Professor Wassily Leontief in the late 1930s. The main purpose of the input-output method is to establish a tessellated input-output table and a system of linear equations.

An input–output table shows monetary interactions or exchanges between the economic sectors and therefore their interdependence. The rows of an IO table describe the distribution of a sector's output throughout the economy, while the columns describe the inputs required by a particular sector to produce its output (Miller, Blair, 1985). The system of linear equations also describes the distribution of a sector's output throughout the economy mathematically, i.e., sales to processing sectors as inter-inputs or to consumers as final demand.

Studies of energy-related problems, based on IO models, date back to the late 1960s. Some experts extended input-output analysis to energy and related environmental issues. In the 1970s and 1980s some energy analyzers further expanded this technology (Miller, Blair,

1985). In recent years much attention has been given to these issues. Some of these studies perform sensitivity analyses on one or more social and economic factors (Kainuma, et al., 2000; Kim, 2002; Lee, Lin, 2001; Paul, Bhattacharya, 2004; Yabe, 2004). Other researchers primarily analyzed the impact of certain economic activities or governmental policies on energy consumption, such as the impact of international trade, and the effects of certain policy reforms or frameworks (Bach, et al., 2002; Christodoulakis, et al., 2000; Sánchez-Ch. Óliz, Duarte, 2004).

Above shows the basic form of the input-output model. When more than one region is taken into account, the basic IO model needs to be extended to encompass multiregion modeling.

Many-region IO models include the interregional IO (IRIO) model and a set of simplified models. The first presentation of the IRIO model is in Isard (Miller and Blair, 1985). The basic IRIO model has high requirements for statistical data, which needs a complete IRIO tableau. Without a precise statistical system, a large-scale investigation using sizeable manpower and material resources will be needed to generate such a data set. What is more, sometimes the reliability of the data from such an investigation is not guaranteed. Up to recently, just a few countries, such as Japan and the Netherlands, have been able to generate complete IRIO tableaux (Miller, Blair, 1985; Zhang, Li, 2004).

Because of the problems of data availability, it is very difficult and complex to directly employ the basic IRIO model. Therefore, various simplified models have emerged, which include: multiregional IO (MRIO) model, also called the column coefficient model, the Leontief model and Pool-Approach model. Among these models, the MRIO model is the acknowledged mainstream simplified IRIO model. Compared with other simplified models, it has advantages such as low datum requirements and high precision results (Liu, Nobuhiro, 2002; Zhang, Li, 2004). The current MRIO tableaux for China was produced with an MRIO model. Thus, an MRIO model has been identified as highly suitable for this study. The basic equation in the MRIO model would be:

$$X = (I - CA)^{-1}CY \tag{3-1}$$

Where: X and Y denotes the total output vector and final demand vector respectively; A denotes the technology matrix; C denotes the interregional trade matrix.

3.1.3 China Energy Demand Analysis System–CEDAS

To facilitate the computational process, we adopted Visual Basic language to develop the computer software, and corresponding software structure and interface are shown in Figs. 3-1 and 3-2 respectively.

3.1.4 Scenario Analysis of Energy Requirement in 2020

3.1.4.1 Model Size

In this study, China is divided into eight regions as shown in Fig. 3-3 and Table 3-1. The Northwest and Southwest regions constitute the west region, which is in line with the extension laid out in the state West Development Strategy. Tibet and Taiwan are not taken into account in this study. In each region, four sectors are considered, i.e., agriculture, manufacturing, construction and service (including freight transport, communication, commerce, catering and non-material production service).

3.1.4.2 Scenarios

Considering actual conditions in China, with recognition of the major driving forces on

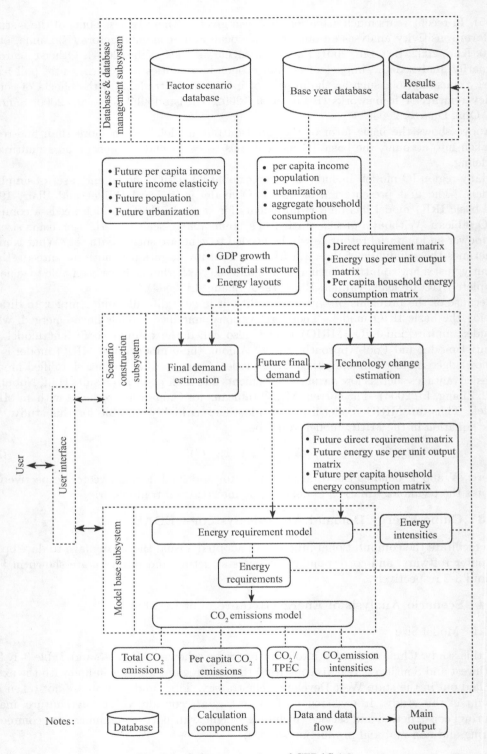

Fig. 3-1 Software structure of CEDAS 1.0

3.1 Forecasting of Energy Requirements in 2020

Fig. 3-2 Main interface of CEDAS

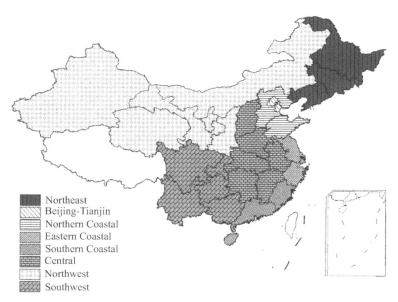

Fig. 3-3 Map of China with the eight economic regions

Table 3-1 Description of region division

Region	Province/ municipality
Northeast	Heilongjiang, Jilin, Liaoning
Beijing-Tianjin	Beijing, Tianjin
Northern Coastal	Hebei, Shandong
Eastern Coastal	Jiangsu, Shanghai, Zhejiang
Southern Coastal	Fujian, Guangdong, Hainan
Central	Shanxi, Henan, Anhui, Hubei, Hunan, Jiangxi
Northwest	Inner Mongolia, Shaanxi, Ningxia, Gansu, Qinghai, Xinjiang
Southwest	Sichuan, Chongqing, Guangxi, Yunnan, Guizhou, Tibet

energy consumption, i.e. GDP, population, urbanization and technology, five scenarios are, identified, as shown in Table 3-2.

Table 3-2 Scenario description

Scenarios	Scenario description
Scenario **BAU** (Business-As-Usual)	Assume that in the coming years up to 2020, China's economy could maintain its current growth rate and realize a relatively high growth rate; *per capita* income achieves the preset "well-off", "developed" society objectives; population and urbanization rate grow at a medium speed; technology advances at a medium rate and achieves the National Energy-Saving Layout in each analysis year.
Scenario **L** (Low economic growth)	Assume that various challenges and risks constrain economic development and urbanization advancement.
Scenario **H** (High economic growth)	Assume a higher economy growth rate on the base of scenario BAU
Scenario **HP** (High economic growth + High Population)	Assume a higher population growth rate on the base of scenario H
Scenario **HT** (High economic growth + High Technology)	Assume technology achieve greater improvement on the base of scenario H

3.1.4.3 Setting of the Major Driving Force Scenarios

(1) Economy scenarios.

In this study, three modes of economic growth are considered based on an investigation of the PRC State Council Development Center led by researcher Li (2005), as shown in Table 3-3.

Table 3-3 Forecasting of national economic growth (%)

Scenario	Year 2005–2010	Year 2010–2015	Year 2015–2020
Economy-base	8.1	7.5	6.8
Economy-low	7.5	5.8	4.8
Economy-high	8.5	8.2	7.7

This investigation also forecasts the future industrial structure. It shows that the ratio of the three industries would be 10.6:54.2:35.2 in year 2010; and 7.0:52.6:40.4 in year 2020.

Referring to the method adopted by the Workgroup of the State Statistics Bureau that estimated the future regional shares in GDP (Workgroup of the State Statistics Bureau, 2004), and based on the weighted average of regional GDP growth in the last seven years (1998–2004), the regional shares in GDP for year 2010 and 2020 could be estimated. Combining the results of regional shares with the base-year regional GDP, plus the forecasts of national GDP growth shown in Table 3-3, the growth rates of regional GDP could be obtained. Finally, taking the future population and urbanization rates into account, with the functions describing how regional household income grows with regional GDP (which are fitted using historical data), the future *per capita* income for each region could be estimated.

(2) Population scenarios.

Here, based on the forecasts of Jiang (1998) and with our own adjustments, two population growth scenarios for each region for year 2010 and 2020 are established, as shown in Tables 3-4 and 3-5. Here the national average for total fertility rate is 2.0 in the medium-growth scenario, and 2.1 in the high-growth scenario.

(3) Urbanization scenarios.

Here we combined the findings of the Institute of Geographic Sciences and Natural Resources Research of the Chinese Academy of Sciences (Liu, et al., 2003), with the population

3.1 Forecasting of Energy Requirements in 2020

Table 3-4 Forecast of regional population (in billions)

	Scenario	NE	BT	NC	EC	SC	C	NW	SW
2010	Medium	0.119	0.025	0.174	0.154	0.132	0.391	0.126	0.264
	High	0.121	0.025	0.176	0.156	0.134	0.395	0.128	0.267
2020	Medium	0.128	0.027	0.186	0.171	0.147	0.413	0.135	0.277
	High	0.130	0.027	0.189	0.174	0.149	0.420	0.138	0.282

scenarios set-out above, plus our own adjustment, to obtain the medium-urbanization scenario. The relatively low annual advancing rates of urbanization during period 1997–2003 are set as the low scenario.

Table 3-5 Forecast of urbanization rate (%)

	Scenario	NE	BT	NC	EC	SC	C	NW	SW
2010	Medium	61.67	79.62	50.61	61.2	69.41	41.83	41.12	36.74
	Low	50.54	78.85	21.81	37.41	23.59	25.51	42.53	23.21
2020	Medium	68.99	83.83	64.71	73.86	80.89	53.67	49.22	47.20
	Low	56.54	83.02	27.89	45.15	27.49	32.73	50.91	29.82

(4) Technology scenarios.
• Energy end-use efficiency improvement.

Because of lack of data, currently only the improvements of energy end-use efficiency are considered in the technology scenario. Here, two sets of technology scenarios are established, as shown in Table 3-6.

Table 3-6 Technology scenarios

Scenario	Description
Medium	The improvements of national average energy end-use efficiency achieved the goal of National Energy-Saving Layouts (Workgroup of the Academy of Macroeconomic Research of the State Development Planning Commission, 1999; National Development and Reform Commission, 2004), assuming uniform rates of technology improvement for all regions.
High	The improvements of national average energy end-use efficiency achieve a 5% greater improvement than the goal of National Energy-Saving Layouts, assuming uniform rates of technology improvement for all regions.

• Ratio of renewable energy over primary energy.

This study utilizes the forecasts of the Academy of Macroeconomic Research Workgroup of State Development Planning Commission (1999a) (Table 3-7).

Table 3-7 Forecast of ratio of renewable energy over primary energy (%)

Ratios	2010	2020
Hydro power	9.3	10.3
Nuclear power	2.3	3.64

3.1.4.4 Data

The base year data of regional GDP, population, *per capita* income of urban and rural residents, and urbanization rates are obtained through arranging and integrating the corresponding provincial data of year 1997 from the *China Statistical Yearbook* (National Bureau of Statistics, 1998).

The base year matrices of technology coefficients, final demand and interregional trade coefficients are obtained by adjusting the eight-sector basic matrix from the multi-regional input-output tables for China, 1997 (the State Information Center of China, 2005). From the multi-regional input-output data we could also identify regional-urban and rural-household aggregate consumptions, which are combined with regional population and urbanization rates to obtain the base-year regional *per capita* consumption by urban and rural residents.

The base-year data of energy consumption per unit output and residential energy consumption *per capita* were obtained as follows: adjusting and integrating the energy balance of each province or municipality from the *China Energy Statistical Yearbook* (Department of Industrial and Transportation Statistics, 2001), total sector energy consumption, as well as aggregate residential energy consumptions of urban and rural residents for each region, could be obtained. The base year data of energy consumption per unit output could be obtained by dividing total sector energy consumption with the corresponding sector total output from the multi-regional input-output tables. Base-year residential energy consumption *per capita* could be obtained by dividing regional aggregate residential energy consumptions of urban and rural residents with the corresponding base-year regional urban and rural population, respectively.

The future urban and rural income elasticities are acquired through a modification of the work of Hubacek and Sun (2001).

3.1.5 Results

In this section, the results for energy requirements for year 2010 and 2020 are discussed.
(1) Energy requirements.
Regional energy requirements for year 2010 and 2020 are presented in Figs. 3-4 and 3-5, respectively.

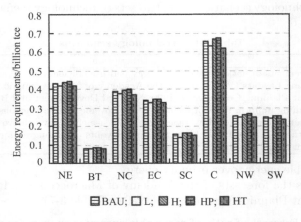

Fig. 3-4 Regional energy requirements for year 2010

Results show that, for year 2010, total national energy requirements will be 2.444–2.635 billion tce, regional requirements will be 0.415–0.442 billion tce for Northeast, 0.079–0.084 billion tce for Beijing-Tianjin, 0.369–0.400 billion tce for Northern Coastal, 0.325–0.348 billion tce for Eastern Coastal, 0.139–0.164 billion tce for Southern Coastal, 0.616–0.675 billion tce for Central, 0.246–0.264 billion tce for Northwest, 0.237–0.257 billion tce for Southwest; for year 2020, total national energy requirements will be 2.840–3.873 billion tce, regional requirements will be 0.462–0.631 billion tce for Northeast, 0.098–0.136 billion tce for Beijing-Tianjin, 0.458–0.624 billion tce for Northern Coastal, 0.404–0.554 billion tce for Eastern Coastal, 0.187–0.256 billion tce for Southern Coastal, 0.671–0.910 billion tce for Central, 0.288–0.393 billion tce for Northwest, 0.272–0.369 billion tce for Southwest.

From Figs. 3-4 and 3-5 it can be seen that, in both analyzed years the Central region occupies the largest share of total national energy requirements, whose share would be 25.22%–25.75% in year 2010 and slightly fall to 22.55%–23.63% in year 2020. The smallest share was occupied by Beijing-Tianjin in both analyzed years in all scenarios.

As for the annual growth rate of energy requirements during the period 1997–2020, in all scenarios, except scenario HT, the highest one corresponds to Eastern Coastal (3.67%–

3.1 Forecasting of Energy Requirements in 2020

5.11%), closely followed by the Southern Coastal (3.62%–5.03%). In scenario HT, the highest annual growth rate appears in Southern Coastal (4.47%), closely followed by Eastern Coastal (4.45%). In all scenarios the lowest annual growth rate would be 2.15%–3.50% of Southwest.

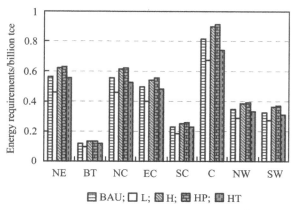

Fig. 3-5 Regional energy requirements for year 2020

In both analyzed years, all regions reach their maximum energy requirements in scenario HP. Other factors being constant, when the national average total fertility rate rises from 2.0 to 2.1, i.e., scenario HP versus H, in year 2010 regional energy requirements will increase by 0.8 (Beijing-Tianjin)–6.37 (Central) million tce; in year 2020 regional energy requirements will increase by at least 2.04 (Beijing-Tianjin)–13.65 (Central) million tce. On the basis of scenario BAU, sensitivity analysis of population was performed to assess how 1% growth of population in each region respectively will affect energy requirements. The results are shown in Table 3-8. It can be seen from Table 3-8 that population growth in each region will all bring about an obvious increase of intra-regional energy requirements. Through the interregional trade relationship, population growth in one region will not only raise energy requirements inside this region but also drive up requirements in other regions. The three regions whose population growth will have the largest outer-regional effects are Eastern Coastal, Central and Southern Coastal. Compared to other regions, 1% of population growth in Central will cause the highest increase of total national energy requirements, i.e., 0.24%

Table 3-8 Impacts of 1% growth of regional population on energy requirements (%)

	Population-growth-region	Variation inside the region	Total variation of the other regions	Total national variation
2010	NE	0.77	0.02	0.15
	BT	0.73	0.02	0.04
	NC	0.57	0.04	0.12
	EC	0.69	0.08	0.16
	SC	0.70	0.06	0.10
	C	0.71	0.07	0.24
	NW	0.73	0.03	0.10
	SW	0.80	0.03	0.11
2020	NE	0.82	0.02	0.15
	BT	0.83	0.02	0.05
	NC	0.62	0.03	0.13
	EC	0.77	0.07	0.17
	SC	0.68	0.04	0.08
	C	0.75	0.06	0.22
	NW	0.79	0.03	0.11
	SW	0.84	0.03	0.10

in year 2010 and 0.22% in year 2020, respectively, followed by those in Eastern Coastal and Northeast.

The growth of population will drive up the total energy requirements in two aspects. On one hand, the increase of population will directly raise aggregate residential energy requirements. On the other hand, total production energy requirements will also go up in order to satisfy extra residential consumption for both energy and non-energy goods incurred by population growth. With the development of economy and the improvement of income levels, per capita household demand for various goods will go up accordingly. Therefore, the increase rate of energy requirements driven by the same extent of population growth will augment correspondingly. It could also be seen from Table 3-8 that 1% growth of regional population in year 2020 will generally cause a higher increase of rates of energy requirements than that in year 2010, either inside the region or at the national level. In order to alleviate the pressure on resources and environment driven by population growth, it is important to actively improve sector energy end-use efficiency, including that of the residential sector. However, it would be a hard and long-term task to obtain at the national level an obvious higher improvement of energy end-use efficiency than the potentials projected in the current energy-saving layouts. Therefore, at least during the important period when China is taking efforts to build a well-rounded, economically vibrant society, the basic state policy of family planning should be seriously carried out in each region.

It can also be seen from Figs. 3-4 and 3-5 that, in both analyzed years for all the regions, regional energy requirements in scenario L and HT are lower than those in other scenarios. However, the low energy requirements in scenario L are obtained at the cost of GDP growth. Compared with scenario BAU, during the period 1997–2020, the annual growth rate of regional energy requirements in scenario L just reduced 0.84% (Southwest)–0.94% (Beijing-Tianjin), while that of regional GDP in this scenario reduced 0.96% (Southern Coastal)–0.98%(Central). Such a scenario is not desired.

During the period 1997–2020, compared with scenario BAU, in scenario HT the annual growth rate of GDP for all regions increases about 0.45%, while that of regional energy requirements will decline 0.05% (Southern Coastal)–0.41% (Central), or only go up 0.01% (Beijing-Tianjin). On the basis of scenario BAU, sensitivity analysis of technology was performed to assess how 1% higher improvement of energy end-use efficiency from each region respectively will affect energy requirements. The results are shown in Table 3-9. It can be seen from Table 3-9 that technology improvements in each region could all generate apparent intra-regional energy saving effects. Such effects are especially evident in Central region. The intra-regional energy-saving effects incurred by 1% higher improvement of energy end-use efficiency in this region are about 71% and 78% higher than those in Beijing-Tianjin in 2010 and 2020, respectively. The national energy-saving effects created by technology improvements in Central are also the highest, which would be 0.41% and 0.32% in year 2010 and 2020, respectively, obviously higher than the 0.03%–0.21% and 0.03%–0.18% brought by the other regions.

Table 3-9 Impacts of 1% higher regional technology improvements than those in BAU scenario

	Technology-vary-region	Variation inside the region/%	Total national variation/%
	NE	−1.02	−0.17
	BT	−0.95	−0.03
	NC	−1.40	−0.21
	EC	−1.18	−0.16
2010	SC	−1.03	−0.06
	C	−1.62	−0.41
	NW	−1.27	−0.13
	SW	−1.48	−0.14

3.1 Forecasting of Energy Requirements in 2020

Continued

Technology-vary-region		Variation inside the region/%	Total national variation/%
2020	NE	−0.81	−0.13
	BT	−0.76	−0.03
	NC	−1.13	−0.18
	EC	−0.93	−0.13
	SC	−0.81	−0.05
	C	−1.36	−0.32
	NW	−1.08	−0.11
	SW	−1.24	−0.12

Figs. 3-6–3-11 illustrate the requirements for coal, crude oil, and natural gas in each region for the two analyzed years, respectively.

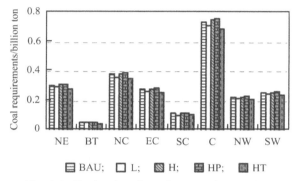

Fig. 3-6 Regional coal requirements for year 2010

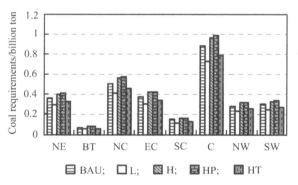

Fig. 3-7 Regional coal requirements for year 2020

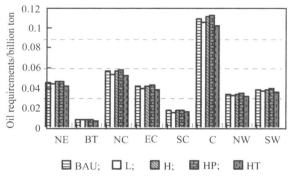

Fig. 3-8 Regional crude oil requirements for year 2010

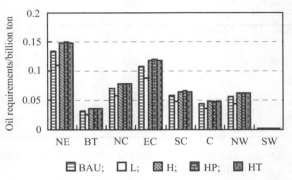

Fig. 3-9 Regional crude oil requirements for year 2020

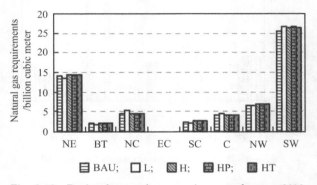

Fig. 3-10 Regional natural gas requirements for year 2010

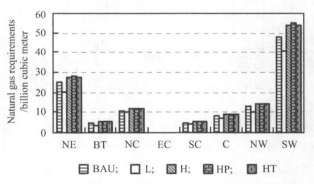

Fig. 3-11 Regional natural gas requirements for year 2020

The results show that for both analyzed years in all scenarios the highest coal requirements appear in region Central, followed by Northern Coastal and Northeast. During the period 1997–2020, the highest annual growth rate of coal requirements corresponds to Eastern Coastal (3.04%–4.46%), followed by Northern Coastal (2.83%–4.25%) and Southern Coastal (2.78%–4.20%). For both analyzed years, all regions reach their highest coal requirements in scenario HP.

As far as crude oil is concerned, for both analyzed years in all scenarios the highest requirements appear in region Northeast, followed by Eastern Coastal and Northern Coastal. During the period 1997–2020, the highest annual growth rate of crude oil requirements corresponds to Eastern Coastal (5.09%–6.55%), followed by Southern Coastal (4.70%–6.10%)

and Northern Coastal (4.49%–5.91%). For both analyzed years, all regions reach their highest requirements in scenario HP.

When it comes to natural gas, for both analyzed years, in all scenarios, the highest natural gas requirements appear in region Southwest, followed by Northeast. During the period 1997–2020, the highest annual growth rate of natural gas requirements corresponds to Beijing-Tianjin (9.69%–11.43%), followed by Eastern Coastal (9.27%–10.69%) and Northern Coastal (9.15%–9.97%). In year 2010, the highest natural gas requirements in region Northern Coastal, Eastern Coastal, and Central appear in scenario L. The other five regions reach their highest requirements in scenario HP. In year 2020, all regions reach their highest requirements in scenario HP.

According to the forecasts by the Workgroup of the Academy of Macroeconomic Research of the State Development Planning Commission (Workgroup of the Academy of Macroeconomic Research of the State Development Planning Commission, 1999), the natural gas production in Northwest is possible to reach 27.21 billion cubic meters and 41.13 billion cubic meters in year 2010 and 2020, respectively. In such a case, up to 2020 in all scenarios, the Northwest could have a relatively sufficient margin available to transfer to other regions. The margin would be 20.26–20.74 billion cubic meters in year 2010 and 26.58–30.77 billion cubic meters in year 2020, respectively.

(2) Energy intensities.

Figs. 3-12 and 3-13 illustrate the results of regional energy intensities for year 2010 and 2020 respectively. It can be seen that, under the assumption of uniform rates of technology improvement for all regions, up to year 2020 there will be distinct differences among regional energy intensities. In all scenarios in the two analyzed years, the regions whose energy intensities are higher than the corresponding national average include Northeast, Northern Coastal, Central and Northwest. The highest energy intensity corresponds to Northeast, closely followed by Northwest. In all scenarios in the two analyzed years, the regions whose energy intensities are both lower than the corresponding national averages include Beijing-Tianjin, Eastern Coastal and Southern Coastal. The lowest energy intensity corresponds to Southern Coastal.

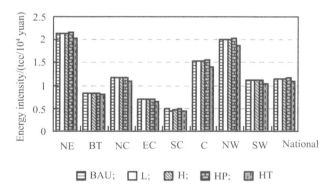

Fig. 3-12 Regional energy intensity for year 2010

The scenario analysis results of regional natural gas requirements show that, up to 2020, in all scenarios Northwest could have a relatively sufficient margin available to transfer to other regions. However because of the inter-dependencies among sectors, the increase of demand for the natural gas in Northwest will directly or indirectly drive up demand for other goods, especially the manufacturing goods in this region. The above results of intensities show that the energy intensity of Northwest will be the second highest in the two analyzed years, slightly lower than that of Northeast. Therefore, while shortening the supply-demand gap in the importing region, increasing imports from this region will on the whole raise the total national energy requirements. On the basis of scenario BAU, sensitivity analysis of trade coefficients was performed to assess how 1% increase of imports of manufacturing goods from Northwest by each of the other regions respectively will affect energy requirements. The results are shown in Table 3-10. It can be seen from Table 3-10 that importing Northwest manufacturing goods by any of the other regions will drive up energy requirements not only inside Northwest but also at the national level. Moreover, different importing regions will bring about relatively highly different variations. The highest increase of energy requirements inside Northwest are caused by imports of region Eastern Coastal (0.14% in 2010 and 0.20% in 2020), followed by those of Central. The highest increase of energy requirements at the national level are also caused by imports of region Eastern Coastal (0.14% in 2010 and 0.19% in 2020), followed by those of Southern Coastal and Central.

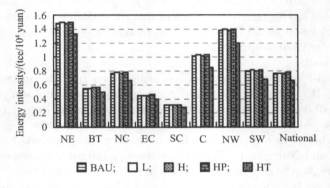

Fig. 3-13 Regional energy intensity for year 2020

Table 3-10 Impacts of 1% increase of imports of manufacturing goods from Northwest on energy requirements

Importing region		Variation of energy requirements/%		National
		Importing region	Northwest	
2010	NE	−0.49	0.85	0.01
	BT	−0.53	0.48	0.03
	NC	−0.33	1.01	0.05
	EC	−0.43	1.99	0.14
	SC	−0.45	1.21	0.09
	C	−0.48	1.89	0.07
	NW	−0.52	0.93	0.05
2020	NE	−0.55	0.96	0.01
	BT	−0.65	0.67	0.04
	NC	−0.38	1.28	0.07
	EC	−0.51	2.71	0.20
	SC	−0.44	1.28	0.10
	C	−0.54	2.11	0.09
	NW	−0.60	1.04	0.05

If accelerating the technology improvements in Northwest on the base of scenario BAU, i.e., suppose this region achieves a 1% greater improvement of energy end-use efficiency, re-performing the above sensitivity analysis of trade coefficients, the results are shown in Table 3-11. It can be seen from Table 3-11 that, if the technology improvement in Northwest region is accelerated, the extra increase of national energy requirements caused by increased demand from Eastern Coastal could be explicitly reduced, and those caused by increased demand from the other regions could be eliminated.

Compared with the results shown in Table 3-10, for variations at the national level, the increased rate of energy requirements cause by imports of Eastern Coastal will decline from 0.14% to 0.01% in year 2010, and from 0.20% to 0.09% in year 2020.

Therefore, accelerating improvements of energy end-use efficiency in Northwest is in favor of facilitating interregional energy transfers, i.e., on the one hand to make good use of the abundant resources in Northwest to satisfy the demands of importing regions and shorten their demand-supply gaps, on the other hand to avoid causing extra pressure on the national energy requirements.

Table 3-11 Impacts of 1% increase of imports of manufacturing goods from Northwest on energy requirements (Accelerating the technology improvement in Northwest)

Importing region		Variation of energy requirements/%		National
		Importing region	Northwest	
2010	NE	−0.49	−0.43	−0.12
	BT	−0.53	−0.79	−0.09
	NC	−0.33	−0.27	−0.08
	EC	−0.43	0.69	0.01
	SC	−0.45	−0.07	−0.04
	C	−0.48	0.59	−0.06
	NW	−0.52	−0.35	−0.08
2020	NE	−0.55	−0.13	−0.10
	BT	−0.65	−0.42	−0.07
	NC	−0.38	0.19	−0.04
	EC	−0.51	1.60	0.09
	SC	−0.44	0.19	−0.01
	C	−0.54	1.01	−0.03
	NW	−0.60	−0.05	−0.06

3.1.6 Policy Implications

Summarizing the above scenario and sensitivity analysis results of the model, the following policy recommendations are proposed:

(1) Continuing effort should be taken to advance the improvement of energy end-use efficiency in each region, especially to accelerate the improvement in Central and Northwest as much as possible.

As shown in the above results of the analysis, in the case of high economic growth, if the energy end-use efficiency of all sectors in all regions could achieve 5% improvement (scenario HT *versus* scenario BAU), during the period of 1997–2020, when the annual growth rates of regional GDP rise approximately 0.45%, the regional annual growth rate of energy requirement will decline 0.05% (Southern Coastal)–0.41% (Central), or go up 0.01% (Beijing-Tianjin). Moreover, the sensitivity analysis of technology shows that technology improvement in each region could generate "apparent" intraregional energy saving effects. Therefore, in order to alleviate the pressure caused by high economic growth on energy and the environment, energy end-use efficiency in all regions should be actively improved.

Model results also show that Central occupies the largest share in total national energy requirements in both analyzed years. What is more, sensitivity analysis results of population show that 1% increase of population in this region will induce the largest rise of national energy requirements in both analyzed years. While the sensitivity analysis of technology shows that both the intra-regional and national energy-saving effect brought by 1%, improvement of energy end-use efficiency in Central are greater than those of the other regions. Especially, the national energy-saving effect brought about from technology improvement in Central, are clearly higher than those brought about by the other regions. Therefore, it would be of great benefit to accelerate the improvement of energy end-use efficiency in Central.

Scenario analysis of regional natural gas requirements shows that, up to year 2020 in all scenarios, Northwest could have relatively sufficient margin available to transfer to other regions. However, the results of intensities show that under the assumption of uniform rates of technology improvement for all regions, the energy intensity of the Northwest region will be second highest in the two analyzed years. In such case, the sensitivity analysis of trade coefficients show that 1% increase of imports of manufacturing goods from the Northwest region by any of the other regions will drive-up the total national energy requirements. If the technology improvement in the Northwest region is accelerated, the extra increase of national energy requirements caused by increased demand from Eastern Coastal could be explicitly reduced, and those caused by an increased demand from the other regions could be eliminated.

In order to facilitate energy saving at the national level and the effectiveness of inter-regional energy transfers, the efficiency improvement in regions Central and Northwest should be accelerated as much as possible.

(2) During the important period when China is making efforts to build a well-rounded, economically vibrant society, the basic state policy of family planning should be enforced in each region.

The results show that in both years analyzed, all the regions reach their highest energy requirement in scenario HP. In the case of high economic growth, if the national average of the total fertility rate increases from 2.0 to 2.1 (scenario HP *versus* H), the energy requirement will increase some 1.08% in year 2010, and the increase rate will mount-up to approximately 1.77% in year 2020. Results of the sensitivity analysis of population show that, with the growth of the economy and improvements of *per capita* income, *per capita* household demand for various energy or nonenergy goods will go up accordingly, thereby the driving-force of population growth to energy requirements will increase correspondingly. Population growth in one region will not only significantly affect energy requirements of the region itself, but also drive-up energy requirements of the other regions.

Therefore, in order to alleviate the pressure on resources and the environment incurred by population growth, on one hand it is important to actively improve efficiency of household energy-use so as to slow down the increase of direct *per capita* residential energy requirements. The household consumption pattern should also be guided to less energy-intensive and environmentally friendly goods. The energy end-use efficiency of production sectors should also be actively improved so as to slow down the increase of production energy requirements indirectly incurred by population growth.

On the other hand, it would be a hard and long-term task to obtain on the national-level a more obvious improvement of energy end-use efficiency than the potentials projected in the current energy-saving layouts; what is more, the effect of energy substitution is not likely to be seen before 2020 (Zhou, Yukio, 1996). Therefore, in order to create a favorable population environment for the important period when China is making efforts to build a well-rounded, economically vibrant society, in the near future the basic state policy of family planning should be seriously followed in each region.

3.2 Economic Analysis of Coal Supply System

China is one of the countries in the world which takes coal as its principal energy. Coal plays a critical role in social and economic development in China. It is not only a main source of fuel in many industries, but also an important raw material of the chemical industry and civil energy. Coal also has been an important export commodity. In 2002, coal production accounted for 70% of primary energy production, and coal consumption accounted for 66% of

3.2 Economic Analysis of Coal Supply System 87

primary energy consumption in China. In the predictable future, coal will have an important influence on China's highly developed economy.

After the coal shortage in 1980s, town and village mines experienced fast development, which resulted in the increasing of the production of coal. China had been the greatest coal producer in 1990s, and the amount of output arrived at its summit in 1997, namely 1.397 billion tons. The coal industry went into decline after that. The amount of output was 1.045 billion tons in 1999, 980 million tons in 2000, 1.16 billion tons in 2001, and 1.38 billion tons in 2002. As a consequence of increased investment, the amount of output reached 1.67 billion tons in 2003. China produced 1.956 billion tons of coal in 2004, an increase of 13.2%. Among them, state-owned coal mines produced 1.24 billion tons, an increase of 13.1%, and town and village mines produced 719 million tons, an increase of 13.4%. Both state-owned and town and village mines play an important role in China.

3.2.1 System Dynamics Model and Data for Coal Production and Supply

3.2.1.1 Flow Diagram of Coal Production and Supply

Through research on reserves, investment in mining and washing of newly-built coal mines, and key state mines production capacity situations of the past, we draw a flow diagram of national coal production and supply system (Fig. 3-14). Coal production system is a dynamic system. Production capacity of key state mines increases with the putting into production of the newly-built mines, and reduces as a result of scrapping. The system is influenced by external variables such as coal demand and national macro economic policy. Variable index is listed in Table 3-12.

Table 3-12 Variable index

No.	Variable name	Detail	Unit	Note
1	ARS	Available reserves for constructing mines	Billion tons	Ordinary variable
2	CD	Coal demand	Thousand tons	Ordinary variable
3	Cmc	Coefficient of mine construction	Null	Parameter
4	ERS	Explored reserves	Billion tons	Ordinary variable
5	ERsm	Extraction rate in state-owned mines	Null	Parameter
6	ERtv	Extraction rate in town or village owned mines	Null	Parameter
7	GPI	Geological prospecting investment	Thousand RMB yuan	Ordinary variable
8	IARS	Increased available reserves for mine construction	Billion tons	Ordinary variable
9	ICGP	Investment coefficient in geological prospecting	Null	Control variable
10	ICmw	Investment coefficient of mining and washing of coal	Null	Control variable
11	IPCsm	Increased production capacity of state-owned mines	Thousand tons	Ordinary variable
12	Ism	Investment in state-owned mine construction	Thousand RMB yuan	Ordinary variable
13	MERS	Mining-employed reserves	Thousand tons	Ordinary variable
14	MERsm	Mining-employed reserves by state-owned mines	Thousand tons	Ordinary variable
15	MRS	Mining reserves	Billion tons	Ordinary variable
16	MERtv	Mining-employed reserves by town or village mines	Thousand tons	Ordinary variable
17	PCcm	Production capacity of constructing mines	Thousand tons	Ordinary variable
18	PCnsp	Production capacity of new starting project	Thousand tons	Ordinary variable
19	PCsm	Production capacity of state-owned mines	Million tons	State variable
20	Psm	Production of state-owned mines	Thousand tons	Ordinary variable
21	Ptv	Production of town or village owned mines	Thousand tons	Ordinary variable
22	SPCsm	Scrapped production capacity of state-owned mines	Thousand tons	Ordinary variable

3.2.1.2 System equations

As shown in Fig. 3-14, we describe the system as the following equations.

(1) Production capacity of state-owned mines:

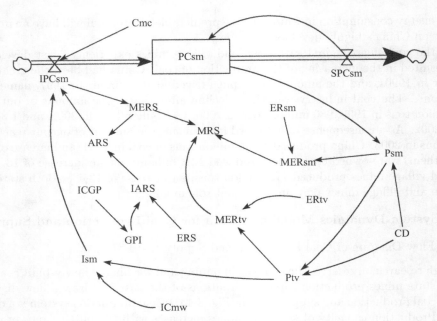

Fig. 3-14 Flow diagram of coal supply

$PCsm[i] = PCsm[i-1] + (IPCsm[i] - SPCsm[i])/1000$

(2) Available reserves for constructing mines:

$ARS[i] = ARS[i-1] + IARS[i] - MERS[i]/1\,000\,000$

(3) Production capacity of constructing mines:

$PCcm[i] = PCcm[i-1] + PCnsp[i] - IPCsm[i]$

(4) Increased available reserves for mine construction:

$IARS[i] = ((GPI[i] - 43157)/261.8)/1\,000\,000$

(5) Investment in state-owned mine construction:

$Ism[i] = 88\,834\,596.978 ICmw[i] + 47.535\,PCcm[i] - 3\,912\,234.835$

(6) Production capacity of new starting projects:

$Ism = 29.18 PCcm + 43.44 PCnsp$

(7) Increased production capacity of state-owned mines:

$IPCsm[i] = 0.4 PCnsp[i-10] + 0.3 PCnsp[i-9] + 0.2 PCnsp[i-8] + 0.1 PCnsp[i-7]$

(8) Mining reserves:

$MRS[i] = MRS[i-1] + IPCsm[i] \times Cmc[i]/1\,000\,000 - (MERsm[i] + MERtv[i])/1\,000\,000$

(9) Mining-employed reserves:

$MERsm[i] = Psm[i]/ERsm[i]$

$MERtv[i] = Ptv[i]/ERsm[i]$

3.2 Economic Analysis of Coal Supply System

3.2.1.3 Applications

(1) Model checking.

The application of this model is used to verify how the model explains the essential features of coal production and supply in China. In other words, we need to make sure whether the model dynamically reflects the relationship between the variables, and to what extent the model fits actual situations.

We take into account two important variables, production of state-owned mines and investment in mine construction. The comparison between simulated values and historical observations of production of state-owned mines during 1990s is shown in Fig. 3-15.

We adapt the method of computing relative errors as follows:

$$error = \frac{|simulation \cdot data - historical \cdot data|}{historical \cdot data} \times 100\%$$

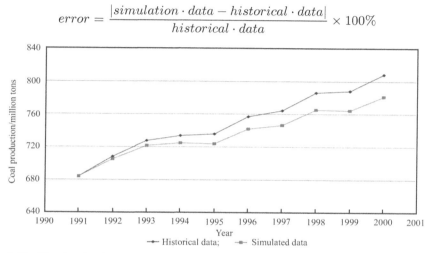

Fig. 3-15 Simulated values versus historical observations of production of state-owned mines

Simulated values versus historical observations along with simulated errors are listed in Table 3-13. Obviously, simulated results reflect the steady growth of production of state-owned mines in 1990s. And the errors between simulation and real values are less than 3.4%.

Table 3-13 **Simulated errors of production of state-owned mines**

Year	Historical value /million tons	Simulative value /million tons	Error/%
1991	684.16	684.16	0.00
1992	708.07	705.07	0.42
1993	727.26	721.26	0.83
1994	733.48	724.48	1.23
1995	735.58	723.58	1.63
1996	756.82	741.82	1.98
1997	764.30	746.30	2.36
1998	786.27	765.27	2.67
1999	788.31	764.31	3.04
2000	808.34	781.34	3.34

Similarly, the comparison between simulated values and historical observations of investment in mine construction during 1990s is shown in Fig. 3-16. Simulated values versus historical observations along with simulated errors are listed in Table 3-14. Simulated results reflect the fluctuation of investment in mine construction to a certain extent. And the errors between simulation and real values vary from 3% to 23%.

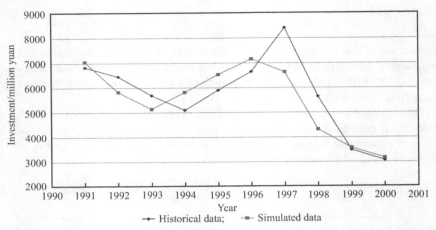

Fig. 3-16 Simulation values versus historical observations of investment in mine construction

Table 3-14 Simulated errors of investment in mine construction

Year	Historical value /million yuan	Simulated value/million RMB yuan	Errors/%
1991	6832.64	7050.5	3.19
1992	6457.08	5825.26	9.78
1993	5696.12	5145.20	9.67
1994	5099.78	5821.77	14.16
1995	5905.21	6543.05	10.80
1996	6656.72	7165.09	7.64
1997	8416.26	6641.21	21.09
1998	5648.84	4321.88	23.49
1999	3483.39	3568.64	2.45
2000	3067.78	3164.86	3.16

(2) Programming objectives.

Meeting coal requirements is our goal. For sustainable supply, we should try to control the system to the following objectives:

A. The production of town or village owned mines shouldn't exceed that of 2003, i.e., 600 million tons.

B. There should be 34 million tons of increased production capacity of state-owned mines in 2020 and 350 million tons of production capacity of constructing mines between 2015 and 2020 to ensure there is enough production capacity after 2020.

C. Available reserves for constructing mines must meet reserve requirements of new starting projects.

D. Lighten the over-load production status of state-owned mines.

(3) Scenarios.

There are two control variables in the model, investment coefficient of mining and washing of coal ($ICmw$) and geologic prospecting investment factor ($ICGP$).

① Investment coefficient of mining and washing of coal ($ICmw$).

Based on the investments in mining and washing of coal between 1991 and 2000, we conduct six possible scenarios for investment coefficient of mining and washing of coal ($ICmw$) as follows:

• Extra-low investment (C1): the investment of mining and washing of coal get to the lowest level in the history, namely the least possible investment is given to the mining and washing of coal. In this case, $ICmw=0.005$.

• Low investment (C2): the investment in mining and washing of coal takes the average of those in 1990s, i.e., $ICmw=0.015$.

3.2 Economic Analysis of Coal Supply System   91

• Middle investment (C3): the investment in mining and washing of coal almost takes the maximal of those in 1990s, i.e., $ICmw$=0.03.
• Higher investment (C4): the investment in mining and washing of coal is 1.3 times to that in scenario C3, i.e., $ICmw$=0.04.
• Much higher investment (C5): the investment in mining and washing of coal is 2 times to that in scenario C3, i.e., $ICmw$=0.06.
• Extra-high investment (C6): mining and washing of coal gets special attention from the whole country. The investment in it reaches 3 times to that in scenario C3, i.e., $ICmw$=0.09.
② Geologic prospecting investment factor ($ICGP$).

Based on investments in the eighth five-year plan and the ninth five-year plan in China, we conduct three scenarios for geologic prospecting investment factor ($ICGP$) as follows:
• Low investment (D1): Geological prospecting investment continues to be neglected for a long time, and is kept to the average of that in the period of the ninth five-year plan, i.e., $ICGP$=0.5.
• Middle investment (D2): Geological prospecting investment is equal to the average of that in the period of the eighth five-year plan, i.e., $ICGP$=1.
• High investment (D3): This scenario fits the situation that there is little available reserve for constructing mines. $ICGP$=2.

3.2.2 Investment Influence on the Capacity of Coal Supply

Various simulations based on above scenarios have been done. The simulations show that scenarios C4 and C5 of investment coefficient of mining and washing of coal can meet pre-designed objectives A and B; scenario C6 can meet objectives A, B and D; scenarios D2 and D3 of geologic prospecting investment can meet objective C.

Let us focus on investments in mining and washing of coal after 2000. When the investment keeps to the level of the 1990s, even if with the highest investment in 1990s, the programmed objectives cannot been reached. When investments in mining and washing of coal are 130% greater than the highest investment in 1990s, production capacity of constructing mines might gradually increase and exceed 350 million tons in 2010; the productions of town or village owned mines might be decreased after a peak. The over-load production status of state-owned mines can not be lightened till 2020 except in scenario C6. It means lightening the over-load production status of state-owned mines is almost impossible under regular investment policies.

Among all scenarios for geologic prospecting investment, the scenario D2 is satisfied and economical. In scenario D2, the available reserves would approximately get to 8.6 billion tons every year, which can meet the requirements of China's economic goal on coal in 2020.

For the programmed objectives, the feasible scenarios are those that combined D2 with C4, C5 and C6. Parts of simulation results of these scenarios are listed as Table 3-15.

3.2.3 Conclusions and Remarks

From the simulations and related analysis, we conclude as follows:
(1) Insufficient investment in mine construction between 1990 and 2000 resulted in coal production of town or village owned mines rising. Town or village owned mines usually were small and poorly equipped, so there were problems such as waste because of lower extraction rate, insecurity and pollution because of poor production conditions, and irregular market behavior. As the Chinese government has recognized the serious consequence of insufficient investment in mine construction at the beginning of this century, mine construction is paid more attention to and gets more investment. This has resulted in effective control of production of town or village owned mines and will result in a decrease of less than 600 million tons after 2010.

Table 3-15   Simulated results of certain feasible scenarios

Scenarios	Year	Investment in state-owned mine construction /million RMB yuan	Production capacity of state-owned mines /million tons	Production of state-owned mines /million tons	Production of town or village owned mines /million tons	Production capacity of constructing mines /million tons	Available reserves for constructing mines /billion tons
(C4,D2)	2000	3160	780	780	510	160	161.4
	2005	14220	820	1030	740	440	131.0
	2010	14290	870	1090	740	450	152.0
	2015	12530	1170	1460	500	360	143.0
	2020	13490	1240	1550	570	410	152.6
(C5,D2)	2000	3160	780	780	510	160	161.4
	2005	17870	820	1030	740	560	106.3
	2010	17630	900	1120	700	550	126.3
	2015	14790	1290	1620	350	440	103.4
	2020	15570	1380	1720	400	480	111.6
(C6,D2)	2000	3160	780	780	510	160	161.4
	2005	23360	820	1030	740	750	69.1
	2010	22650	940	1180	650	710	87.7
	2015	18180	1480	1480	490	550	44.0
	2020	18680	1590	1590	530	570	49.9

(2) Because of long periods for construction of mines, the problems of insufficient mine construction investment during the 1990s started to appear in the early 2000s. That is, overload production of state-owned mines and too much production of town or village owned mines. At the beginning of this century, the Chinese government clearly recognized this problem and corrected it immediately. From 2001 to 2005, mine construction investment increased rapidly, but it is impossible to change the status of the over-load in production of state-owned mines until 2020. Having missed the best window of opportunity for investment, it is not easy to correct it in a decade.

(3) Simulations show that geological prospecting during the ninth five-year-plan did not receive enough attention, and suggest the level of geological prospecting investment during the eighth five-year-plan is more suitable.

(4) If the investment coefficient of mining and washing of coal is between 4% and 9%, the coal production of town and village owned mines might be limited to a certain level. But the status of over-load production of state-owned mines might not be changed in 2020, and it could be changed after 2020 because production capacity of constructing mines is enough in 2020. If the investment coefficient of mining and washing of coal is greater than 9%, the over-load production status of state-owned mines could be solved before 2020.

To sum up, coal requirements increased rapidly from 2002. It caused state-owned mines to increase their production and lead to an over-load status. On the other hand, town or village owned mines had an opportunity to increase their production, which resulted in a worse status for the coal industry in China for a long period. Periods of geological prospecting and mine construction are long, the delay of investment moments in 1990s caused a serious impact on the coal system. In order to ensure coal productions and supplies to meet coal requirements caused by high economic development in China, investment in mining and washing of coal and geological prospecting must be increased immediately.

3.3   Summary

This chapter undertook forecasts on China's energy supply and requirements for year 2020.

As for energy requirements, this chapter established a model for energy requirements scenario analysis based on multi-regional input-output method. Considering actual conditions in China, with recognition of the major driving forces on energy consumption, five scenarios are identified, representing different growth paths of technology and economy. With the data

3.3 Summary

of year 1997, and dividing China into eight economic regions, scenario analysis of energy requirements was performed for each region.

Results show that, for year 2010, total national energy requirements will be 2.444–2.635 billion tce, and regional requirements will be 0.415–0.442 billion tce for Northeast, 0.079–0.084 billion tce for Beijing-Tianjin, 0.369–0.400 billion tce for Northern Coastal, 0.325–0.348 billion tce for Eastern Coastal, 0.139–0.164 billion tce for Southern Coastal, 0.616–0.675 billion tce for Central, 0.246–0.264 billion tce for Northwest, 0.237–0.257 billion tce for Southwest; for year 2020, total national energy requirements will be 2.840–3.873 billion tce, regional requirements will be 0.462–0.631 billion tce for Northeast, 0.098–0.136 billion tce for Beijing-Tianjin, 0.458–0.624 billion tce for Northern Coastal, 0.404–0.554 billion tce for Eastern Coastal, 0.187–0.256 billion tce for Southern Coastal, 0.671–0.910 billion tce for Central, 0.288–0.393 billion tce for Northwest, 0.272–0.369 billion tce for Southwest.

The sensitivity analysis of technology shows that technology improvements in each region could generate "apparent" intraregional energy saving effects. Therefore, in order to alleviate the pressure caused by high economic growth on energy and the environment, energy end-use efficiency in all regions should be actively improved. Model results also show that Central occupies the largest share in total national energy requirements and CO_2 emissions in both analyzed years. What is more, sensitivity analysis results of population show that 1% increase of population in this region will induce the largest rise of national energy requirements in both analyzed years. The sensitivity analysis of technology shows that both the intra-regional and national energy-saving effect brought by 1% improvement of energy end-use efficiency in Central are greater than those of the other regions. Especially the national energy-saving effect brought about from technology improvement in Central, is clearly higher than those brought about by the other regions. Scenario analysis of regional natural gas requirements shows that Northwest could have a relatively sufficient margin available to transfer to other regions, which will be 20.26–20.74 billion cubic meters and 26.58–30.77 billion cubic meters for year 2010 and 2020, respectively. However, the results of intensities show that under the assumption of uniform rates of technology improvement for all regions, the energy intensities of the Northwest region will be second highest in the two analyzed years. In such case, the sensitivity analysis of trade coefficients shows that 1% increase of imports of manufacturing goods from the Northwest region by any of the other regions will drive-up the total national energy requirements. Therefore, in order to facilitate energy saving at the national level and the effectiveness of inter-regional energy transfers, the efficiency improvement in regions Central and Northwest should be accelerated as much as possible.

In the case of high economic growth, if the national average of the total fertility rate increases from 2.0 to 2.1 (scenario HP versus H), the energy requirements and CO_2 emissions will increase some 1.08% in year 2010, and the increase rate will mount-up to approximately 1.77% in year 2020. Population growth in one region will not only significantly affect energy requirements of the region itself, but also drive-up energy requirements of the other regions. It would be a hard and long-term task to obtain on the national-level a more obvious improvement of energy end-use efficiency than the potentials projected in the current energy-saving layouts; what is more, the effect of energy substitution is not likely to be seen before 2020 (Zhou, Yukio, 1996). Therefore, in order to create a favorable population environment for the important period when China is making efforts to build a well-rounded, economically vibrant society, in the near future the basic state policy of family planning should be seriously followed in each region.

At the energy supply aspect, by using system dynamics, this chapter focuses on the forecasting of coal supply until 2020. The conclusions show that the amount of coal produced by state-owned mines will be between 1.55 billion tons and 1.72 billion tons in 2020, and town or village mines will produce coal from 400 million tons to 570 million tons. Production capacity of constructing mines will be between 410 million tons and 570 million tons, and available reserves for constructing mines will be between 49.9 billion tons and 152.6 billion tons.

Simulations show that, because of long periods of construction for mines, the problems of insufficient mine construction investment during 1990s started to appear in the early 2000s. Because of the same reason, the status of over-load production of state-owned mines might not be changed in a long time. Simulations show that geological prospecting during the ninth five-year-plan did not receive enough attention, and suggest the level of geological prospecting investment during the eighth five-year-plan is more suitable. If the investment coefficient of mining and washing of coal is greater than 9%, the over-load production status of state-owned mines could be solved before 2020. State-owned mines, which must be complemented by town or village mines, can not meet the rapidly increased coal requirements. Town or village mines, whose production have low safety degree, must be reformed.

To sum up, coal requirements increased rapidly from 2002. It caused state-owned mines to increase their production and lead to an over-load status. On the other hand, town or village owned mines get an opportunity to increase their production, which results in a worse status of coal industry in China for a long period. Periods for geological prospecting and mine construction are long, and the delay of investment moments in 1990s caused a serious impact on coal system. In order to ensure coal production and supplies meet coal requirements caused by high economic development in China, investment for mining and washing of coal and geological prospecting must be increased immediately.

CHAPTER 4

Fluctuation in Oil Markets and Policy Study

As the consumption of petroleum and its products permeates through all sectors of industry, the status of petroleum in the national economy and security is gradually increasing. Nowadays, Chinese domestic production of petroleum cannot satisfy growing demand, thus imports are rapidly increasing. The increasing import dependency threatens the domestic petroleum market; even national security. Against this background, the policy makers put more emphasis on the fluctuation of oil prices and the causes. Thus,
- What are the regularities behind the fluctuations?
- What is the relationship between international and domestic crude oil prices?
- To what extent does the fluctuation of oil prices impact on China?
- How can we predict future oil prices?
- How can China minimize the international market risks and by what kind of policies?
- What are China's strategies to cope with the hot competition in international oil markets?

In this chapter, we put forward some empirical studies on these problems and proffer policy suggestions.

4.1 The Characteristics of International Petroleum Price Fluctuation

Since the international petroleum prices increased rapidly in 2003, petroleum prices became a hot issue internationally. First, this section introduces the relationship between petroleum prices and the global economy. Next, we describe the three historical petroleum crises and features of the present oil prices. Furthermore, based upon the above-mentioned analysis, we discuss the long-term tendency and short-term fluctuation. Finally, we pose some suggestions about how to deal with international oil price fluctuations.

4.1.1 International Petroleum Prices and Their Impact on the Global Economy

Nearly all fluctuation in oil prices causes great impact on the global economy. Since 2003, international petroleum prices have drastically increased. According to the World Bank, a one-year-long 10 US$/barrel increment decrease about 0.5% to the world economy, among which 0.75% to the third world. Jiao (2005) found that increasing international petroleum prices have a direct impact on China's economy, that is, 50% increment of oil prices would bring about 1.468%, 0.106%, 0.206%, 0.137% and 0.636% decline in exports, investment, consumption, actual GDP and RMB exchange rate, respectively, while 1.468% and 0.675% increments to foreign currency expenditure and general price standards by GDP deflator. According to the total import volume, namely, 70 million tons of crude oil, every one US$ increment of oil price, would cost 4.3 billion US$ or more to import that same amount. As international crude oil prices increase 20%, export will decrease 0.841%; as international crude oil prices increase 50%, export will decrease 1.468%.

The speculators around the world are buying petroleum at high prices because of the globally increasing prices and the massive demand in Asia and the Pacific. In May 2004, Morgen Stanley and Bank of Germany set aside 932 million US$ to purchase half of the total crude oil production of Canadian Royal, that is, 36 million barrels. In August, Morgen Stanley invested 775 million US$ to buy potential production capability of 24 million barrels in four years of another oil producer (Wang, et al., 2004). Against this background, the tendency of growing demand for crude oil is inevitable to be a target of many speculators and bankers. Thus, how to make the best use of limited petroleum resources and design suitable strategies to evade the risks, are crucially important nowadays.

4.1.2 Oil Price Level Characteristics of Three Oil Crises

4.1.2.1 Wavelet Model for Oil Price Level Characteristics Analysis

Oil price level is the range of oil price volatility over a long period, describing the long-term trend and general level of oil prices. The present oil price level must be analyzed and the future one must be estimated in order to work out long-term national oil strategies, which can thus acutely consider the effect of oil price volatility. Oil price level's change is the structure change of oil price series in time series analysis, namely the inherent change in oil price time series. There are numerous approaches and applications regarding structure changes of time series. Most methods test structure changes of time series based on linear models, where firstly every structure of time series is fitted with linear models, and then the final position of break points at which one structure changes to another structure are decided according to the parameter change of the fitted linear model. However, there is a serious deficiency in linear models, which cannot correctly divide different structures and positions of break points for non-linear time series.

Because of complex related factors, oil price level takes on highly non-linear, even chaotic characteristics. Therefore oil price time series can be considered as a non-linear system and a method sensible in non-linear structure changes and accurate in break point detection is expected for the structure change and break point approach of the oil price. Currently, as a strong non-linear analysis tool, wavelet is widely applied in many fields of non-linear science. When the time series structure changes, the singularity usually changes in the break point. The singular point appears because one structure changes to another different structure and this process leads to discontinuity. Wavelet analysis can effectively detect the local singularity from different scales, namely signals suddenly change in some points, causing discontinuity or derivative discontinuity, showing the structure changes induced by accidents. Lipschitz index is usually used to measure the signal singularity. Oil price time series can be taken as numeral signals. The oil price times series will show great singularity when price level changes are evident.

4.1.2.2 Oil Price Level Characteristics of Three Oil Crises

The monthly data of refiner acquisition cost of imported crude oil (RACI) in the United States from 1970-01 to 2005-02 is adopted for the analysis of oil price level characteristics of three oil crises. The RACI monthly data is shown in Fig. 4-1.

The complex Morlet wavelet is used to perform singularity analysis in 64 scales for RACI and the calculated Lipschitz index is shown in Fig. 4-2. The Lipschitz index in the position of the circle is smaller than 1, which is a singular point. Because the head and tail data is dealt in a wavelet transform, leading to certain errors in actual computation, the head and tail singular points are neglected.

The position of singular points in actual curve is shown in Fig. 4-3, the vertical dotted lines indicate the oil price is a singular point, where the Lipschitz index is smaller than 1.

Important oil price related events will change the inherent structure in oil price and make it jump from one price level to another. In the transition period of two oil price structures, different factors interact and make oil prices violently fluctuate. In these periods, singular points will frequently appear. Many close singular points indicate a changeable period

4.1 The Characteristics of International Petroleum Price Fluctuation

affected by related events, so we consider close singular point as one period. Four different periods according to the singular points are shown in Table 4-1.

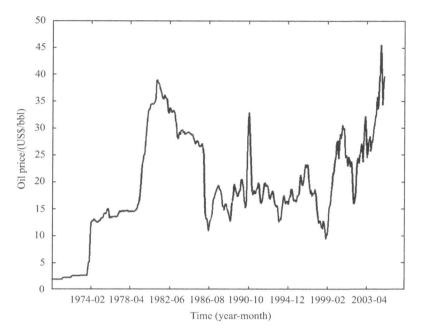

Fig. 4-1 RACI monthly data
Data source: Energy Information Administration.

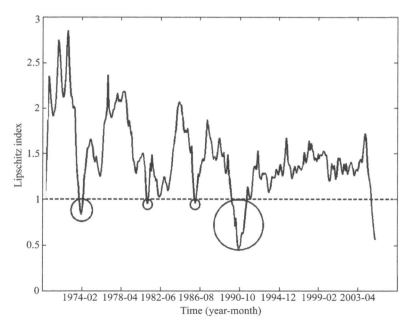

Fig. 4-2 Lipschitz index of RACI

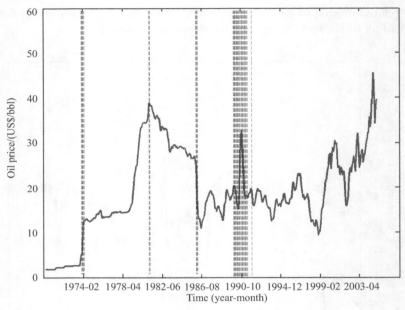

Fig. 4-3 Singular points of RACI

Table 4-1 Four different periods corresponding to singular points

Order	Period
1	From 1973-11 to 1974-03
2	From 1981-01 to 1981-02
3	From 1986-01 to 1986-03
4	From 1989-12 to 1991-11

The three oil crises corresponding to the oil price singularity detection are shown in Tables 4-1 and 4-2, which show the accuracy of wavelet singularity detection. From Tables 4-1 and 4-2, the periods of singular points are close to corresponding historical events, which indicate that sudden events lead to the structure change and singular points of oil price.

Table 4-2 Oil price and history events at singular points

Order	Oil price	History events
1	Rise from several US$/bbl to more than 10US$/bbl	The first oil crisis (1973–1974), the west countries, economies decay, US GDP dropped by 4.7% and EU GDP fell by 2.5%
2	Rise from more than 10US$/bbl to more than 20US$/bbl	The second oil crisis (1979–1980), Iran-Iraq War, main west countries, economies seriously decay
3	Fall from more than 20US$/bbl to more than 10US$/bbl	Oil demand decreased because of the last two oil crises and oil price fell sharply (1986)
4	Violently fluctuates between more than 10US$/bbl and more than 30US$/bbl	The third oil crisis (1990), Gulf War

Through the analysis of singular points in three oil crises, we find that the singular point has the ability of capturing oil price level's violent fluctuations. The properties of singular points when oil price level changes are shown below:

(1) The singular points of oil price time series indicate sharp changes of oil price, namely the structure of oil price which appears inherent changes from one price level to another different level. Taking Table 4-3 and Fig. 4-4 for example, and divided by the four singular points, historical oil price can be separated into five different periods evidently. Oil price

4.1 The Characteristics of International Petroleum Price Fluctuation

fluctuates smoothly inside the same period but there is great difference between different periods for average oil price level, which shows different oil price structures.

Table 4-3 Mean and standard deviation of oil price in different periods (US$/bbl)

Order	Period	Mean	Mean change	Standard deviation
I	From 1970-01 to 1973-10	2.26	—	0.41
II	From 1974-04 to 1980-12	18.00	15.74	7.47
III	From 1981-03 to 1985-12	30.91	12.91	3.56
IV	From 1986-04 to 1989-11	15.99	14.92	2.40
V	From 1991-12 to 2005-02	21.16	5.17	6.96

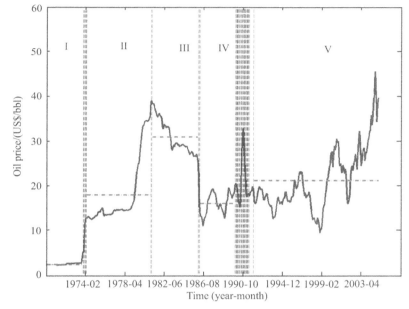

Fig. 4-4 Oil price level in different periods

From Table 4-3, the average difference of oil price is above 10US$/bbl between period I, II, III and IV and the one between IV and V is also above 5US$/bbl. The standard deviation is quite small in I, II and IV, which indicates the stability of oil price, while the one in II and V is greater, which indicates the oil price fluctuates more violently in these two periods. However, there are no singular points in the two periods, which indicates the oil price changes smoothly, that is, the price level in the whole period doesn't jump sharply as shown in Fig. 4-4.

(2) The important events closely related to oil price usually correspond to the singular points. The appearance of important events will lead to violent fluctuation of the oil price and evidently affect the oil price trend and cause the singular points. At the same time, singular points indicate the change of oil price structure. If oil price changes are inherent, then there must be corresponding events. Three oil crises are closely related to singular points as shown in Table 4-2.

(3) There are more singular points in the periods of complex and lasting events. The related factors are complicated and the oil price changes rather violent. For example, in the 1990 Gulf War, the international oil market was changeable and the singular points in this period are numerous and the oil price reached the highest level in history.

In the three oil crises, oil price levels appear to change and there are singular points. According to He et al. (2005), the three oil crises are described as follows:

(1) The first crisis (1973–1974).

The fourth Middle East War broke out in 1973-10. In order to hit Israel and its supporters, the Arab countries of OPEC declared the claim of pricing rights of crude oil in December and raised the basic crude oil price from 3.01US$/bbl to 10.65US$/bbl. The oil price rose by more than double.

(2) The second oil crisis (1979–1980).

There was a serious political crisis in the world's second largest oil export country, Iran, in 1978. The Shah of Iran, close to the United States, collapsed and the second oil crisis broke out. The oil production was seriously affected because of the Iran-Iraq War at the same time, from 5.8 million bbl/d to below 1 million bbl/d, which broke the unstable equilibrium in the international oil market. As the sharp decrease of production, there was a gap of 5.6 million bbl/d in the international market. The oil price rose sharply from 1979, from 13US$/bbl to 34US$/bbl in 1980.

(3) The third crisis (1990).

Iraq was under sanctions after its intrusion into Kuwait in 1990-08, which interrupted its crude oil supply; and international oil prices rose sharply to 42US$/bbl.

4.1.3 The Oil Price Level Characteristics in 2005

4.1.3.1 The Singularity before the Change of Oil Price Level

When oil price level changes, i.e., before singular points appear, oil price and relevant Lipschitz index appear to follow certain rules, which can prove an important reference for us to forecast singular points and further oil price trends. Lipschitz index before singular points is shown in Table 4-4 ($-1, -2, -3, -4, -5$ indicate separately 1 month, 2 months, 3 months, 4 months and 5 months before singular points).

Table 4-4 Lipshitz index 5 months before singular points

Order \ Month	-5	-4	-3	-2	-1
1	1.6436	1.4103	1.2436	1.1857	1.0301
2	1.4833	1.3962	1.3472	1.1242	1.0055
3	1.5053	1.5858	1.3156	1.2099	1.0887
4	1.4869	1.2716	1.2304	1.141	1.0455

From Table 4-4, we can conclude the Lipschitz index shows monotonic decreases evidently before singular points appear. At the same time, oil price will evidently change before singular points appear, as shown in Table 4-5. At singular point 1, the oil price rose by 57.14%, from 2.59US$/bbl, some two months before, to 4.07US$/bbl one month before, which indicates that the oil price may go on rising after this singular point. At singular point 2, the oil price rose by only 0.02US$/bbl between five months and four months prior, while up to 0.54US$/bbl between two months and one month prior. This evident change also indicates that the oil price may go on rising after this singular point. At singular point 3, located at the beginning, the oil price fluctuated around 26.6US$/bbl, but suddenly fell by 0.91US$/bbl between two months and one month prior, which indicates that the oil price may go on falling after this singular point. Similarly, at singular point 4, the oil price rose sharply, which indicates that the oil price may go on rising after this singular point. Combined with a real change of oil price and Lipschitz index, we can understand the precedent phenomenon before singular points and judge the appearance of singular points and the trend in any future oil price.

4.1.3.2 The Oil Price Level Characteristics in 2005

Singular points did not appear in 2001, which indicates that the oil price went on rising

4.1 The Characteristics of International Petroleum Price Fluctuation

Table 4-5 Oil price 5 months before singular points

Order \ Month	−5	−4	−3	−2	−1
1	2.59	2.59	2.59	2.59	4.07
2	34.44	34.46	34.63	35.09	35.63
3	26.61	26.56	26.79	27.12	26.21
4	17.99	17.23	17.62	18.29	18.32

continuously; although it changed smoothly. Compared to the three oil crises, the degree of change is much smaller. Meanwhile, the Iraq War with the U.S. did not lead to a singular point in 2003, which indicates the defense against an oil crisis is more strengthened with the construction and precision of an oil strategy store and realization of financial futures mechanism. However, the recent high oil price rising should not be neglected, because the oil price always changes in one direction before singular points appear. If the oil price cannot be controlled, another oil crisis may happen.

The crude oil futures price went over 60US$/bbl post 2003, which is the highest price in history since crude oil futures traded in 1983. However, the crude oil price continually rises, even greater than 70US$/bbl in some periods. There have been three oil crises since 1947. The oil price rises in the recent two years may be evidence of a fourth oil crisis and governments should be put on notice.

There has been great speculation leading to this oil price rise, but there are always some (natural) oil supply and demand changes which cannot be neglected. After exploitation for hundreds of years, there have been no large oil field finds in the past 30 years. Some 80% high production areas were discovered before 1972 (Zhao, 2005). Presently, oil fields are in a phase of extensive exploitation or decline, such as the Daqing Oil Field in China and the North Sea Oil Field in Britain. Because the North Sea Oil Field is in decline, Britain has to change from an oil export country to oil import country. Global oil consumption increases every year. In part, this is due to the economy of China and India developing at high speed, and thereby the global oil consumption has increased. However, the daily production of OPEC is larger than the official quota, over 2 million bbl, and most OPEC members have reached the production limit. However, the crude oil supply cannot increase over a short period (Zhao, 2005). Because the oil store of oil producing countries is in decline and oil demand increased sharply, the imbalance of demand and supply will lead to higher oil prices.

According to Wen (2005), oil demand raised by 1.26%, from 53 million bbl per day in 1972 to 77.3 million bbl in 2002. Interestingly, the oil supply rose by only 1.17% from 54 million bbl per day in 1972 to 76.6 million bbl in 2002. At the same time, the experience of the past 30 years indicates no matter how oil price fluctuates, oil demand increases all the time. Even in 1973, the oil price rose from 3US$/bbl to 12US$/bbl, while the oil demand dropped a little from 57.3 million bbl/d in 1973, to 56.4 million bbl/d in 1974 and 56.2 million bbl/d in 1975. However, it went back sharply to 59.5 million bbl/d in 1976. In the period that the oil price rose so much, oil demand did not decrease, which indicates oil consumption is stable and there are no substitutions. Because of the stability, even if oil price rises substantially, oil consumption is difficult to reduce.

To summarize, the present oil price level can not be treated optimistically. Depending on the oil supply and demand, and short-term substitution, the future oil price level will no doubt keep at its present high level. The low oil price era was in the past.

4.1.4 The Long-Term Tendency of International Crude Oil Prices

We take the Daqing crude oil which is produced in the largest oil field (Daqing) in China as a domestic representative and Brent crude oil as the international comparator. Because China coalesces its domestic market with the international market in 1998, we choose the

FOB price time series dating from Jan. 1997 to Dec. 2004, both of which are taken from the U.S. EIA website. Fig. 4-5 illustrates the tendency of the spot FOB Daqing and Brent crude oil prices.

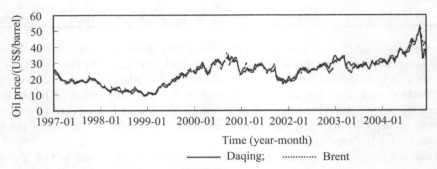

Fig. 4-5 The tendencies of international and domestic crude oil prices

China became a net import country of crude oil starting in 1993 and of petroleum since 1996. The imported crude oil makes up a third of the total domestic consumption. China cannot retreat from the international market because the domestic prices are deeply influenced by international prices. We find that in Fig. 4-5 the price tendencies of international and domestic crude oil are basically similar. Generally speaking, during this period, they both experienced two periodic circles of increasing and declining, with the average price of second circle being higher.

Since 1997, the lowest domestic price is 9.5US$/barrel on Dec. 25, 1998, 0.3US$/barrel higher than the lowest international price of 9.20US$/barrel on Dec. 11, 1998; the highest domestic price is 49.54US$/barrel on Oct. 15, 2004, 1.94US$/barrel lower than the highest international one which is 51.48US$/barrel on Oct. 29, 2004; at the same time the average price of international and domestic are almost equivalent.

According to Fig. 4-5, we divide the price tendency into four phases:

The first is the declining stage, from January 1997 to December 1998. Brent price declined from 23.40US$/barrel in January 1997 to less than 10US$/barrel in December 1998. In the same period, the Chinese crude oil price also declined from 23.90US$/barrel to less than 10US$/barrel; the lowest was 9.20US$/barrel and 9.50US$/barrel, respectively. The main reason for the decline was insufficient demand. During this period, Asia suffered from a financial crisis and slow economic growth. Some countries, such as Indonesia, Malaysia, the Philippines and others, recorded approximately 8% negative growth; Japan recorded 2.5% negative growth. After the crisis, the demand for petroleum products decreased about 3% in Asia. For the global economy, economic growth slowed down from 3.2% in 1997 to 2% in 1998.

The second is the oscillating and upswing stage, from January 1999 to September 2000. Brent price started to increase successively, rising from 10US$/barrel at the beginning of 1999 to 26US$/barrel at the end of the year; and breaking through 30US$/barrel in August 2000; the highest was 37.61US$/barrel. In 1999, the U.S. economy began to grow rapidly, and at the same time the Asian economy miraculously revived from the financial crisis, which stimulated the increase in demand for petroleum products. OPEC and non-OPEC countries maneuvered smoothly through the treaty by cutting 2.5 million barrels/day of their production in March 1999, which accounted for 3.3% of the total production in the world and almost used the surplus of supply. From that point on, international crude oil prices have increased by a large margin. The main reason for price rises during that period was the impact of factors associated with demand and supply.

The third stage is the oscillating and dropping stage, from October 2000 to December 2001. The world economy growth slowed down, and there was a slump in demand for oil, especially as a result of "9/11"; several days after crude oil prices quickly reaching more than 30US\$/barrel, the price dropped rapidly to 19.87US\$/barrel. The margin of decrease in 2001 was about 40%. OPEC price was less than 22US\$/barrel (the lower bound of OPEC Price Zone) over three consecutive days in 2001, and a low again in October 2001 when it reached 19.07US\$/barrel. The main reason for the price change in this stage is: weak demand and market dynamics where speculation played a major role in the violent aftermath of "9/11".

The fourth stage is the oscillating and rising stage, from January 2002 to December 2004. OPEC cut its production three times by an accumulated 1.5 million barrels/day in 2001. Russia, Mexico, Norway and other non-OPEC countries also reduced their crude oil exports. This had the effect of stabilizing the crude oil market; and the price began to rise at the end of the year 2001. OPEC (except Iraq) cut production to a total of 22.69 million barrels/day until February 2002 that resulted in the lowest production output in the last ten years. In addition, the world economy began to revive, which increased the demand for petroleum, and funds speculated balefully in the futures market, which made the oil price rise rapidly. During the Iraq war, the crude oil price reached 34.18US\$/barrel. After that, the price dropped by a small margin, but the price began to rise slowly from May 2003. Thereafter, from April 2004, the price increased dramatically, and led to panic in the world. The determinants of the trend in this period are: the demand, supply, war and funds speculation, along with devaluation of the U.S. dollar that accelerated the price rising.

In summary, the main factors which affect crude oil prices include fundamental factors akin to production and demand of petroleum, as well as non-fundamental factors like some major events, e.g., "9/11" and the Iraq war. Furthermore, global speculation power plays an important role in pushing the prices higher. Speculation on crude oil markets amplifies the fluctuations of oil prices.

After all, it is fundamental factors resembling production and demand of petroleum that decide the long-run tendency of crude oil prices. Because of its non-renewable feature, the supply elasticity of crude oil is not flexible, so that the main factors which affect crude oil prices are the status of the world economy, if there are no newly found large-scale oil fields, and major technology innovation. As the world economy increases, the demands for crude oil will gradually increase, so that the world may confront insufficient supply in the future.

4.1.5 The Short-Term Fluctuations of International and Domestic Crude Oil Prices

4.1.5.1 The Crude Oil Price Tendency in 2005

Since 2005, the domestic crude oil prices keep rapidly increasing and achieving new high records. Daqing broke through 60US\$/barrel in Sept. 2005 and Brent in Aug. 2005. Daqing prices stayed only a short time at this price, and then fell down below 60US\$/barrel, while Brent remained at that price for a long time. According to an IMF report, the international crude oil price would keep on increasing: the average price would remain 54.23US\$/barrel that year, and increase to 61.75US\$/barrel the next year, which are higher than a previous prediction in April, that is, 46.5 and 43.75, respectively.

Geopolitics is another factor that affects oil prices. The political situation in world key oil production areas and terrorism that destroy oil facilities accelerate the increase of oil prices. Furthermore, crude oil will play an even more important part in the world economy, which will boost prices up further. In the first two decades, petroleum remains the most important primary energy worldwide. It is predicted that in the 2030s, petroleum will make up 40% of all energy consumption and that daily global demand will increase from the present 84.30

million barrels to 120 billion barrels. At the same time, the oil reserves are limited. As a consequence, the future prices will still have room for further increase, i.e the average prices of world petroleum will possibly be greater than 48US$/barrel, and the low price epoch will never return. As for global distribution of oil reserves, the OPEC region has the richest reserves, which take up 80% of all oil reserves, thus OPEC is still one of the most influential powers in oil pricing.

Nearly all nations concerned with oil pricing influence the fluctuation of oil prices, due to the direct relationship with oil production countries' international balance, national economic growth, and economic interests of oil importing countries. As the transfer of petroleum supply and demand develops, OPEC may have greater power to control the market and prices.

Furthermore the crude oil supply and demand has features of rigidity and unsubstitutability. The demand for energy in China and India is increasing rapidly, while that of industrialized countries like U.S. Japan, European countries is increasing gradually too. The strong demand for petroleum brings about supply risks, which offers a good reason for speculation. As for the short-term tendency, some evidence suggests that international oil prices can hardly fall down dramatically, and that high oil prices will remain (Wang, 2005).

4.1.5.2 The Price Fluctuations of International and Domestic Crude Oil

Price fluctuations of international and domestic crude oil can reflect directly the price risks. Fig. 4-6 illustrates the fluctuations of the log-log returns of Daqing and Brent crude oil prices.

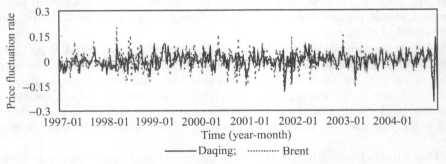

Fig. 4-6 Crude oil price fluctuation

Fig. 4-6 shows us that although domestic oil prices share a similar tendency with the international ones, there exist some differences between them. During the observation period, the biggest upside fluctuation of international prices is 19.86%, which happened on Mar. 27, 1998; at the same time that of domestic ones is 10.4%; the biggest downside fluctuations are 15.58% for Brent during the Iraq war and 10.79% for Daqing. The "9/11" tragedy encouraged malicious speculations in oil, which drove the prices down. Daqing prices suffered an 18.76% decline and Brent 14.2%. The biggest upside and downside fluctuations are 24.89% on Nov. 12, 2004 and 13.42% on Nov. 26, 2004, respectively, which are both greater than what happened to international ones. Table 4-6 illustrates that although the mean values and standard deviations of domestic prices is less than those of international ones, the fluctuations are usually higher than international ones under certain situations such as "9/11", war, and high prices in 2004. Thus, the domestic market is more fragile than international ones. The plausible reason lays in the domestic reference pricing mechanisms, that is, Tapirs for light crude oil, Minus for intermediate I, etc. Those references are all Middle East or South Asian crude oils, which represent only regional supply and demand, unlike the global references, e.g., Brent and WTI. That is why domestic market is much more vulnerable to speculations and manipulations in international markets.

Table 4-6 Descriptive statistics of oil price fluctuation

	Minimum	Maximum	Average	Std.
Daqing	−0.2489	0.1342	0.0005	0.04292
Brent	−0.1558	0.1986	0.0008	0.05299

Compared with the situation a few years ago, the magnitude of oil price fluctuations is not so large, that is 6.7% and 8.6% for the largest upside international and domestic ones respectively, and 1.1% and 1.2% for average ones. So the prices, both international and domestic ones, are steadily rising in 2005.

4.1.6 Concluding Remarks and Policy Suggestions

4.1.6.1 The Implications and Conclusions of International Oil Price Fluctuation

In this section, we analyze the fluctuations of international oil prices and relative issues. Based on the study, we draw the following conclusions:

(1) The relationship between the world economy and international oil price fluctuations becomes intensively correlated.

In the industrialization society, demand for energy is increasing rapidly. Petroleum becomes the most important primary energy as a substitute for coal. That is why the world economy and international oil price fluctuations become intensively correlated nowadays. Every persistent change of oil prices will influence the world economy.

(2) The import dependency of China on petroleum is increasing every year. China's economic growth relies on international oil prices.

As China's economy growth accelerates (as in recent years), the consumption of petroleum is, accordingly, increasing rapidly. The import dependency of China on petroleum is increasing, in that domestic production is far from sufficient for growth in demand, which can only be satisfied by more imports. With the international oil prices getting higher, the cost of import is rising accordingly, which gradually become the obvious burden on China's economy.

(3) The three oil crises are accompanied by salient changes of oil prices.

Historically, there were three identifiable serious petroleum crises, all of which brought about disastrous outcomes for society and the economy. The common feature of the three crises is that they were accompanied by salient changes of oil prices, especially those changes of jumping from a relatively lower to higher standard. Relative research results illustrate that the prediction of this price jumping can be a superior indicator for a petroleum crisis and a useful tool for making preparations for it.

(4) The complex interactions of fundamental and non-fundamental factors make the oil prices hard to be predicted.

Fundamental factors, including production and consumption of petroleum, largely decide the long-term tendency of oil prices. Non-fundamental factors, e.g., "9/11" incident, Iraq war, speculation, oil field explosion, strikes, etc., mainly influence the short-term fluctuation. The complex interactions of fundamental and non-fundamental factors make the oil prices hard to be predicted.

(5) Under normal circumstances domestic oil price fluctuations are less in magnitude than international ones, but in outburst incidents the former will suffer a greater fluctuation than the latter.

At present China's current pricing mechanism is deeply influenced by the reference oil prices, especially those of Middle East and South Asia whose production and trading volumes are far less than Brent and WTI. That is the main reason why China's domestic oil prices are more fragile than international ones. This vulnerability makes domestic prices easy to be influenced or even manipulated.

(6) The abnormal changes in these two years may implicate the beginning of the fourth petroleum crisis.

Since 2003 the oil price keeps increasing rapidly. Recently the prices stably remain about 60US$/barrel. The current situation is similar to the former three petroleum crises. In 2004 and 2005 many incidents happened to boost the oil prices upwards such as terrorism attacks, strikes, hurricanes, etc. As for fundamental factors, oil prices increase because world economy recovers from the financial crisis, which stimulates world demand for petroleum. OPEC production capability is reaching to their thresholds but the supply can hardly satisfy growing demand. Furthermore, the disasters and speculations accelerate the increase speed. If this tendency persists, the fourth petroleum crisis is drawing near to damage world economy. In the past China could satisfy its demand for petroleum by means of its own production, so that the former three crises had no impact on China's economy. But against the background of increasing oil import dependency and growing demand for more petroleum, China's economy is exposed to high volatilities and risks of high oil prices.

4.1.6.2 Policy Suggestions to Evade International Market Risks

To stabilize economic growth and protect national security, China should fully realize the importance of stipulating and enforcing the correct petroleum strategy so as to minimize the damage of potential risks to its economy. We propose the following suggestions:

(1) Put more emphasis on stipulating petroleum strategy to prevent the potential oil price risks.

In the past, China could satisfy its demand for petroleum by its own production; and imported petroleum would make up a relatively small proportion in gross energy consumption. Under such circumstances, the importance of stipulating a petroleum strategy to prevent the potential oil price risks was not an urgent task. With hot global competition for limited oil reserves to satisfy domestic demand, especially with fast growing consumption and oil importation, China realizes the importance of the right strategy and puts forward foresighted policies, e.g., circulation economy and economized economy, etc. To better evade oil price risks, China should put more emphasis on stipulating long-term petroleum strategy by means of quantitative or qualitative scientific research, minimizing the potential oil price risks and ensuring a smooth and stable economic growth.

(2) Put more emphasis on oil exploitation to increase the self-support capability during a crisis.

The import risks make up a large proportion of the total oil price risks; international price changes will bring about import costs. Thus, another way to minimize oil price risks is to produce more domestic petroleum instead of importing foreign oil when the prices are high. As a consequence, we should put more emphasis on oil exploitation to increase the self-support capability and to make better understanding of domestic oil reserves; besides, we should improve the current techniques of oil exploitation; when necessary, China should reserve some oil wells for future usage like some other countries do.

(3) Disperse import risks and protect import security by broadening import channels.

In the past, China usually imported petroleum via only a few channels, that is, mainly by oil tankers, which might be very vulnerable to military blockage, and local geopolitics. Thus there are two ways to disperse import risks and protect import security by broadening import channels: firstly, China can import petroleum from more countries, especially those outside of OPEC, e.g., Russia, South American and Asian countries, to disperse import risks; secondly, China can choose more sea routes, e.g., those in the Indian Ocean, and various ways besides oil tankers to import, e.g., the oil pipelines connecting China to Russia and Middle East.

(4) Take so-called "international cooperation" strategies more positively.

Threatened by insufficient oil supply and a potential oil crisis, China should invest more funds in so-called "international cooperation" strategies, that is, more oversea investment and cooperation on oil exploitation and higher strategic status in international competition, so as to protect our interests and security. The implementation of these strategies implies that first of all, China should invest more in foreign oil companies to improve our competition capability; secondly, China should foster a closer relationship and more co-operation with different countries with rich reserves; thirdly, China should learn more advanced technologies from competition and cooperation with foreign energy industrials; fourthly, China should be more flexible in the way of cooperation, e.g., we can invest solely or jointly in all parts of oil exploitation, production, refinement and sales, etc.

4.2 Analysis of the Co-Movement between Chinese and International Crude Oil Prices

Production and consumption are separate spatially in the world, that is, the primary petroleum production countries, e.g., OPEC countries, only consume a very small proportion, while the primary consumption countries, e.g., OECD countries, own far less oil reserves than OPEC. Furthermore, the complex political and religious situations in the Middle East where there is rich oil reserve make the oil price fluctuations more drastic.

China began to depend on net importing of crude oil in 1993, and to rely on net importing petroleum products in 1996. Now, more than a third of all consumption is derived from the international market. Chinese crude oil price suffered greatly from large imports from the international market after the 1990s. In addition, China changed its petroleum pricing mechanism after 1998, and partly or totally released the power of regulating petroleum products' pricing. In this section, by studying the co-movement between Chinese and international crude oil prices, we quantitatively explore the effect of the Chinese petroleum pricing mechanism reform, identify problems, and analyze the impact of international crude oil price on the Chinese oil price. What are the effects of these reforms? Are there any problems in these reforms? By means of time series analysis, we will study quantitatively the interaction of domestic and international crude oil prices in order to answer these questions.

4.2.1 The Long-Term Relationship between International and Domestic Crude Oil Prices

The time frame of the analysis is from January 1997 to the end of 2004. Two sets of data series were obtained from the web of the U.S. Department of Energy and Energy Information Administration. To simplify the acquisition of stationary series, we choose a logarithm series identified as Logarithm of Brent (LBR) price and Logarithm of Daqing (LDQ) price.

To test co-integration between international and domestic crude oil prices, it requires a certain stochastic structure of the individual time series (Engle, et al., 1987). First of all, we focus on first-order non-stationary integrated processes (Dickey, et al., 1979, 1981; Phillips, et al., 1988), $I(1)$ processes, which require first differences to become stationary. Thus, to test for the presence of stochastic non-stationary characteristics in the data used here, it is necessary to investigate the order of integration of the individual time series preceding any other tests; this is known as unit root tests. The results of stationary testing on logarithmic prices illustrate that there exists a long-term co-integration relation between international and domestic crude oil prices in that under critical value at the 1% level. Both of them are non- stationary while these non-stationary series are differenced 1 time to become stationary, then it is said to be integrated of 1 order: i.e., $I(1)$. Thus, the test shows that the presence of co-integration might exist.

Trace test statistic (Johansen, 1988) shows that there exists at least one integrated relationship between logarithmic international and domestic crude oil prices. The co-integrated relationship, namely, long-term equilibrium relationship between them is defined as:

$$\ln pd = 1.017 \ln pb - 0.057 \qquad (4\text{-}1)$$

Where: $\ln pb$ and $\ln pd$ are the natural logarithmic international and domestic crude oil prices.

We adopt the Wald test to the coefficient of $\ln pb$. The result shows probability $p = 0.681 > 0.05$, so we accept the null hypothesis, which means the difference of LDQ and LBR is approximately a constant, its value is -0.057.

$$\ln pd - \ln pb \approx -0.057 \qquad (4\text{-}2)$$

It also means the ratio of LDQ to LBR is approximately a constant, its value is 0.945.

$$pd \approx 0.945 pb \qquad (4\text{-}3)$$

The result may imply that LDQ is happening after the change of LBR, that is, LDQ fully follows LBR.

4.2.2 The Short-Term Relationship between International and Domestic Crude Oil Prices

4.2.2.1 Granger Causality Analysis

The forgoing results illustrate that there exist a long-term co-integration relationship and approximate ratio relationship between the international and domestic crude oil prices. But it is not enough to understand the sequence of mutual impacts. What is the driving force? The Granger causality test is applicable for this task.

The Granger causality test was carried out to test for causality between Chinese and international crude oil prices. The long-term causality (level) and short-term (first difference) causality of these variables were tested. The results of the Granger causality analysis are reported in Table 4-7.

Table 4-7 The Granger causality test for Chinese and international crude oil prices

Null hypothesis	Obs.	F-statistic	Probability
lags 3			
LBR does not Granger cause LDQ	357	51.0419	0.00000
LDQ does not Granger cause LBR		6.55812	0.00025
lags 5			
LBR does not Granger cause LDQ	355	29.9265	0.00000
LDQ does not Granger cause LBR			0.00214
lags 3			
ΔLBR does not Granger cause ΔLDQ	356	48.1278	0.00000
ΔLDQ does not Granger cause ΔLBR		2.13419	0.09562
lags 5			
ΔLBR does not Granger cause ΔLDQ	354	28.5730	0.00000
ΔLDQ does not Granger cause ΔLBR			0.34369

Results show that there are bilateral Granger causes between LDQ and LBR both in lags 3 and 5. But in the short-term, there is only one directional Granger cause, LBR is

4.2 Analysis of the Co-Movement between Chinese and International Crude Oil Prices

LDQ Granger cause, and LDQ is not LBR Granger cause. This underlines the fact that an international crude oil price shock can be related immediately to China, but the international market makes a response only when it finds the demand is changed when Chinese crude oil price fluctuates. So, in the formation of international crude oil pricing, China still has a passive role; it does not play an active part. It might be argued therefore, that this is an unfavorable and unfair situation for China. This is especially pertinent as more than a third of China's consumption comes from the international crude oil market.

4.2.2.2 The Short-term Mutual Impacts of International and Domestic Crude Oil Prices

Although the Granger causality test can be used to explain the sequence (or directions) of the long-term causality (level) and short-term (first difference) causality of the prices, it can not give a quantitative relation between them, while Vector Error Correction Model (VECM) (Li, 2000) can estimate the dynamic influences. According to LR rule (Yi, 2002), a VECM with 3-week lags, which is restricted with one co-integrating vector, was estimated. Applying the OLS regression, the VECM model can be listed below in Table 4-8.

Table 4-8   LDQ's vector error correction model estimate

Variants: $\Delta \ln pd$ Variant	Coeffient	std	t-statistics	
C	0.0002	0.0018	0.1299	0.8967
$ecm(-1)$	**−0.0602**	0.0337	−1.7893	**0.0743**
$\Delta \ln pd(-1)$	**0.1160**	0.0601	1.9303	**0.0543**
$\Delta \ln pd(-2)$	**−0.0978**	0.0593	−1.6501	**0.0997**
$\Delta \ln pd(-3)$	−0.0313	0.0485	−0.6458	0.5188
$\Delta \ln pb(-1)$	**0.3690**	0.0444	8.3040	**0.0000**
$\Delta \ln pb(-2)$	0.0180	0.0473	0.3817	0.7029
$\Delta \ln pb(-3)$	0.0391	0.0453	0.8637	0.3883
S.E. of regression	0.0360	Akaike info criterion		−3.7896
Sum squared resid	0.5047	Schwarz criterion		−3.7093
Log likelihood	760.2380	F-statistic		25.1450
Durbin-Watson stat	1.9322	Prob(F-statistic)		0.0000

LDQ is not significant to its own variables lagged 1,2, or 3 terms at 5% level, but variables lagged 1,2 terms is significant at 10% level; LBR's lagged 1 term is significant at 1% level. From quantity, LDQ changes 0.369% when LBR(1) changes 1%, LDQ only changes 0.12% when LDQ(1) changes 1%.

The results of VECM are illustrated in Table 4-9.

Table 4-9   LBR's vector error correction model estimate

Dependent: $\Delta \ln pb$ Variable	Coefficient	S.E.	t-statistic	Probability
C	0.0012	0.0026	0.4741	0.6357
$ecm(-1)$	0.2139	0.0479	4.4617	0.0000
$\Delta \ln pd(-1)$	−0.0913	0.0855	−1.0679	0.2862
$\Delta \ln pd(-2)$	−0.0203	0.0844	−0.2405	0.8101
$\Delta \ln pd(-3)$	**−0.1766**	0.0691	−2.5559	**0.0110**
$\Delta \ln pb(-1)$	**0.2296**	0.0633	3.6295	**0.0003**
$\Delta \ln pb(-2)$	**0.1148**	0.0673	1.7063	**0.0887**
$\Delta \ln pb(-3)$	0.0793	0.0644	1.2310	0.2191
S.E. of regression	0.0513	Akaike info criterion		−3.0833
Sum squared resid	1.0227	Schwarz criterion		−3.0030
Log likelihood	620.0282	F-statistic		4.0297
Durbin-Watson stat	1.9929	Prob(F-statistic)		0.0003

LBR is significant to the change of LDQ (3 weeks) at 2% level, and LBR changes 0.1766% when LDQ (3 weeks) changes 1%. LBR (1 weeks) is significant at 1% level, LBR (2 weeks) is significant at 10%. VEC model results show that LBR change 1%, 1 term (week) later LDQ changes 0.369%, while LDQ changes 1%, 3 terms (weeks) later LBR only changes 0.1766%. So, the impact of LBR on LDQ is bigger and faster, while the impact of LDQ on LBR is smaller and slower.

4.2.3 The Dynamic Impact of International Crude Oil Prices on Their Chinese Counterpart

4.2.3.1 Impulse Response Analysis

Using the estimated VECM, the impulse response analysis (Lutkepohl, et al., 1992) was performed to study the impact of Chinese and international crude oil price shock on its counterpart. Both LDQ and LBR's one standard deviation have approximately the same impact on its counterpart. The results of the impulse response over 52 weeks are presented in Fig. 4-7.

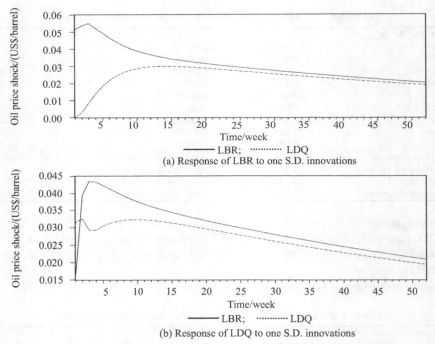

Fig. 4-7 Impact to one SD deviation shock in oil price

Both LDQ and LBR's one standard deviation have approximately the same impact on its counterpart. Both have a small positive margin, but the impact of LBR on LDQ has more intensity in the short-term. The effect is also more significant; while the impact of LDQ on LBR is more steady and slow, and the effect is smaller too. Specifically, the addition of one standard deviation to LBR results, in about two weeks later, in an 80% change of LBR that will be transmitted to LDQ. There is a small positive margin from the third week to the fifth week after which the LDQ begins to drop. But the impact of LBR on LDQ did not disappear totally; one year later, there was still approximately a 37% effect. The impact of LBR's one standard deviation on itself would maintain less than two weeks after which the impact had a small increase, and then it decreased sharply some seven weeks later. The speed of descending tends to decelerate; one year later LBR rose about 0.02US$/barrel.

4.2 Analysis of the Co-Movement between Chinese and International Crude Oil Prices

The response of LBR to one deviation of LDQ is comparatively small. The reaction in the first week is small; since the second week, the LBR raised quickly. This kind of trend was maintained up to the tenth week. Later, the trend decelerated, and closely mirrored the reaction of LDQ to LBR. By the 52nd week, the impact of LDQ on LBR was almost the same with LBR on LDQ.

4.2.3.2 Variance Decomposition (VDC) Analysis

Variance Decomposition (VDC) analysis (Lutkepohl, et al., 1992) provides a tool of analysis to determine the relative importance of Chinese and international crude oil price shock in explaining the volatility of its counterpart. The results of the VDC over ten weeks are presented in Table 4-10 and Fig. 4-8.

Table 4-10 Variance decomposition of LDQ

Period	S.E.	LBR	LDQ
1	0.0356	21.7920	78.2080
2	0.0624	47.1829	52.8171
3	0.0815	56.1406	43.8595
4	0.0968	59.7456	40.2544
5	0.1099	61.0958	38.9043
6	0.1215	61.5216	38.4784
7	0.1319	61.5171	38.4829
8	0.1414	61.3012	38.6988
9	0.1500	60.9864	39.0137
10	0.1580	60.6342	39.3658

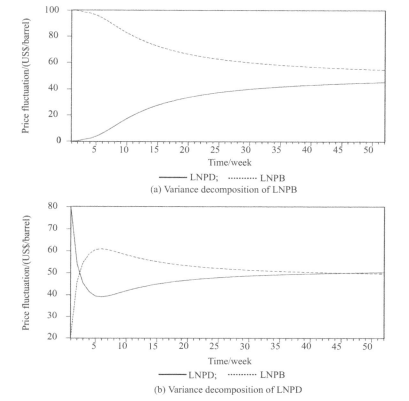

Fig. 4-8 Variance decomposition for international/Chinese crude oil price

The results of VDC analysis suggest the part of China's price in explaining its own variance from a 78.208% decrease to 39.366% after ten weeks. It also explains from the first week to the third week, the part of international prices in explaining the variance of China's price increase rapidly from 21.8% to 56.1%. However, China's 43.86% after five weeks suggests that the change tends to be smoother. Both explain about 50% variance of the Chinese price as show in Table 4-11. The major part of variance in international crude oil price comes from its own shocks. At the beginning, the estimated error of international crude oil price completely comes from its own, along with time of change, the construable part of itself dropped slowly, the role of Chinese price gradually augments. Up to the tenth week, the contribution rate of Chinese crude oil price is 14.8%. One year later, the situation tends to stabilize, and the contribution rate of Chinese crude oil price is about 40%. The result of VDC analysis shows that Chinese crude oil price reacts to the international crude oil price quickly, and it responds intensively, greatly, but the reaction of international crude oil price to its counterpart is slower, steadier and the influence is less.

Table 4-11 Variance decomposition of LBR

Period	S.E.	LBR	LDQ
1	0.0514	100.0000	0.0000
2	0.0744	99.8416	0.1584
3	0.0929	99.0651	0.9349
4	0.1074	97.6215	2.3785
5	0.1196	95.7280	4.2720
6	0.1302	93.5945	6.4055
7	0.1397	91.3803	8.6197
8	0.1482	89.1956	10.8044
9	0.1561	87.1099	12.8901
10	0.1634	85.1615	14.8385

4.2.4 Conclusions

According to the analysis, we can draw the following conclusions:

(1) There is a long-run co-integration relation between Chinese and international crude oil prices.

The elasticity of Chinese price to the international one is 1.017; that is to say, when the international price changes 1%, the Chinese price will change an average of 1.017%. Over the longer-term-run, the value is near to one. The Wald test result shows that the relationship between Chinese price and international price is approximately a ratio so their relative price almost has no change.

(2) International crude oil price affects the Chinese price whether over the short-term or long-term. While the Chinese price only affects the international price over the long-term, it does not affect the international one in the short-term.

The conclusions drawn above explain that whatever the variety, the Chinese crude oil price follows international price very closely. This suggests that China only connects price level with the international price level, but does not connect its pricing mechanism with its international counterpart. There are two possible reasons: first, China is passively tracking the international price, and China's price does not reflect the change of demand and supply in the domestic market. Second, China has let loose its pricing. Crude oil pricing is not rigidly regularized by the government, even though the degree of openness in the Chinese oil market is minimal. Also, the market is monopolized by three big state-owned petroleum companies, resulting in lack of competition and risk. For example, they often transfer all or most of the price risk to domestic petroleum consumers. This leads to the expression: "Buy rise and don't buy fall" in petroleum importing.

4.2 Analysis of the Co-Movement between Chinese and International Crude Oil Prices

Because of Chinese pricing mechanism and petroleum market conditions, China only connects its crude oil price level with international prices, rather than its pricing mechanism. The existent problem is that Chinese crude oil price passively follows international prices. China has no independent "offer system", therefore it cannot reflect the relationship between the supply and demand of local markets. Moreover, it also cannot transfer the variety of the Chinese petroleum market to international crude oil market with the form of price signal, and participate in the process of international crude oil price formation.

(3) There exists a positive feedback in Chinese crude oil price to the change of international crude oil price.

The rise of international crude oil price will cause the Chinese price to rise quickly. However, the Chinese crude oil price will drop a little soon after; and the range is very limited, but it cannot eliminate completely its preceding rise. In this view, the risk of the international crude oil price has already influenced the Chinese market; so China should build up appropriate mechanisms to avoid this kind of risk. Now, there are many public opinions about a strategy of building up petroleum storage. While a storage strategy is not suitable as a normal mechanism to avoid market risk, because of its high cost, it is suitable to be used as a mechanism to avoid risk, for example, when petroleum supply is disrupted due to war or accidents. To avoid market price risk, China should use market methods, such as establishing a petroleum future market or forward trade system, thereby establishing a Chinese independent quota system. This will allow the development of a market power to disperse the price risk. It will also allow participation of investment in international future markets, influence international crude oil price, and lower Chinese importing costs by holding a proper future position in the international crude oil future market.

(4) From the analysis of decomposition, the variance of forecast errors, we find that Chinese crude oil price had almost no impact on its international counterpart in the short run period. However, there is still a certain amount of impact over the longer period. China should continuously enlarge this kind of influence.

Along with the development of the Chinese economy, Chinese crude oil reduction will increase the dependence on the international crude oil markets.

China is a great consumer in international crude oil markets, only the international crude oil price reflects more about Chinese crude oil demanding variety. Such price is just reasonable, and it is just fair to Chinese crude oil users. This requires Chinese crude oil manufacturers and users to participate in the formation of the international crude oil price actively, but not just reflect the international crude oil price to Chinese market passively.

Without pricing power, China suffers greatly from high international oil prices, e.g., the so-called "Asia Premium".

According to an expert of China International petroleum Chemicals Co. Ltd., Saudi Light crude oil prices in North-east Asia are usually 3 US$ higher than those in Europe. It is strange that sometimes it would be cheaper to purchase Saudi oil from U.S. than from Saudi Arabia directly. Asia Premium is the result of lacking pricing power in North-east Asian Countries. Middle East crude oil main is purchased by Asian, European countries and U.S. But the prices for Middle East crude oil in Europe usually coalesce with British IPE Brent oil prices, and the prices in the United States with NYMEX WTI prices. Because there is no successful crude oil futures market in Asia, the prices for Middle East crude oil in Asian countries usually coalesce with Dubai and Amen crude oil prices, which only reflect the supply and demand in Northeast region, in PLATT's price quote system.

Since Asian countries lack pricing power in global petroleum markets, can China participate into the world markets and affect international prices? Can China become a regional or even global pricing center?

It is generally recognized that China should set up its own crude oil futures market in order to become a regional or even global pricing center. Thus, China should build up a good spot market and the producers, consumers, wholesalers and speculators in China should participate into market activities.

The current situation in China is that there is no mature oil market system but a primary trading system with a rigid pricing mechanism in China. Furthermore, because of relaxation of domestic crude oil prices, the prices for the main petroleum products–gasoline and diesel oil, etc., are still under strict regulations. Governmental participation in price formation goes against the healthy development of China's oil market. In order to fully develop oil market mechanism in China, the authority should prevent its direct interference into the pricing mechanism as far as possible and influence the prices mainly by indirect ways, e.g., taxation to guide and regulate the market to form true market prices.

In 2004, Shanghai Futures Exchange put forward a fuel oil futures contract. The curtain of crude oil futures is undraping a little. As an important energy futures contract, fuel oil futures contracts exerted a price discovering function perfectly during the previous year. The perfect functioning of this contract will offer a lot of useful experiences for other petroleum futures contracts, that is, it is just a matter of time until crude oil futures would be put forward (Li, 2005).

Furthermore, a popular view, or so-called "China factor", states that the dramatic increasing oil prices are incurred by rapidly growing demand for petroleum in China. The research results put forward strong evidence that this view is completely wrong.

According to the study in this sector, international crude oil prices are exogenous to China's domestic prices. Thus, the changes in domestic prices in the short run will not incur changes of international ones, i.e., China has no power to influence global prices. As a consequence, the changes in China's demand will have a lagged and quantitatively small impact on international prices. "The rising high prices are caused by China's demand" is nothing but a lie.

So we conclude that the "China factor" does have a certain impact on global oil markets, but the high oil prices are still the result of the comprehensive interactions of fundamental and nonfundamental factors.

4.3 Analyses of the Features of the Changes in China's Crude Oil and Its Products

The main reason why petroleum prices are the global focus is that they will exert substantial influence on the social economy and residential welfare. Those influences are largely implemented by the oil products of crude oil. Therefore, the study of crude oil price fluctuations concentrates on the features of price fluctuations of oil products and its pricing behavior. If crude oil prices fluctuate and the prices of oil products fail to adjust accordingly, the profits and performance of all industries will suffer great loss. As China's three biggest oil companies come into foreign stock markets, it is unfair to domestic consumers if this kind of adjustment is too big in magnitude, which implies a large proportion of consumer interests will be transferred to the hands of these three companies, that is, to the oversea investors of these companies; on the other hand, it will do great harm to China's oil industry and endanger their taxation if the adjustment is too small in magnitude, and the profits of China's oil industry will suffer great loss.

Under normal circumstances, there exist some price differences between the fluctuations of crude oil as a primary raw material and those of its oil products. If the price differences deviate from the normal scope, distorted prices will not serve as a useful tool to adjust relocation and distribution of resources, which may result in the chances for arbitrage of crude oil and oil products in financial markets, in that the speculators can buy underestimated contracts and then sell them at high prices. Therefore, it is crucially important to study

4.3.1 The Price Correlation Analysis of Oil Products and Crude Oil

The intimate relationship between the oil products and crude oil in the same industrial chain requires that the price correlation analysis of oil products and crude oil should be included in the study of prices for oil products. Related research shows that there exists a long-term difference (logarithmic price) relationship between them. In this section, we will study the relationship between the oil products and crude oil, and then compare the results with the international ones, thus providing valuable information to predict and manage price risks for domestic oil companies and oil pricing reforms by better understanding of their relationship.

4.3.1.1 Data

In this section, the data are taken from the articles entitled as "Recent International Key Crude Oil Spot Prices" and "Recent Tendency of Domestic Oil Products" which were published in *International Petroleum Economics*. The price time series of domestic oil products are monthly prices dating from Sept., 2001 to Mar., 2003, while that of international prices are taken from the United States EIA website, dating from Sept., 2001 to Sept., 2003. In this section, we apply six time series for international and domestic crude oil, gasoline and diesel oil respectively. To facilitate our research, we pretreated these times series using logarithmic methods, which can be defined as Table 4-12.

Table 4-12 Variables and their definitions

Variable	Definition
$\ln ic$	Logarithmic prices of Brent crude oil
$\ln ig$	Logarithmic prices of international gasoline
$\ln id$	Logarithmic prices of international diesel oil
$\ln nc$	Logarithmic prices of domestic Daqing crude oil
$\ln ng$	Logarithmic wholesale prices of domestic $90^{\#}$ gasoline average
$\ln nd$	Logarithmic wholesale prices of domestic $0^{\#}$ gasoline average

4.3.1.2 Analysis of the Price Correlation Between Oil Products and Crude Oil

The intimate relationship between oil products and crude oil can be directly reflected by the strong correlation between them. Table 4-13 illustrates the correlation coefficients between them.

Table 4-13 The correlation coefficient matrix between all oil products

Variable	International gasoline	International diesel oil	International crude oil	Domestic gasoline	Domestic diesel oil	Domestic crude oil
International gasoline	1					
International diesel oil	0.912	1				
International crude oil	0.973	0.927	1			
Domestic gasoline	0.842	0.878	0.853	1		
Domestic diesel oil	0.789	0.876	0.800	0.980	1	
Domestic crude oil	0.955	0.928	0.943	0.887	0.856	1

The correlation coefficients between international crude oil and oil products are high, e.g., that between the international crude oil and gasoline is 0.973, while that between the international crude oil and diesel oil is 0.927 which is relative lower. Comparably, they all are significant at 1% level, which implies a very strong correlation between international crude oil and oil products. Fig. 4-9 illustrates the same tendency of all three prices.

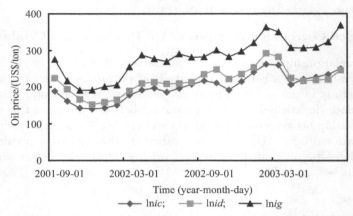

Fig. 4-9 Crude oil price and gasoline/diesel oil price in international market

The correlation coefficients between domestic Daqing crude oil and oil products are high, too, e.g., that between the domestic Daqing crude oil and gasoline is 0.887, while that between the international crude oil and diesel oil is 0.856 which is relative lower than international cases. They all are significant at 1% level, which is influenced by current domestic pricing mechanism, that is, the crude oil prices now are fully coalesced with international prices, while prices for the oil products are still decided by National Development and Reform Commission (NDRC). Usually the prices for oil products are rigid and lagged, compared with crude oil prices.

The correlation coefficients between domestic gasoline and diesel oil and international ones are 0.842 and 0.876, respectively, which are lower than the correlation coefficient between domestic crude oil and international one, 0.943, which implies the domestic crude oil can quickly and promptly reflect international market changes.

The correlation coefficients between domestic gasoline and the international and domestic crude oil are higher than those between domestic diesel oil and international and domestic crude oil, which implies that the adjustment of domestic gasoline is relatively quicker than international crude oil price changes.

4.3.2 The Ratio (Logarithmic Difference) Analysis of Oil Products and Crude Oil Prices

The above-mentioned strong correlation between the oil products and crude oil demonstrates there may exist some kind of quantitatively and stable relationship between them. There may be logarithmic difference relationship:

$$\ln p_c - \ln p_f = \alpha \qquad (4\text{-}4)$$

Or ratio relationship:

$$p_c = \exp(\alpha) \cdot p_f \qquad (4\text{-}5)$$

We compute the average ratios of international gasoline and diesel oil to crude oil during all periods, which is 1.399 and 1.162 for gasoline and diesel oil respectively. We treat the average ratios as the stable ratio in the period and apply Eq. (4-5) (namely, take $\exp(\alpha) = 1.399$ and 1.162) to compute the prices according to average ratios, then compare them with the actual prices. The results are listed in Figs. 4-10 and 4-11.

The average ratios of domestic gasoline and diesel oil to crude oil are 2.094 and 1.907.

4.3 Analyses of the Features of the Changes in China's Crude Oil and Its Products

According to Eq. (4-5), the comparisons between the actual prices and simulated prices are illustrated in Figs. 4-12 and 4-13. Table 4-14 lists the simulated errors.

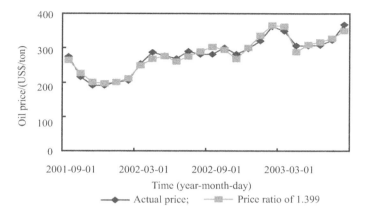

Fig. 4-10 The comparison between the actual prices of international gasoline and simulated prices according to the ratios

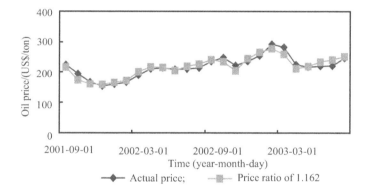

Fig. 4-11 The comparison between the actual prices of international diesel oil and simulated prices according to the ratios

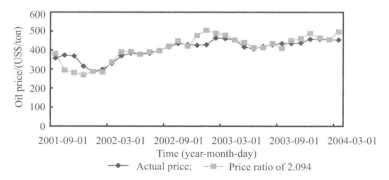

Fig. 4-12 The comparison between the actual prices of domestic gasoline and simulated prices according to the ratios

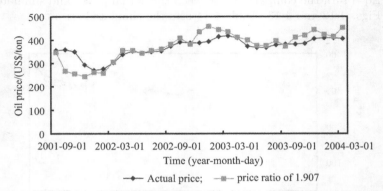

Fig. 4-13 The comparison between the actual prices of domestic diesel oil and simulated prices according to the ratios

Table 4-14 The simulated errors according to the comparisons

Variable	Minimum	Maximum	Average	Std.
Domestic gasoline	−0.1801	0.2362	−0.0076	0.0835
Domestic diesel oil	−0.1694	0.2663	−0.0100	0.0947
International gasoline	−0.1026	0.1184	−0.0037	0.0615
International diesel oil	−0.0913	0.1068	−0.0029	0.0543

The maximum simulated errors of international cases are far less than those of domestic ones, that is, those of international diesel oil and gasoline are 0.118% and 0.107% than those of domestic ones (0.236% and 0.266% respectively); the average absolute simulated errors of international cases are far less than those of domestic ones, too, that is 0.0029% and 0.0037%, respectively, which is less than 0.010% and 0.0076%; furthermore, the std. results also imply the similar conclusions. Hereinafter we will apply more rigid statistics to test the response of domestic price oil products.

Unit root test shows that $\ln ig$, $\ln id$, $\ln ic$, $\ln ng$, $\ln nd$, $\ln nc$ are $I(1)$. We test the cointegration between international and domestic crude oil prices and the international and domestic oil products, whose results illustrate there exists the following relation:

$$\ln ig = 1.071 \ln ic + 0.405, \quad \hat{\beta} = 1.071 \qquad (4\text{-}6)$$

$$\ln id = 0.975 \ln ic + 0.286, \quad \hat{\beta} = 0.975 \qquad (4\text{-}7)$$

$$\ln ng = 0.784 \ln nc + 1.868, \quad \hat{\beta} = 0.784 \qquad (4\text{-}8)$$

$$\ln nd = 0.716 \ln nc + 2.118, \quad \hat{\beta} = 0.716 \qquad (4\text{-}9)$$

The coefficient test of $\hat{\beta}$ shows that there exists stable spread, that is, 0.405 US$/ton between international gasoline and crude oil prices, which implies that international gasoline and crude oil prices will change approximately according to a fixed ratio of 1.499. Similar results can be found in international diesel oil and crude oil prices, whose fixed ratio is 1.331.

There is no fixed ratio, or say, no stable spread between domestic oil products and crude oil price, which implies that the domestic prices of oil products can not fully reflect the fluctuations of domestic crude oil price. The reason for this mainly lies in the current domestic pricing mechanism. Generally speaking, the commodity prices in a fully competitive market can fully reflect the costs of this commodity and are sensitive to the changes in costs. On the contrary, the commodity in a monopoly market is not sensitive to the changes in costs, in that the monopoly producer(s) can easily ensure his (their) income and profit by increasing

4.3.3 The Dynamic Response Relation between Oil Products and Crude Oil Prices

4.3.3.1 Asymmetric Vector Error Correction Model (VECM) Estimates

According to microeconomics, cost is one of the most important components of commodity price. The costs of oil products mainly are the prices of crude oil, so that the response of oil products price to crude oil prices can be an indicator to measure the price reasonability. In this section, we study the dynamic response relation between China's 90# gasoline and 0# diesel oil prices to Daqing crude oil prices. International researchers on this issue mainly study the asymmetric response, namely the different response of prices to the upward and downward changes in its costs, or oil products prices to crude oil prices. First of all, using the Asymmetric Vector Error Correction Model (AVECM) and long-term relationship between them, we analyze the different response of China's 90# gasoline and 0# diesel oil prices to the upward and downward changes in Daqing crude oil prices; then, using the Wald test to test the symmetric response of China's 90# gasoline and 0# diesel oil prices to the upward and downward changes of Daqing crude oil prices; finally, using the cumulative adjustment function (Bettendorf, et al., 2003), we obtain the fully track of this response.

Bacon's model (Bacon, 1991) is one of the earliest models to explain the gasoline price's response to input costs. Bacon noticed that the partial adjustment model cannot explain and evaluate the asymmetric responses to two independent volatilities of target prices, unless the sample can be properly divided into sub-samples during the observation period, which is just impossible beforehand. Thereby, Bacon put forward a squared partial adjustment model, by means of which he found that the British gasoline prices' response to upward changes of crude oil is greater than that of downward changes (Bacon, 1991).

The shortcoming of Bacon's model is that, after crude oil price fluctuation, the adjustments of gasoline prices to a new equilibrium are equal to each other in fixed ratios of current prices to long-term equilibrium prices. To overcome this flaw, Borenstein et al. (1997) put forward an asymmetric model.

In this section we choose the optimum lagged period m, n (Robert, et al., 2003) according to AIC rules, thus the asymmetric model can be defined as:

The first stage estimates model

$$\Delta y_t = \theta_0 ecm_{t-1} + \sum_{k=0}^{m} \theta_k^+ \Delta c_{t-k}^+ + \sum_{k=0}^{n} \theta_k^- \Delta c_{t-k}^- \qquad (4\text{-}10)$$

Where: $ecm_{t-1} = y_{t-1} - \hat{\psi}_0 - \hat{\psi}_1 c_{t-1}$ stands for error correction. Model (4-10) is an Asymmetric Vector Error Correction Model (VECM).

In the second stage we test the symmetric price response, that is, test $H_0 : \theta_k^+ = \theta_k^-$, if $k \leqslant \min(m, n)$; $\theta_k^+ = 0$ if $k > m$ or $\theta_k^- = 0$ if $k > n$. To reject H_0 implies that there exist different responses to upward and downward changes of crude oil cost. If $\theta_k^+ > \theta_k^-$ is significantly valid, then we can say that there exist stronger responses to upward crude oil costs and downward ones.

Based on the AVECM model, Bettendorf et al. (2003) study the one-period changes spot prices (upward or downward) and the adjustment at the i−th period. Taking the Kirchgasser model (1992) as an example, one change in t−th period (e.g., a declining cost) and the adjustment can be defined as:

$$\frac{\partial \Delta y_t}{\partial \Delta c_t^-} = \theta_1^- \tag{4-11}$$

The adjustment at $(t+1)$-th period can be defined as the prompt response at t-th period plus error correction:

$$\frac{\partial \Delta y_{t+1}}{\partial \Delta c_t^-} = \theta_1^- + \hat{\psi}_1(\frac{\partial y_t}{\partial \Delta c_t^-} - \frac{\partial y_t^*}{\partial \Delta_t^-}) = \theta_2^- + \theta_0(\theta_1^- - \hat{\psi}_1) \tag{4-12}$$

The following adjustment can be obtained by this method.

In our study, using the AVECM model defined in Eq. (4-10), we analyze the symmetric responses of Chinese $90^{\#}$ gasoline prices and $0^{\#}$ diesel oil prices to the changes of crude oil. Using the Bettendorf et al. (2003) model we compute the cumulative adjustment of all periods of oil products to negative or positive fluctuations of crude oil prices. The oil products prices we used in the section are the average wholesale price, rather than the retail prices which is due to two reasons: one, the retail prices are different nationwide and hard to obtain; two, the wholesale prices can reflect fully the current pricing mechanism and its responses to market changes.

4.3.3.2 The Empirical Study of Dynamic Response of Oil Products Prices to Crude Oil Prices

The first stage: estimates of long-term equilibrium prices of gasoline and diesel oil.

The results of the unit root test show that all the three series are $I(1)$. Testing the co-integration relation between these series, we find that there exists at lease one co-integration function at 1% significant level between domestic gasoline and crude oil prices, and at 5% significant level between domestic diesel oil and crude oil prices.

The corresponding functions are

$$\ln ng = 0.784 \ln nc + 1.868 \tag{4-13}$$

Error correction: $ecm_{t-1} = \ln ng - 0.784 \ln nc - 1.868$.

$$\ln nd = 0.716 \ln nc + 2.118 \tag{4-14}$$

Error correction: $ecm_{t-1} = \ln nd - 0.716 \ln nc - 2.118$.

The second stage: according to AIC rule, we obtain the 0 lagged period for positive crude oil price fluctuations and 2 for negative ones. By means of OLS, we eliminate insignificant variables, and the results are listed in Table 4-15.

Table 4-15 The OLS results of $90^{\#}$ lead-free gasoline prices

Dependent: $\Delta \ln ng$ Variable	Coefficient	Std.	t-statistics	Probability
ecm	-0.5784	0.1230	-4.7025	0.0001
$\Delta \ln nc_t^+$	0.4115	0.1002	4.1051	0.0004
$\Delta \ln nc_{t-1}^-$	0.3513	0.2272	1.5461	0.1352
$\Delta \ln nc_{t-2}^-$	0.2272	0.1091	2.0833	0.0480
R-squared	0.770237	Mean dependent var		0.007441
Adjusted R-squared	0.741516	S.D. dependent var		0.055836
S.E. of regression	0.028388	Akaike info criterion		-4.154163
Sum squared resid	0.019341	Schwarz criterion		-3.963848
Log likelihood	62.15828	F-statistic		26.81844
Durbin-Watson stat	1.713082	Prob(F-statistic)		0.000000

4.3 Analyses of the Features of the Changes in China's Crude Oil and Its Products

Where the statistics of Wald test $\chi^2(3)=19.584$, significant probability $p = 0.0002$, thus $H_0 : \theta_0^+ = 0, \theta_1^- = 0, \theta_2^- = 0$ is rejected, namely, the responses of Chinese 90# gasoline prices are asymmetric to positive and negative crude oil price fluctuations. The gasoline prices can respond quickly to the positive increments of crude oil costs, whose responses are significantly greater than the negative ones. But the lagged 1 and 2 period(s) (months) responses of gasoline prices to negative crude oil costs are significantly greater than those of positive ones, which the latter one is insignificant. We can conclude from the significance of these coefficients that the gasoline prices can respond promptly, strongly and shortly to the positive increments of crude oil costs, but slowly and persistently to the negative ones. The conclusion is consistent with related international research results. The reasons for this phenomenon are: on one hand, due to the price rigidity; on the other hand, due to the current monopoly oil market. The monopoly oil companies have no intention and driving force to respond to market fluctuations since they can easily transfer their own production costs to consumers when the prices are increasing and maintain high prices. But because of the low energy efficiency, there are still large margins for energy-saving and substitute energy, so that it is impossible to maintain monopoly high profits in the long run and oil companies are resistant to adjust the changes in the markets.

The OLS results of 0# diesel oil prices are listed in Table 4-16.

Table 4-16 The OLS results of 0# diesel oil prices

Dependent: $\Delta \ln nd$ Variable	Coefficient	Std.	t-statistics	Probability
ecm	−0.5743	0.1201	−4.7802	0.0001
$\Delta \ln nc_t^+$	0.3836	0.0917	4.1822	0.0003
$\Delta \ln nc_{t-1}^-$	0.3796	0.2006	1.8925	0.0705
$\Delta \ln nc_{t-2}^-$	0.2484	0.0956	2.5983	0.0158
R-squared	0.827232	Mean dependent var		0.005463
Adjusted R-squared	0.805636	S.D. dependent var		0.055005
S.E. of regression	0.024250	Akaike info criterion		−4.469237
Sum squared resid	0.014113	Schwarz criterion		−4.278922
Log likelihood	66.56932	F-statistic		38.30486
Durbin-Watson stat	1.696873	Prob(F-statistic)		0.000000

Where the statistics of Wald test $\chi^2(3)=20.994$, significant probability $p = 0.0001$. Therefore $H_0 : \theta_0^+ = 0, \theta_1^- = 0, \theta_2^- = 0$ is rejected, namely, the responses of Chinese 0# diesel oil prices are asymmetric to positive and negative crude oil price fluctuations. The gasoline prices can respond quickly but shortly to the positive increments of crude oil costs, while slowly and persistently to the negative ones.

According to the definition of cumulative adjustment function by Bettendorf et al. (2003), we compute the cumulative adjustment functions of Chinese 90# gasoline prices and 0# diesel oil prices by means of AVECM, namely, compute the adjustment when the prices go up or down 1 US$/ton, whose results are illustrated in Figs. 4-14 and 4-15.

1 US$/ton of crude oil price increment will incur the 0.412 US$/ton increment to 90# gasoline oil prices this period (month), while 1 US$/ton of price decrement will bring about only 0 US$/ton. But in the second month, there are strong responses of gasoline prices, i.e., the total accumulated adjustment reaches 0.804 US$/ton, among which the positive accumulated adjustment reaches 0.627 US$/ton. In the third month, the positive accumulated adjustment reaches 0.718 US$/ton; while the negative accumulated adjustment reaches 1.019 US$/ton, which is greater than the volatility of crude oil costs. In the fourth month, the positive accumulated adjustment keeps increasing and then converges to 0.784 US$/ton; while the negative accumulated adjustment is declining rapidly to 0.883 US$/ton at first, then keeps declining until it converges to 0.784 US$/ton, which is the long-term equilibrium.

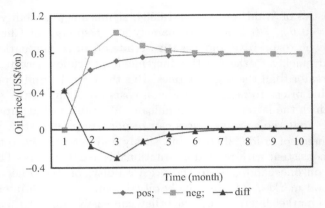

Fig. 4-14 90$^{\#}$ diesel oil accumulative adjustment function

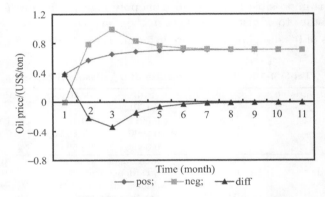

Fig. 4-15 0$^{\#}$ diesel oil accumulative adjustment function

1US$/ton of crude oil price increment will incur the 0.384US$/ton to 0$^{\#}$ diesel oil prices this period (month), while 1US$/ton of price decrement will bring about only 0US$/ton. But in the second month, there are strong responses of diesel oil prices, i.e., the positive accumulated adjustment reaches 0.575US$/ton and the negative accumulated adjustment reaches 0.791US$/ton. In the third month, the spread between the positive and negative accumulated adjustment is enlarged, but the negative accumulated adjustment reaches 0.996 US$/ton, which is less than the volatility of crude oil costs, namely, 1US$/ton. In the fourth month, the spread between the positive and negative accumulated adjustment is getting smaller, until it converges to 0 and the final adjustment reaches its long-term equilibrium.

The forgoing results illustrate that China's gasoline and diesel oil prices have an instant response to increasing crude oil prices, and can adjust quantitatively to long-term equilibriums. As for the decreasing crude oil prices, China's gasoline and diesel oil prices have no response in the first month. After the first month, the decreasing crude oil prices will have obviously strong impacts, which are greater than the accumulative amount of adjustment in continuous two months on the prices for oil products. After the fourth month, negative cumulative adjustment will fall down significantly until it converges to zero. Because during all of the adjustment period, only the negative cumulative adjustment in the third month is greater than negative volatility of crude oil prices, that is, there is an over-reaction effect, which will be seriously unfavorable to gasoline producers. Otherwise, this stage is favorable to producers, but not to consumers. Generally speaking, accumulative adjustment in all months is basically below the volatility of crude oil prices, so that distortion exists in the formation of China's gasoline and diesel oil prices.

4.3.4 Concluding Remarks and Policy Suggestions

(1) There exists distortions of gasoline and diesel prices, among which the latter suffers more distortions than the former.

In the long run, there exists distortions of gasoline and diesel l prices, while $90^{\#}$ gasoline is much more adjustable (about 78.4%) to the cost of domestic crude oil prices than $0^{\#}$ diesel oil (about 71.6%), which means that there exists distortions to gasoline and diesel prices.

(2) The reactions of gasoline and diesel prices to the cost of crude oil are asymmetric.

Concretely speaking, the increasing cost of crude oil will bring about the rapid changes of gasoline and diesel prices, whilst the decreasing cost of crude oil will bring about the slow and lagged changes of gasoline and diesel l prices.

(3) The negative cumulative reactions of gasoline and diesel oil prices are greater than the positive ones.

According to the long-term and short-term adjustment, the negative cumulative reactions of gasoline and diesel oil prices from the second month on become greater than the positive ones in our study, until both of them reaches long-term equilibrium.

The main reason for these results lies in the present domestic pricing strategies stipulated in 2001, that is domestic gasoline and diesel oil prices follow the prices in Singapore, Rotterdam and New York markets. The retail price is government guided, namely, the government will decide and announce the retail prices, while the concrete prices will go up/down 8%. This mechanism actually encourages the speculators to buy in and store the oil product, because there would be a lagged effect that the speculators can precisely predict the future price tendency and make a profit on it.

The solution to this question is a new pricing mechanism where the price is decided by market supply and demand. By means of taxation and economics, the government instructs the formation of prices but just gives a specific one. Because the two companies (China Petroleum and China Petroleum Chemical) still are in charge of the production, import and wholesale of gasoline and diesel oil, a total free market pricing mechanism is still a long way off, although a reform of it is inevitable. First of all, the government should notice the shortcomings of lagged pricing in the current mechanism and make intermediate prices can reflect the real time changes in both international and domestic markets. Secondly the government should redesign the import, wholesale and retail sectors to foster a healthier petroleum circulating system and governmental supervising policy. Finally based on the foregoing two steps the government will then implement the total market mechanism.

4.4 The Impact of Rising International Crude Oil Price on China's Economy

After two big oil crises in the 1970s and 1980s, it is generally believed that increases of crude oil prices contribute to inflation, whilst bringing down demand for consumption, slowing down economic growth, and resulting in economic depression. The main reason is that crude oil and its products are the principal raw materials for industrial production. High energy prices have influenced national economies since the year 2004; moreover, the problem of the fiscal impact of oil price fluctuation on every country's economy has become a global focus of attention.

As fundamental energy and raw material, the fluctuations of petroleum prices will have great impact on the macro-economic situations worldwide. In this chapter, we established a model based on Computable General Equilibrium (CGE) which reflects the macroeconomic operation of China; thus, we applied this model to simulate the different scenarios, i.e., the different degrees of rising oil prices, and their impact on China's macro-economy (e.g., GDP, gross investment, consumption, import and export); at the same time, we analyzed the oil risks and uncertainties by integrating the changes of the technology advancement of crude oil extraction and refinement sectors, and ratios of transportation consumption. The results show that the rising crude oil prices will impose negative impacts on China's GDP, gross investment, consumption, etc., while technology advancement has positive impacts on controlling oil price risks. The research also put forward evidence to prove that ratios of transportation consumption to total consumption have no significant effects in oil price risks. From a beneficial point of view, rural residents will suffer more in loss of social benefits from rising oil prices. Finally, some applicable policy suggestions are proposed for China's energy strategy and policy-making for improving China's petroleum price security.

China became a crude oil net importing country in 1996. The overall tendency for crude oil imports, and its level of dependence on international oil markets gradually increased year-to-year; with the exception being only during the Asian Financial Crisis in 1998. China's crude oil imports totaled 120 million tons in 2004, causing its dependence on international oil to increase to around 42%. It has been estimated that the import of oil by China will reach 230 million tons; increasing its dependence on international oil some 60%. The Chinese crude oil price has basically realized co-movement with international crude oil price after the reform and adoption of the "oil pricing mechanism" in 1998. Any fluctuation in crude oil prices in international markets will inevitably influence Chinese domestic prices; and in-turn, will influence the Chinese economy. Such relational pricing mechanisms and impacts can be represented using a CGE model that reflects a dynamic Chinese economy. Fig. 4-16 represents the simulation model's output of impact on the Chinese economy of crude oil price rising in international crude oil markets.

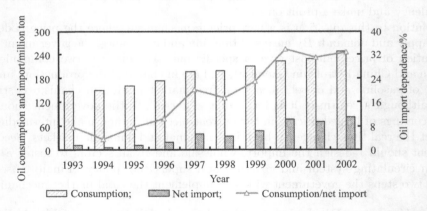

Fig. 4-16 Chinese oil consumption, net import and degree of dependence on international oil

4.4.1 Computable General Equilibrium (CGE) Model of Oil Prices Fluctuation

4.4.1.1 The Computable General Equilibrium Mode

The theory underpinning the Computable General Equilibrium model originated from the publication of the book *An Introduction to Political Economics* (Walras) in 1874. He advanced the theoretical model of General Equilibrium. Subsequently, the existence, unique-ness, and stabilization of the model's solution were successively proven by Arrow in 1951, and Arrow and Debru in 1954. Scarf provided an integrated convergence algorithm to com-pute fixed points in 1967, which made it possible to compute equilibrium prices. It is Scarf's

4.4 The Impact of Rising International Crude Oil Price on China's Economy

work that makes the General Equilibrium model a practical application model from its pure theoretical framework.

The basic idea of General Equilibrium theory is: according to the principle of maximizing profit or minimizing cost, producers make input decisions under resource constraints. Producers then determine optimal supply, according to the principle of maximizing utility. Consumers make optimal expenditure decisions under budget constraints that ultimately determines optimal demand. Equilibrium price makes optimal supply equal to optimal demand. This result in resources having the most rational use, consumers getting satisfaction, and the economy reaching a stable equilibrium state.

Early-stage CGE models include Johansen (1960) and Harberger (1962) who built a two-sectors CGE model; Shoven and Whalley (1973, 1974) built a CGE model on Tax Reform for developed countries. Global Trade CGE models have also been developed (Shoven, Whalley, 1992; McFarland, et al., 2004). Almost all countries now have various CGE models to explore relevant policy development (Bye, Åvitsland, 2003; Dellink, et al., 2004; Willenbockel, 2004; Das, Alavalapati, et al., 2005; Babiker, 2005). The merit of the CGE model is that it embodies market mechanisms and policy instruments which play an important role in identifying price incentive mechanisms. Moreover, it can describe inter-dependence factors that exist in production, demand and international trade. When an economy suffers an unexpected shock, we can study the impact of the shock on gross economics, economic structure, relative prices, and so forth. The CGE model is an appropriate tool for studying various impacts when an economy suffers unexpected international oil price shocks.

Literature on Chinese CGE models has increased in recent years. CGE models specifically using Chinese characters include: DRCCGE developed by The Development Research Center of the State Council (DRC) (Zhai, Li, 1997; Li, Wang, et al., 2000), PRCGEM developed by Chinese Social Science Academy (Zheng, Fan, et al., 1999 ; Wang, Shen, 2001), CNAGE model (Solveig Glomsrd, Wei, 2001), Sulfur-Tax CGE Model for China (Wu, Xuan, 2002); and applied research on analyzing CO_2 control in China using a CGE model (Zhang, 2000). These models are mainly used in policy simulation, such as international trade, tax, or the environment (for example, Sulfur-Tax or CO_2 tax). Hou and Xuan (2003) developed a one-sector CGE model to analyze the impact of international crude oil price shock on the Chinese macro economy. Wei (2002) used the CNAGE model to simulate the impact of international crude oil price rises on the Chinese economy.

In this chapter we discuss the building of our CGE model that was developed by drawing upon the above literature. Application of the model allowed simulation of the impact of international crude oil price occurring with differing shocks on several Chinese economic indexes; together with the impact on import/export/value added at the industry level. We also simulated the action of crude oil mining, petroleum and chemical, and the transportation sectors occurring at low/medium/high technological advances on resistive effects of oil price risk.

4.4.1.2 Model Structure

Our CGE model is composed of six blocks; they are prices block, production block, revenue and consumption block, investment block and closure block. For stressing energy and energy related sectors, we consider twelve sectors (Table 4-17): they are Agriculture (AGRI), Heavy Industry (HIND), Light Industry (LIND), Petroleum and Chemistry (PECH), Construction (CONS), Transport and Post (TRAN), Commerce (COMM), Nonmaterial (NONM), Coal Mining (COAL), Crude Oil Mining (OIL), Natural Gas (GAS) and Electricity (ELEC). Also considered are four kinds of agents: they are rural residents, city residents, enterprises and Government. Additionally, two kinds of factors are considered: Labor and Capital. To compute the impact on rural and city residents caused by oil prices fluctuation, we added the welfare block into our CGE model. The welfare block was not used to contribute to a final solution, but to compute residents' welfare (upon final execution of the model).

Table 4-17 Sectors in model corresponding to sectors in input-output table

Sectors in model	40 sectors	124 sectors
01 Agriculture	01	001—005
02 Heavy Industry	04—05, 13—26	009—013, 048—085, 087—089
03 Light Industry	06—10	014—035
04 Petroleum and Chemical	11—12	036—047
05 Construction	27	090
06 Transportation	28—29	091—099
07 Commerce	30—31	100—101
08 Nonmaterial	32—40	102—124
09 Coal Mining	02	006
10 Oil Mining	03	007
11 Natural Gas Mining	03	008
12 Electricity	24	086

(1) Prices block.

Prices block comprises definitions of various prices in the CGE model; included is the Armington assumption (Armington, 1969). The assumption is made that Chinese international trade contributions to the global trade are only small; and Chinese products and the domestic price do not influence international price; and China only accepts international prices. Compound commodity prices means that imported product prices are compounded by the product prices supplied for the domestic market. This supposes that they satisfy the constant elasticity substitute function, which indicates that consumers minimize their expenditure, and that producers maximize their profit. Value added price is a compound commodity price that is taken off indirect taxes and median inputs. Price index is defined by GDP deflator, calculated by nominal GDP divided by *real* GDP.

(2) Production block.

In our model we argue the relevance of a 2-nested production function. First, every sector uses labor and capital to produce one's value added (according to CES function). Subsequently, the value added and its median input to produce gross production is considered, according to the Leontief function. Technologies are represented by production functions which exhibit constant elasticity of substitution. Technological progress (both embodied and disembodied) is taken as exogenous to the model. For every sector listed in Table 4-17, production in each time period is represented as a constant elasticity of substitution (CES) value added function of capital and labor inputs, where the elasticity of substitution can vary between zero and infinity. The outputs of the other manufacturing industries also enter the production-function specifications as fixed-factor components of the national input-output matrix for the thirteen producer goods. It is assumed that in each time period, producers maximize profits in a competitive market environment. Treating output and input prices as parameters, profit maximization, based on the described production technology, yields output supply and factor demands for each production sector and factor market in the model.

(3) Revenue and consumption block.

This expresses rural and city residents' demand for commodities and reserves. Residents' incomes come from labor income, capital income, and transfer payments from government, enterprise and rest of the world to residents. It represents disposable income, after taking off income tax. Residents' savings is residents' disposable income multiplied by marginal propensity to save; the consumption of residents to commodities is described by Extended Linear Expenditure System (ELES). Government income comes from enterprises' direct

4.4 The Impact of Rising International Crude Oil Price on China's Economy

taxes and indirect taxes, residents' income taxes, import taxes, etc. "Total saving" is composed of residents' savings, government saving, enterprise saving and foreign net saving.

(4) Trade block.

Producers and consumers select a group of compound commodities compounded by imports and domestic products according to the Armington assumption. Import demand function is obtained by minimizing CES cost function; export demand function is obtained by maximizing CET profit function.

(5) Investment block.

Total investment equals total savings. This supposes that every sector's investment is a fixed proportion of the total investment.

(6) Closure block (as shown in Fig. 4-17).

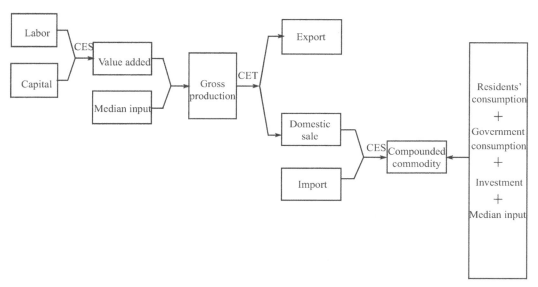

Fig. 4-17 Commodities' flow path

Model closure conditions include market equilibrium and macro equilibrium. There are four market equilibriums in the model; they are commodity market equilibrium, factor markets equilibrium, capital market equilibrium and foreign exchange market equilibrium.

Commodity market equilibrium indicates that every kind of compound commodities supply equals domestic demand for this kind of commodity. Domestic aggregate supply comes from domestic enterprise production and import. Domestic aggregate demand is divided into two parts: one is other sectors median input demand; another is government and residents' final demand and investment.

Factor market equilibrium means that the aggregate demand for every kind of factor equals its supply in every sector. We suppose factor market can make full adjustment by the factor's relative price in every sector when it suffers an external shock. Such as in the labor market, labor aggregate supply is a given exogenous variable, every sector's relative wage is an endogenous variable, and labor supply allocation in every sector is decided by relative wage.

Capital market equilibrium means total savings equals total investment.

Foreign exchange market equilibrium means the balance of foreign exchange receipts and expenditures. Foreign exchange expenditures are composed of import expenditures and the transfers from capital income to the rest of the world. Foreign exchange receipts are composed of export receipts, the transfer from the rest of the world to residents, to government, and foreign net savings.

(7) Welfare block.

We use Hicks equivalent variation to measure the impacts on residents' welfare when oil price rises. On the basis of every kind of commodity's price before implementing a policy, Hicks equivalent variation measures utility change with payment function (Varian, 1992). In this chapter we measure the change of residents' welfare before and after the change of international crude oil price, on the basis of commodity's price before the oil price rising.

4.4.1.3 Data Used in the CGE Model

The model is calibrated to a 1997 data set with these data coming from a variety of sources. Data on the following were obtained: income and expenditures for each of the income categories; the amount of imports and exports in each of the traded sectors; use of labor and capital by each of the producing sectors as well as their level of output; investment by sector; government revenues and expenditures. These data mainly come from the Chinese Input-Output Table (1997), *Chinese Statistical Yearbook* (1998; 1999; 2000), and *Chinese Financial Yearbook* (1999), amongst other sources.

Apart from a large amount of original data, many parameters are also used in the model. These include the supply and demand elasticity for commodities, production elasticity of factors, trade substitute elasticity of commodities, plus others. These parameters are sets of coefficients, which are related to technological aspects of the model. Some of them are estimated using traditional statistical inference; some are collected from the existing literature (Wu, Xuan, 2002; Zhang, 2001). All parameters, however, have been calibrated to a 1997 base. A number of data adjustments are necessary to impose a general equilibrium structure on the economy. Basically this required us to eliminate all inconsistencies in the social accounting matrix (SAM) and fit all production and utility parameters so that the model replicates the actual 1997 data (Ballard, et al., 1985).

4.4.2 Scenarios Analysis of Oil Price Fluctuations

4.4.2.1 The Impact of International Crude Oil Price Rises on the Chinese Economy

We label the state before any international oil price rises as "base state". Using the previously described model, we empirically studied how the Chinese economy would be affected when international crude oil price rises 5%, 10%, 20%, 40%, 50%, 100%, respectively. Figs. 4-18–4-39 show the results. They are the impact on *real* GDP, total investment, consumption of rural/city residents and government, RMB exchange rate, GDP deflator, import/export, and value added, respectively.

The rise of international crude oil price causes a reduction in Chinese *real* GDP. Fig 4-18 shows Chinese *real* GDP reduces 0.029%, 0.053%, 0.088%, 0.126%, 0.137%, 0.159% when international crude oil price rises 5%, 10%, 20%, 40%, 50%, 100%, respectively.

The rise of international crude oil price causes production costs to increase; which reduces products' profit; diminishes the return on investment, then cuts down total investment. Chinese total investment will be cut by 0.026% and 0.106% if crude oil price rises 10% and 50%, respectively.

Fig. 4-20 and Fig. 4-21 shows the impact on consumption. First, government consumption reduces but then increases, which reflects the rigidity of government consumption. Second, rural/city residents' consumption decreases with the oil price rise; note that the degree of rural residents' consumption is bigger than that of city residents. The reason for this result is that rural residents' *real* income reduces much more than that of city residents. Rural residents' income mainly comes from sales of agricultural products; however the input cost of chemical fertilizers and pesticides has a bigger range, and the prices of agricultural products only have a small range.

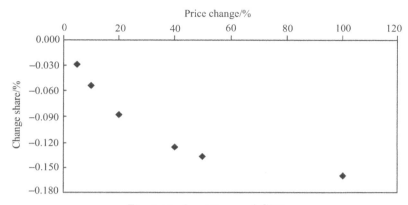

Fig. 4-18 Impact on real GDP

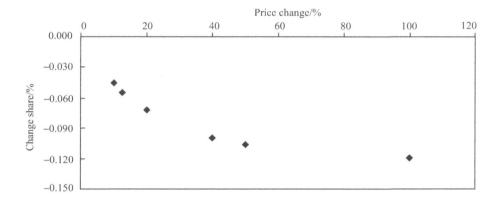

Fig. 4-19 Impact on total investment

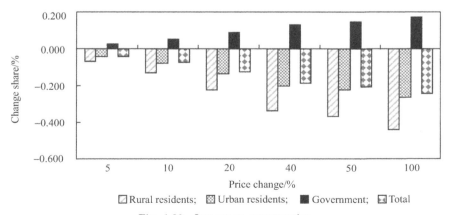

Fig. 4-20 Impact on consumption

The hikes in international crude oil price bring about sliding pressure for the RMB exchange rate (Fig. 4-22). But the range is not great. When crude oil price rises 10%, RMB exchange rate slides 0.226%; while the RMB exchange rate slides 0.755%, when crude oil price rises 100%. In addition, the RMB exchange rate quickens its sliding when oil price goes down by small amounts, after rising a certain degree, sliding velocity tends to become gentle. Chinese imported products become more expensive, and export products become more inexpensive due to the slide of the RMB exchange rate. In turn, this increases pressure on China's foreign exchange payments. However, the range of the RMB exchange rate sliding is moderate so the impact of an oil price hike on the Chinese economy is not serious.

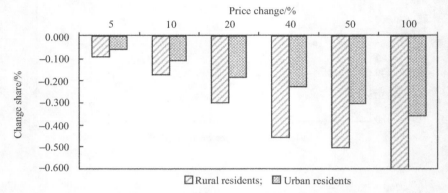

Fig. 4-21 Impact on residents' real income

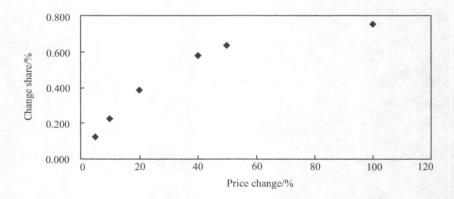

Fig. 4-22 Impact on exchange rate

The Chinese overall price level goes up when international crude oil prices rise; which brings about inflationary pressure. Fig. 4-23 shows the change of overall price level measured by a GDP deflator as oil price rises. The results show that Chinese price level rises 0.4% and 0.8% when international crude oil price rises 20% and 100%. The same is true of *real* GDP and total investment: when international crude oil price has a small rise, the speed of price level rise will accelerate; when the range of international crude oil price rises beyond 50%, the speed of price level rise will slow down.

Chinese import/export cuts down when international crude oil price rises (Fig. 4-24). However, the range of import reduction is greater than that of exports. For example, as international crude oil price rises 20%, imports cut down 1.468%, and exports cut down 0.841%. The result may imply that the impact of oil price on Chinese balance-of-trade payments probably is not as serious as some might envisage. The reason is that the RMB exchange rate depreciates if oil price rises, which makes Chinese exports more competitive, and subsequently weakens Chinese import capacity.

4.4 The Impact of Rising International Crude Oil Price on China's Economy

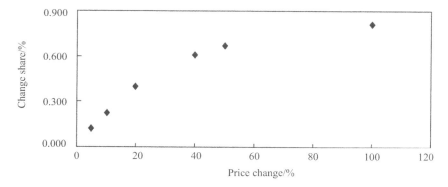

Fig. 4-23 Impact on GDP deflator

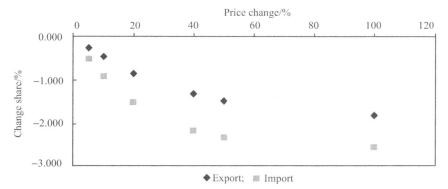

Fig. 4-24 Impact on import/export

Figs. 4-25–4-27 show the results of import, export and value added variation in Primary Industry, Secondary Industry, Tertiary Industry. Import reduces 1.941% and export reduces 1.763% in Primary Industry if crude oil price rises 50%. The reason for this result is probably related to agricultural products' small elasticity of demand, and their high import taxes, which reflects the policy of government protection towards agriculture, and supports on agricultural products export. A major reason of value added reducing in the agricultural sector lays in input cost increasing, while the prices of agricultural products only have a small rise, thus leading to the profit coming down in Primary Industry. With regard to the Secondary Industry, if international crude oil price rises, import/export both decrease, and the range of import is greater than that of export. The difference between Primary Industry and Secondary industry is that value added in the Secondary Industry increases as crude oil prices rise. Import reduces 2.417%, export reduces 1.811% and value added increases 0.291% in Secondary Industry if international crude oil price rises 50%. The reason lies in the rising of domestic crude oil price as international crude oil price is rising, and then oil price risk diffuses along with the industry chain. This is due to crude oil being a primary energy, its downstream being very long, and almost all sectors far and near suffering its price risk. The impact on the petroleum and chemical sector are greatest because of the tight relationship between it and crude oil. All Chinese large-scale petroleum enterprises are integrated upstream-downstream enterprises. About 80% crude oil used by their downstream products are supplied by only 20% of the enterprises' upstream (Chinese degree of dependence on international oil is 20.17%). When there are rises in international crude oil price, these enterprises' upstream products (crude oil) prices also go up; but the price range is greater. Combined with these factors, the profit of Chinese petroleum enterprises does not reduce but, in fact, increase; which is consistent with the fact that the more oil price rises, the more profit petroleum enterprises get.

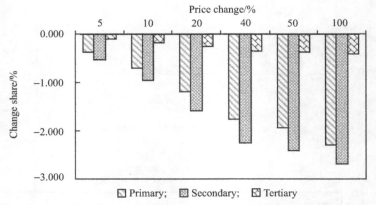

Fig. 4-25 Impact on import based on industry level

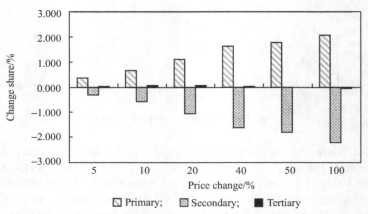

Fig. 4-26 Impact on export based on industry level

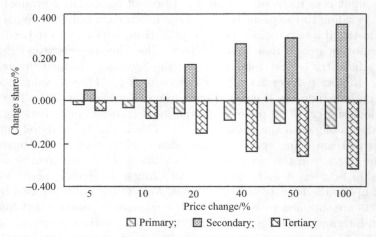

Fig. 4-27 Impact on value added based on industry level

4.4 The Impact of Rising International Crude Oil Price on China's Economy

In the Tertiary Industry, imports reduce as oil price rises. However, value-added increases when the range of oil price rise does not go beyond 40%. Moreover, if the range of oil price rises continues, exports will decline. The reason probably lies in *real* income suffering little impact when crude oil price rises a little, (and Chinese labor is cheap), so countries importing from China do not reduce their imports. However when the range of rises in crude oil price is greater, real income reduces considerably. In addition, Tertiary Industry products have a larger elasticity of demand, thus import reducing begins with the Tertiary Industry, causing reductions in exports. However, whether rises in oil price are small or large, the impact on the Tertiary Industry is much smaller than that on the Primary and Secondary Industry.

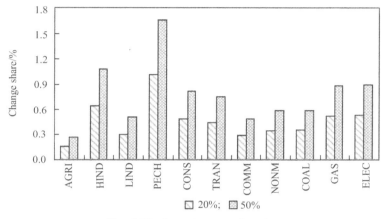

Fig. 4-28 Impact on supply prices

The hike in international crude oil price brings about a rise in China's domestic crude oil price. Domestic crude oil price rises 5.258% and 8.494%, if international crude oil price rises 20% and 50%, respectively (as shown in Fig. 4-28). Fig. 4-29 shows the results of other sectors' products supply price variation caused by international crude oil price rising 20% and 50%. All sectors prices are going up. The price of petroleum and that in the chemical sector has the greatest range, when compared with heavy industry. The agricultural sector has the smallest range. When international crude oil price rises 50%, the range in petroleum and

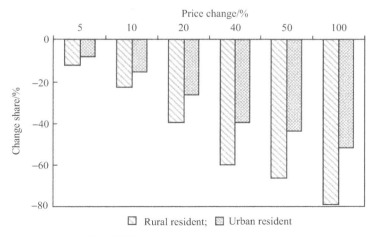

Fig. 4-29 Impact on welfare of resident

chemicals is 1.670%, heavy industry is 1.084%, electricity is 0.898%, construction is 0.814%, and agriculture is 0.263%. The results show that prices go up the most if it has strong relationship with crude oil. Since prices are going up in all sectors, the level of *cost-push* inflation will be raised.

International crude oil price rises make energy become more expensive, which spurs on enterprises to energy conservation. Fig. 4-30 shows the change of energy demand in unit GDP. The rise of international oil price makes crude oil become more expensive; next is natural gas. So the demand for crude oil and natural gas in unit GDP reduces much more. When international oil prices rise 20%, demand for crude oil in unit GDP reduces 1.294%; natural gas reduces 1.159%, coal reduces 0.287%, and electricity reduces 0.083%. The results indicate that China's energy efficiency still has a far way to go on one hand; on the other hand, the rise of oil price will lead to the change of Chinese energy structure. That is to expand the current tendency of substituting crude oil and natural gas with coal and electricity.

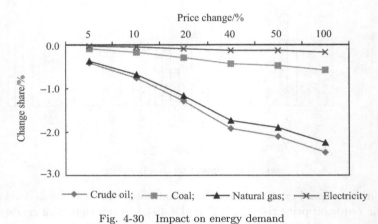

Fig. 4-30 Impact on energy demand

4.4.2.2 Effects of Low/Medium/High Technological Advances in Crude Oil Mining Sector to Resist Oil Price Risk

In economic perspective, other factors removing labor and capital from production function can be regarded as the contribution of technology to output; scale parameter ϕ_i of production function expresses the contribution of technological advances in sector i (in a broad sense) to output. In the light of literature, we call ϕ_i increasing 2% as low technological advance, ϕ_i increasing 4% as median technological advance, ϕ_i increasing 8% as high technological advance.

Figs. 4-31 and 4-32 show the effect of low/medium/high technological advances in the sector of crude oil mining on resisting international oil price risk to China's real GDP and total investment. When there are no technological advances in the sector of crude oil mining, China's real GDP and total investment suffer negative effects resulting from the crude oil price. From the simulated results, we find low technological advances in the sector of crude oil mining has a significant effect on resisting oil price risk to China's *real* GDP (when oil price has a small rise, e.g., 5%, 10%). When international crude oil price rises 10%, low technological advances in the crude oil mining sector can cause China's *real* GDP to reduce less: 0.034%; that accounts for 64% of total effect. However, when oil price has a large rise, the effect of low technological advances in the sector of crude oil mining on resisting oil price risk has a remarkable reduction. When international crude oil price rises 100%, low technological advances in the crude oil mining sector can cause China's *real* GDP to

4.4 The Impact of Rising International Crude Oil Price on China's Economy

reduce only 0.043%; which accounts for 27% of total effect. Medium technological advances in the crude oil mining sector can resist fully the negative effect of oil price increases on China's *real* GDP, so long as the range of oil price rises does not go beyond 10%. It resists half of the negative effect of oil price on China's *real* GDP, when oil price rises 100%. High technological advances in the crude oil mining sector can fully resist the negative effect of increases in oil price on China's *real* GDP even if oil prices rise 100%.

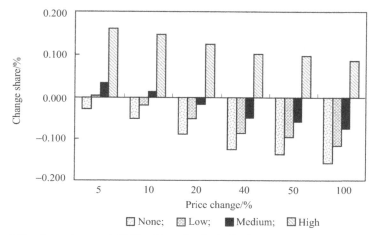

Fig. 4-31 Impact on real GDP with technological advances in crude oil mining sector

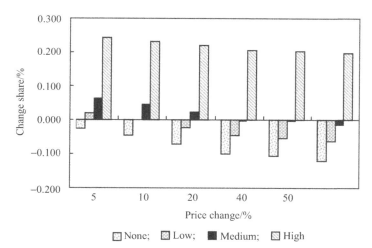

Fig. 4-32 Impact on investment with technological advances in crude oil mining sector

Low technological advances in the crude oil mining sector can fully resist the negative effect of increases in oil price on China's total investment, if the range of oil price rises does not go beyond 10%. Medium technological advances can resist fully the negative effect of oil prices on China's *real* GDP, if the range of oil price rises does not beyond 40%. High technological advances can resist fully the negative effect of oil price on China's *real* GDP, even if oil prices rise 100%. From the relative level, low technological advances can resist 46% of total negative effect on total investment; median technological advances can resist 89%.

According to above results, we can draw the conclusion that technological advances in the crude oil mining sector play an important role in resisting the international crude oil price risk.

4.4.2.3 Effect of Low/Medium/High Technological Advances in Petroleum and Chemical Sectors to Resist Oil Price Risk

Figs. 4-33 and 4-34 show the effect of low/medium/high technological advances in the sectors of petroleum and chemicals on resisting international oil price risk to China's *real* GDP and total investment. Low technological advances can fully resist the negative effect of oil price rise, if the range of oil price rises does not go beyond 20%; medium/high technological advances can fully resist the negative effect of oil price on China's *real* GDP, even if oil price rises 100%. Low/medium/high technological advances in the sectors of petroleum and chemicals can fully resist the negative effect of oil price rises on China's total investment. From a relative level, low technological advances can resist 81% of total negative effects on *real* GDP when oil price rises 50%; resist 69.8% when oil price rises 100%.

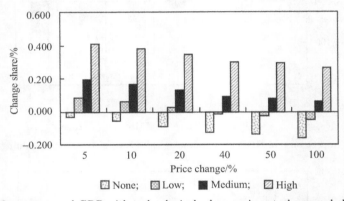

Fig. 4-33 Impact on real GDP with technological advances in petroleum and chemical sector

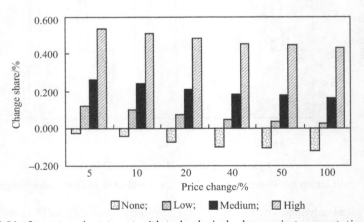

Fig. 4-34 Impact on investment with technological advances in transportation sector

4.4.2.4 Effect of Low/Medium/High Technological Advances in Transportation Sector to Resist Oil Price Risk

Figs. 4-35 and 4-36 show the effect of low/medium/high technological advances in the transportation sector on resisting international oil price risk to Chinese *real* GDP and total investment. Low technological advances can fully resist the negative effect of oil price rises, if the range of oil price rises does not go beyond 20%; medium/high technological advances can fully resist the negative effect of oil price on Chinese *real* GDP, even if oil price rises

100%. Any technological advance in the transportation sector is able to resist fully the negative effect of oil price rises on total investment, and continue the growing tendency. Low technological advance is able to resist 77% of impact of oil price rises on *real* GDP if oil price rises 50%, resist 66.7% of impact if oil price rises 100%.

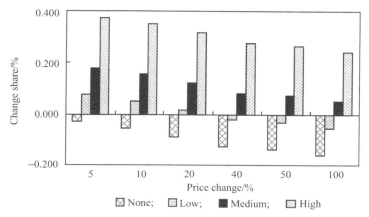

Fig. 4-35 Impact on real GDP with technological advances in transportation sector

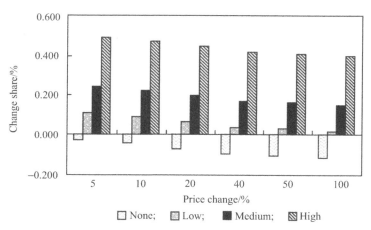

Fig. 4-36 Impact on investment with technological advances in transportation sector

Technological advances play a positive effect in resisting the loss of residents' welfare and reduction of residents' *real* income caused by oil price increases. Figs. 4-37 and 4-38 show the results of rural/city residents' real income and welfare change when international crude oil price rises 5%, 10%, 20%, 40%, 50%, 100%, at the same time low/medium/high technological advances are born in the crude oil mining sector, petroleum and chemical sector and transportation sector. Compared with no technological advances, technological advances effectively restrain residents' real income and welfare loss, and the effect in the transportation sector is a little greater than in the petroleum and chemical sector; they are both greater than the crude oil mining sector. In addition, the effects on rural/urban residents are different; such as when there are no technological advances, and crude oil price rises 100%, rural and city residents' real income reduces 0.6% and 0.36%, welfare loses 0.56% and 0.37%, respectively. However, when there is medium technological advance in the petroleum and chemical sector, their income reduces 0.393% and 0.155%, welfare loses 0.37% and 0.18%, respectively, i.e., technological advances makes rural/city residents' income reduce less 34.5% and 56.9%, welfare lose less 34.5% and 56.9%, and other scenarios have similar results.

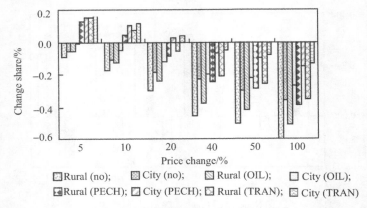

Fig. 4-37 Impact on real income with technological advances in different sectors

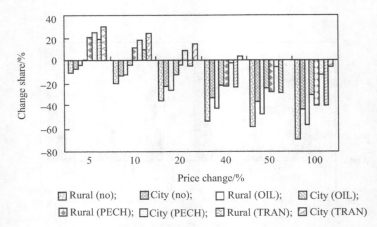

Fig. 4-38 Impact on welfare with technological advances in different sectors

4.4.2.5 The Role of the Transport Sector Consuming Ratio Change in the International Crude Oil Price Hike

The transport sector consumption ratio change is helpful to change the impact of the oil price fluctuation on the real GDP, but the result is not very clear. Specifically the reduction of the transport sector consumption is helpful to alleviate the negative impact on the economy brought by the oil prices hike. Conversely, the increase of the transport sector consumption will worsen the negative impact. However, the effect of both directions is not big enough. As shown in Fig. 4-39, the reduction by 10% of the transport sector consumption at the time that the international crude oil price is rising by 20% will reduce China's real GDP reduction by 0.12%. Conversely, in the oil price rise, if the transport sector consumption increases by 10% it will enhance the reduction of the real GDP by 0.14%. Fig. 4-40 shows China's total investment changing when the transport sector consumption is changing in different international oil price rise degrees. In various degrees of the international oil prices

4.4 The Impact of Rising International Crude Oil Price on China's Economy

rise, the effect of the impact of the total investment brought by the transportation sector consumption proportion changes is significant. The reduction of the consumption can almost remove the negative impact of the oil price rise (except the total investment will reduce a little when the oil price doubles), and increasing the consumption will further reduce the total investment. Besides, the transportation sector consumption proportion change is nearly helpless to the consumption price index, but can release the degree of the RMB exchange rate devaluation.

Compared to the role of technological progress on the resistance to the risk of oil prices, the effect of the change to the transportation sector is much less. We think the main reasons are:

(1) The transportation sector consumption proportion of the total consumption is rather small, about 2.7%.

(2) The transportation sector consumption is terminal behavior, so its impact on the entire economy is not so great as that of the industries in the front-end or mid-range products chain. This conclusion implies that, nowadays, when the scale of China's transportation sector is rather small, to cope with the high oil price to restrict the auto industry development will be not effective enough. We need to work from the source, for instance enhancing vehicle fuel economy.

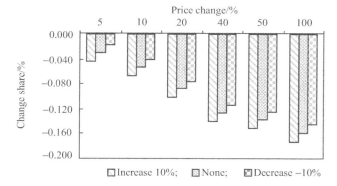

Fig. 4-39 Impact on real GDP with the ratio change of transportation sector consumption/total consumption

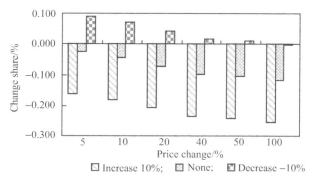

Fig. 4-40 Impact on investment with the ratio change of transportation sector consumption/total consumption

4.4.2.6 The Impact of the Sector Factor Demand by the International Oil Price Rise

The rises of international crude oil price will be an incentive to the crude oil mining sector to increase investment and enlarge production, as Fig. 4-18 and Table 4-19 show.

If the international crude oil price rises by 10%, the capital input of the crude oil mining sector will increase by 16.497% while the labor input will increase by 16.814%. When the international crude oil price rises by 100%, the capital input will increase by 60.179% while the labor input will increase by 61.786%. From the result we can see that the increase degree of the crude oil mining sector's labor input and capital input are almost the same. The impact of demand on natural gas exploration and petrochemical sector capital and labor input are rather great, brought by the oil prices rising, followed by heavy industry and coal mining industries. Because we assume that the labor and capital can be adjusted fully, therefore the increase of labor and capital input of the crude oil mining sector can be transferred from other sectors which are the natural gas exploration sector, petrochemical sector, heavy industry sector and coal mining sector and so on. Except for the crude oil mining sector, other sectors' capital input decreases by various degrees because of the oil price rising. However, besides the crude oil mining sector's labor input increase, the light industry sector, construction sector, nonmaterial sector and electricity sector's labor also increase by a small degree. The main reasons of the natural gas sector's labor and capital input decrease are: ① the substitution between natural gas and crude oil is large, however, compared to the crude oil price rise, the natural gas price rise degree is much smaller, therefore, the expected yield is much bigger than natural gas and as a result funds and labor are attracted to the crude oil sector; ② the scale of China's natural gas industry is rather small, actually, the absolute decrease of the investment is very small. As in the same degree of the oil price rise, the increase (decrease) of capital input is smaller (bigger) than the labor input increase (decrease). Therefore, from this point of view, when the oil price rises, the relative share of labor input will increase a little in each sector. This implies that in each sector, there is a tendency that labor substitute for capital, which may be relevant to the abundance of the cheap labor.

Table 4-18　Relative change of capital input (%)

Price change	AGRI	HIND	LIND	PECH	CONS	TRAN	COMM	NONM	COAL	OIL	GAS	ELEC
5%	−0.156	−0.247	−0.080	−0.605	−0.140	−0.149	−0.161	−0.105	−0.254	8.765	−0.476	−0.123
10%	−0.294	−0.469	−0.156	−1.116	−0.261	−0.280	−0.305	−0.199	−0.476	16.497	−0.883	−0.229
20%	−0.519	−0.834	−0.286	−1.906	−0.454	−0.493	−0.540	−0.352	−0.831	29.038	−1.520	−0.398
40%	−0.806	−1.305	−0.463	−2.848	−0.692	−0.704	−0.840	−0.549	−1.276	44.933	−2.293	−0.607
50%	−0.894	−1.449	−0.519	−3.122	−0.763	−0.846	−0.932	−0.609	−1.409	49.755	−2.521	−0.669
100%	−1.084	−1.763	−0.643	−3.701	−0.916	−1.024	−1.132	−0.739	−1.696	60.179	−3.006	−0.802

Table 4-19　Relative change of labor input (%)

Price change	AGRI	HIND	LIND	PECH	CONS	TRAN	COMM	NONM	COAL	OIL	NGAS	ELEC
5%	−0.012	−0.104	0.063	−0.462	0.004	−0.005	−0.018	0.038	−0.111	8.921	−0.333	0.021
10%	−0.023	−0.199	0.115	−0.848	0.009	−0.009	−0.034	0.072	−0.206	16.814	−0.614	0.041
20%	−0.042	−0.358	0.193	−1.435	0.024	−0.016	−0.062	0.126	−0.356	29.657	−1.047	0.080
40%	−0.067	−0.569	0.280	−2.123	0.048	−0.024	−0.101	0.193	−0.539	44.013	−1.565	0.134
50%	−0.074	−0.634	0.304	−2.321	0.057	−0.027	−0.113	0.213	−0.594	50.993	−1.715	0.152
100%	−0.092	−0.778	0.354	−2.735	0.078	−0.032	−0.140	0.256	−0.710	61.786	−2.033	0.193

4.4.3　Concluding Remarks and Main Policy Suggestions

(1) The direct impact of international crude oil prices on the Chinese economy is to reduce *real* GDP. It is well known that investment, consumption and export are the three major factors contributing to GDP growth. When oil price rises 50%, export reduces 1.468%, total investment reduces 0.106%, consumption reduces 0.206%, which leads to a *real* GDP reduction of 0.137%.

(2) International crude oil price hikes can upset the trade balance. The RMB exchange rate is faced with the sliding pressure if international crude oil price increases. The RMB exchange rate slides 0.636%, if international crude oil price rises 50%. The RMB depreciates,

making imported products more expensive and export products become more inexpensive, which will cut down real income, and then the Chinese international payments position deteriorates.

(3) The rise of international crude oil prices leads to increases in input cost of enterprises, as oil is their fuel or raw material. Consequently the pressure of products' cost has the probability of diffusing along the industrial chain, and which in turn increases inflationary pressure. When international crude oil price rises 50%, China's price level measured by GDP deflator rises 0.675%.

(4) The main indirect impact of international crude oil price on the Chinese economy is the potential risk of exports decreasing. When international crude oil price rises 50%, Chinese exports will drop by 1.468%. On the one hand, Chinese products' exports become less competitive due to increased production costs. Conversely, import capacity reduces in countries that import products from China (due to the difficulty of international balance of payments provoked by oil price increases).

(5) The effect of technological advances on resisting oil price risk is remarkable. Among them the effects of technological advances in the petroleum and chemical sector and transportation sector are bigger than that of the crude oil mining sector. Low technological advances in the petroleum and chemical sector and transportation sector can resist fully the negative effects of oil price increases on China's *real* GDP when the range of oil price hikes does not go beyond 20%; medium technological advances in these two sectors can resist fully the negative effects of an oil price hike on China's *real* GDP when the range of oil price does not go beyond 50%; high technological advances in these two sectors can fully resist the negative effects of oil price increases on China's *real* GDP even if oil price rises 100%.

(6) The impact of an oil price hike on rural/city residents' welfare is different. Generally speaking, oil price increases have negative effects on rural/city residents' welfare; but rural residents lose much more in welfare than city residents. When oil price rises 50%, rural residents' welfare loses 58.912%; city residents' welfare loses 36.731%. Secondly the speed of rural residents' welfare reduction is faster. Finally the positive effect of technological advances brings down city residents' welfare loss faster than that of rural residents' welfare. When oil price rises 50%, low technological advance in the petroleum and chemical sectors can make rural/city residents' *real* income less reduced (34.5%/56.9%), and welfare less reduced (43%/70.6%).

(7) International crude oil price rise causes energy products' prices to rise, and thus causes resources to move from other sectors to the crude oil mining sector. When the oil price rises 50%, the demand for capital and labor in the crude oil mining sector will be increased by 49.755% and 50.993%, respectively. The increase of input factors means that the scale of production expands, which is beneficial to the development of the petroleum industry. Additionally, the hike in international crude oil price reduces the demand for energy in unit GDP. When oil price rises 50%, crude oil demand reduces 2.093%, natural gas reduces 1.890%, coal demand reduces 0.476%, and electricity power demand reduces 0.132%. So oil price increases promotes Chinese energy saving, thereby improving Chinese energy efficiency, which in-turn is beneficial to alleviate China's paradox of oil supply and demand.

4.5 Forecast of International Oil Price

International oil price changes all the way because of numerous complicated factors. How to judge the trend of future oil price is important for policy makers to design strategies. The reasons and the effects of international oil price changes and the importance of oil price forecasts are firstly introduced in this section. Secondly, a long-term forecast method of oil price trend is presented. Then mid-term and short-term forecasts of oil price is commented. Considering the significant impacts of oil futures to oil spot, the mid-term and short-term forecasts of oil price adjusted by futures are advanced. Finally, some policies for high oil price are proposed.

4.5.1 Changeable International Oil Price

4.5.1.1 Complicated Oil Market and Price under Multiple Impacts

When market conditions change, crude oil price usually takes on more violent volatility than other commodities. The historical trend of crude oil spot price of WTI (West Texas Intermediate) from Jan. 2, 1986 to Sep. 29, 2005 is depicted clearly in Fig. 4-41.

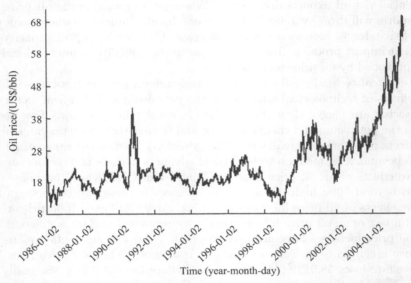

Fig. 4-41 Crude oil spot price of WTI (from Jan. 2, 1986 to Sep. 29, 2005)
(Data source: Energy Information Administration)

No matter whether from a long-term view, or from a short-time manner, the oil price is so unstable that it fluctuates violently frequently. The changeable characteristics of oil price depend on numerous complicated factors including mainly:

(1) World economy.

The economy of main oil-consuming countries and the whole world is the basic driver and decisive factor of oil price. When GDP grows faster, oil price rises; while GDP grows slower, oil price falls. Therefore, GDP decides the basic trend of oil price.

(2) Disaster and climate.

Climate changes oil demand, while disaster destroys production equipment, and interrupts production, and seriously affects oil supply. At the same time, climate and disaster alter people's expectations. These are important decisive factors for oil price. So once climate is unusual or disaster appears, oil price will fluctuate violently.

(3) Political and military actions.

Political and military actions in main oil-production regions is usually affect oil supply, unbalance the demand and supply, and consequently lead to violent fluctuation of oil price. In addition, terrorism causes worries for economic growth and oil supply security.

(4) OPEC policies.

As the most important source of crude oil, OPEC decides oil supply to a great extent. Especially in these years when OPEC is hard on increasing production while oil demand grows at a greater speed, the impact of OPEC on world oil market becomes more and more evident.

(5) Stock.

Stock indicates the relation of oil demand and supply. Stock rises, indicating that oil supply surpasses demand, and oil price may fall. Stock decreases, indicating that oil demand surpasses supply, and oil price may rise.

(6) Speculation.

The capacity of oil futures market is limited. With more and more institutional investors owning large wealth joining oil speculation, the trade of oil futures is manipulated to a great extent. Oil price is seriously affected correspondingly.

4.5.1.2 Social Effects of Oil Price

As one important pole of the world economy, China is more and more related to the economy of the whole world. With the fast economic growth in China, the oil demand rises day to day. Foreign trade dependency on oil increases continuously and the Chinese economy will certainly be affected by international oil price. Large changes of oil price, especially high oil price, cause a series of evident social effects including mainly:

(1) Decreasing economic growth.

Oil price change affects the national economy widely, from production to consumption, from cost to price, and from trade to investment, all of which are affected more or less by oil price change. High oil price seriously impedes economic growth. With international oil price rising, three factors improving economic growth, investment, consumption and exports, decrease certainly. Therefore economic growth falls.

(2) Increasing pressure of inflation.

China is a large oil-consuming country. Most of oil consumption depends on imports. The import amount keeps on rising. If oil price rises, the cost in corresponding industry taking oil as energy or material will increase. At the same time, the cost in refinery sharply rises when international oil price goes high, which leads to the rising cost of oil products. The cost in transport sectors such as road, waterway and air transportation increases correspondingly and so does it in service industry. Consequently, the cost of products consumed by households rises. Wide range of price rises is bound to a strong expectation of inflation, which puts pressure on inflation.

(3) Decreasing international competitive of enterprise.

When international oil price rises, China should pay more dollars for oil import, evidently shrinking trade surplus. What is more, with oil price rises, the cost of corresponding industry increases, and the price of products related to oil increases, causing wider range of rising of enterprise costs. In the international and domestic market where supply is greater than demand, market competition is extremely violent. The product price cannot rise at the same rate as the cost, so it is impossible to transfer the rising cost to downstream enterprise or the consumer, let alone foreign countries. The industry profit is bound to decrease and some enterprises even fall into deficiency, production reduction and stopping. The international competitive of foreign trade enterprise decreases so that the economic activity of the whole society will also decrease.

(4) Deterioration of people's lives.

Domestic petroleum price is directly affected by the rising of international crude oil price, especially in car industry close to people's lives. When gasoline price keeps on rising, the crowding out effect on personal purchase of cars and driving consumption is coming. At the same time short supply causes inconvenience in people's lives.

(5) Deteriorating national finance.

When oil price rises, international financial market will raise the stock and bond value of oil-import countries, while decreasing that of oil-export countries. Because China needs to import lots of oil, if the government gives premiums for oil products, oil price rises will deteriorate the finances of China. With international oil prices rising, premium load will be heavier, putting pressure on government budget balance.

4.5.1.3 Significance of Oil Price Forecast

As the above analysis indicates, oil price fluctuation evidently affects every field in society such as economy, enterprise, life and finance. So it is valuable and significant to forecast future oil price and give advice to the decision makers of government and enterprise.

(1) Benefit for government to design better energy strategy.

Government cannot design the best energy strategy without the forecasting of future oil price. Governments in the world design national energy strategy from security and development in order to meet the era of high prices. Firstly, every country includes oil supply into national security strategy. Secondly, every country makes policy to use and save normal energy within reason, research new clean energy and protect environment, so as to realize economic sustainable growth, social general development, effective utility of resource, and improvement of the environment. If we can forecast future oil price accurately, we will master future scenarios and adopt different strategies according to future oil price. The resource allocation will be optimized and the best energy strategy will be worked out through bettering energy structure and controlling the realization process of energy strategy.

(2) Benefit for government to select better time to purchase oil.

The way of buying oil in China is mainly spot trade. The expenditure of buying oil is directly related to current oil price, so it is important for saving government expenditure to select a suitable period to buy at a low oil price. So far, China doesn't consider the best period selection when buying oil, mainly because future market and oil price cannot be decided accurately. Therefore we often buy oil at high price but not low price, causing great loss. The forecast of oil price can help government to judge future scenarios and decide when to purchase so that it can minimize the offer of oil purchase and save the cost, which is better to improve economic development and people's lives in China.

(3) Benefit for government's macro control and economic market stabilization.

If oil price rises, enterprise goes into deficiency, investment decreases, economy will be impeded seriously. Government can augment premium, lower interests, perform preferential policies in some industries to stimulate investment in order to keep normal economic growth. While low prices will cause excess investment and hot economic growth, so government should adopt corresponding macro control policies to prevent excess investment and realize economic soft-landing. Financial market appears unstable because of oil price fluctuation. Investors speculate on the chance of oil market's change. The stock market may fall into collapse with excess speculation, leading to an unstable society. If oil price is expected to fluctuate violent, government should control the financial market as early as it can. The financial market can be stabilized by adjusting interests and injecting capital. Therefore the oil price forecast can be used as a indicator of social economy and financial market, based on which future scenarios can be estimated beforehand. It is good for the macro control of government and stability of national economic and financial market.

(4) Benefit for enterprise to avoid oil price risk.

For oil-import enterprise, oil price rises will increase import cost, which makes enterprise profit fall. While for oil-export enterprise, it is good time for international trade at high oil price, when the oil at hand should be sold to gain greater profit. Oil price is also related to oil-using enterprise closely. When oil price is low, enterprise should purchase and stock oil in case that future oil price rises and the cost can be decreased. If oil price is expected to go down, the stock can be sold to gain profit and oil is repurchased at low oil price to complement stock. The degree that enterprise avoids risk mostly depends on future oil price, i.e., oil price forecasting. So oil price forecasting is necessary for enterprise to avoid the risk

4.5 Forecast of International Oil Price

brought by oil price change.

(5) Benefit for personal investment and consumption in future.

With the level of people's lives rising, more and more people invest in the stock and bond markets and the amount of investment increases year and year. International oil market more or less affects the profit of most of enterprise, so is directly related to personal investment and profit. In addition, personal purchase for a large commodity such as a car increases evidently, but how to design personal plan to purchase more cheaply, which gains focus from numerous consumers. Oil price forecasting can give advice to personal investment and consumption for decision optimization.

4.5.2 Long-Term Forecast for Oil Price Based on Wavelet Analysis

4.5.2.1 Normal Long-Term Forecast Model for Oil Price

Long-term forecast of oil price means to predict the oil price in future one year or more, which is especially significant in the strategic level of nation and enterprise. The oil strategy of nation and enterprise is quite important in decision, related to the policy, route and guideline for the future and detailed realization. Deviated strategy will lead to great loss. For example, if future oil price keeps high for a long time, government and enterprise should invest in new energy and technology in order to effectively relieve the pressure of high oil price. However, if oil price rises only temporarily, and it will fall back soon, we can put most resources to other urgent sectors. So it does with oil enterprise, which depends on long-term trend of oil price to decide whether a company should be public and what service the company should perform.

In view of the importance of long-term forecasting of oil price, it is deeply studied by many famous scholars on how to improve the accuracy of forecasting. Oil price is affected by politics, economy and psychology and so on. These factors take on highly non-linear even chaotic characteristics which increases the difficulty of forecasting especially the long-term forecast problem. Time series models are the most usual methods to forecast oil price, in which mathematical models are firstly established according to the historical time series of oil price and the parameters are estimated, and then the models are solved to generate forecasts. One of the advantages of the time-series forecast methods applied to oil price is that it can avoid prediction bias of many other factors which also need to be predicted in some multivariate macro-methods. There are only a few time series methods for oil price forecast such as the Holt-Winters and Theta models. Except time series models, there are some multivariate forecast models, which consider factors related to oil price such as GDP and enterprise action. But these multivariate models must estimate all factors before forecast. Therefore, unless all factors can be estimated correctly, the result may be more inaccurate than time series models. Common multivariate models include belief networks, probabilistic models and target zone theory models and so on.

4.5.2.2 Long-Term Forecast Model Based on Wavelet Analysis

Wavelet analysis, as a powerful tool in non-linear science, plays a more and more important role in the forecast fields such as futures and stock, and a lot of methods related with wavelet analysis emerge endlessly. The advantage of wavelet methods over traditional models in dealing with non-linear problems is that it can describe inherent properties of non-linear problems in depth. Since wavelet analysis has so many advantages dealing with non-linear problems, oil price time-series is considered as non-linear time-series in this chapter and a new long-term trend forecasting method is presented here by wavelet multi-scale function to realize the oil price long-term trend prediction in the future.

Oil price is affected by multiple factors including not only long-term regulative factors but also temporary random short-term factors. For example, the seasonality of oil price is the characteristics of climate, one of the long-term factors. When summer comes, households in the United States the first large oil-consuming country, like to drive out for travel, which leads to the oil-consuming climax in summer and oil price will go up correspondingly. When winter comes, heating oil demand also brings another oil-consuming climax and pulls oil price upward. However in other seasons, oil consumption is relatively small and oil price keeps on a lower level. Short-term factors are usually some incidence close to oil such as speculation. They have a short period influence and cause temporary violent fluctuation of oil price but do not affect the long-term trend. The smooth low frequency elements and the violent high frequency elements in oil price represent respectively the effects of long-term and short-term factors. Because long-term factors are stable, the long-term trend of oil usually takes on evident and regulative fluctuation. Wavelet analysis can be applied to distinguish long-term and short-term factors. Fig. 4-42 describes the fundamental principle of wavelet decomposition and reconstruction. The approximation of wavelet decomposition contains the low frequency elements of the signal, while the detail of wavelet decomposition comprises the high frequency elements, respectively corresponding to the long-term trend induced by long-term factors and the short-term shocks induced by short-term factors. Various levels of wavelet decomposition indicate different time scales. The approximation in higher levels represents the oil price movements in longer periods. While the detail in lower level represents the oil price shocks in shorter periods.

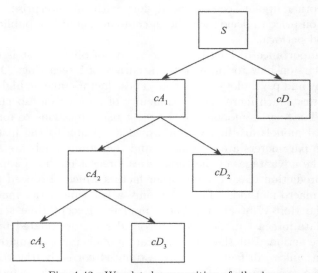

Fig. 4-42 Wavelet decomposition of oil price

Here, S is the signal we will analyze; cA_1, cA_2, cA_3 and cD_1, cD_2, cD_3 is respectively the reconstruction of the approximation and detail of the first, second, third level wavelet decomposition. There exists a relation between the signal and the reconstruction as follows:

$$S = cA_3 + cD_3 + cD_2 + cD_1 \tag{4-15}$$

Because the low frequency elements of the oil price time-series wavelet decomposition indicates the long-term trend of the oil price, first of all, the methods suggested below have wavelet decomposition on the oil price. Secondly, we select approximations and details at some appropriate levels to take part in later predictions, which can avoid the impact of some short-term stochastic factors. Then, for selected approximations and details, the wave of cosine function is adopted to simulate the long-term fluctuation of the oil price,

4.5 Forecast of International Oil Price

which makes results which can fit the long-term trend of the oil price very well. Finally, all predicted results of the selected approximations and details are integrated to compose the whole multi-step forecast of the oil price.

4.5.2.3 Long-Term Forecast in Future 1 Year

The above model is applied to perform 1-year forecast for WTI oil spot price and the result is shown as Fig. 4-43.

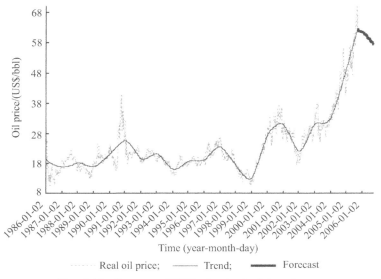

Fig. 4-43 Forecast of WTI oil price in future 1 year

From Fig. 4-43, oil price has evidently gone up since 2004. But forecast result shows oil price will go back gradually from the late half year in 2006. However, compared with past periods of low oil price, oil price will still keep on a high level before the late half year in 2006. Combining macro factors affecting oil price, this forecast is reasonable, because:

(1) World economy develops well, and oil demand rises greatly, while production capacity of oil-producing countries almost reaches the limit, supply and demand relation depends that the era of high oil price has come.

The continuous rising of oil price in these years and in the forecast is dependent on the basic supply and demand relation. Main countries in the world and the global GDP growth rate take on a stable rising trend since 2003. The GDP growth rates of the United States, EU, Japan, China and the world are respectively 3.1, 0.4, 2.7, 9.1 and 3.9 percentages in 2003, while reaches 3.6, 2.2, 4.4, 9.5 and 5 percentages in 2004. Good state of world economy pulls oil demand enormously. The most famous event is the rising economy of China and India in the 21st century. The rising of the economy of China with 1.3 billion population will certainly change the world energy market. Since 2000, Chinese economy grows fast at a rate above 7%. According to underestimated exchange rate at present, average personal GDP in China reaches 1000 dollars in 2003. With higher income, China goes into the era of cars. Car production and consumption grows quickly. According to Xu (2005), 5.15 million cars were sold in China in 2004, growing by 22.64% from 4.4439 million in 2003. Car demand keeping on fast growth is still the main force to pull the petroleum consumption. In 2005, the car stock in China will reach 33 million growing by 16.4% from last year; the motorcycle stock will reach 70 million, growing by 8.5% from last year. Air industry is still developing fast. Domestic air transportation will keep on a growth rate of 15%. Diesel oil for farming and the oil used by construction, mining enterprise and business will keep on fast growth. According to the oil-consuming data provided by Energy Information Administration, Chinese oil demand is 6.3 million barrels/day in 2004, while only 5.5 million barrels/day in 2003, growing by 15%, which accounts for 30% of the global growth. Because China performs a policy of macro control and saving energy, Energy Information Administration estimates that oil demand in China is 6.7 million barrels/day, and the growth rate drops to 7%. ExxonMobil Corporation estimates that Chinese cars will reach 115/thousand persons, but 1 thousand persons only own 5.16 cars in 2003 in China (Chen, 2005). Therefore Chinese households just begin to go into an era of cars. Car sales will grow stably and fast in the future decades and push oil demand to grow continuously.

However, under the circumstances that demand continuously rises, oil supply doesn't increase much. The world oil production rises to a limit. Because the proven oil reserves of non-OPEC countries are limited and the oil industry of the most important oil-producing countries such as Saudi Arabia doesn't open to outside, the oil production capacity is limited. Furthermore, the construction cycle for oil production capacity is long; it spends 3–4 years to form new production capacity. So the growth rate of oil supply will decrease. In the fourth season of 2004, OPEC production is above 29.5 million bbl/day, exceeding the quota by 2.5 million bbl/day, and remaining only 1 million bbl/day of capacity, which indicates that the ability of OPEC to increase production in 2005 is very limited. In addition, according to OPEC's report, the oil supply of non-OPEC countries is 53.7 million bbl/day in 2004, and 55.11 million bbl/day in 2005, with a growth of only 1.4 millon bbl/day (Chen, 2005). Therefore, oil price will certainly keep on a high level driven by large demand and urgent supply needs.

(2) The fierce rising of oil price from 2004 is induced by numerous causal incidents in the short run, and the force driving oil price to continuously rise is weak, so oil price can only stay high at the same level.

These two years, all kinds of causal incidents seriously affecting oil price happened continuously. After the war of the United States and Iraq, terrorism on oil equipment in Iraq always goes on. Most of the derricks and oil pipes are damaged and the oil production and export of Iraq cannot recover to the level before the war. Venezuela, the third largest oil-producing country in OPEC, has unstable politics, which seriously affects the oil production and export. Russian oil corporation Yukos, which had a daily oil production of 1.7 million bbl/day occupying 2% of daily production in the world, was checked because of tax dodging and the oil production dropped down much. Nigeria, the largest oil-producing country in Africa and the seventh oil-exporting country in the world, has a daily oil production of 2.5 million bbl/day, occupying about 3% of daily production in the world. Workers in Nigeria are organized by the labor union to start a strike and this event blocks the oil export, furthermore, it decreases the expected oil supply in the international oil market. So the unstable international oil market was shocked once again and it directly drove oil price to rise. Hurricanes such as Ivan frequently struck the Gulf of Mexico and seriously damaged local oil pipes. Investment funds speculate by all kinds of ways in the oil market. From the early half year of 2005, free capital of about 8000 billion dollars has come into the oil market (Zi, 2005). This capital is from IT and the real estate market. After the bubble of IT and realty broke, a large amount of international capital has no where to go. Therefore it ran into the oil market. Especially under the circumstances that the oil supply is vital and oil price is high, speculators expect that the oil stock of the countries in the world especially the United States in the summer will decrease further, so they go on buying oil futures, which makes oil futures price rise sharply. In Asia, because oil demand increases continuously, and oil stock is little, the speculation in oil futures is more evident. These factors added

4.5 Forecast of International Oil Price

together, the stability of oil supply in the market is doubted. In the sensitive market, oil price is certainly stimulated to go up.

However, the political and military conflicts of oil-producing countries, strike of oil workers, climate events such as hurricanes and speculation of investors are non-basic factors, which are only short time inducement to oil price rising. The long-term trend of oil price finally depends on basic factors, i.e., demand and supply relation. Demand and supply relation, decides that long-term price in future will stay at a high level while cannot go on rising continuously unless the happening of causal incidents like those in these two years.

(3) Oil price will reverse in a long time, affected by international economic environment and high price.

According to "Ten Forecasts in Chinese Economy in 2005" published by Shanghai Security Newspaper and National Information Center, the growth rate of GDP in China will slow down to 8.5%, because the growth rate of fixed capital investment slows down from 29.7% in 2003, 29% in 2004 to 20% in 2005. Real estate growth is the main force of investment growth in 2003. Among 29.7% of total investment growth, one fourth is the result of realty growth. At present, Chinese realty is at an inflexion, and some part of the houses cannot be sold in all big cities. So it is estimated that realty investment in 2005 will drop down to a large extent. The growth of real investment reached the highest historical level in 2003, which rose by 7% and occupied 43% of GDP, so the high growth of GDP in China is mainly pulled by the high growth of fixed capital investment. Therefore the slowdown of fixed capital investment will lead to the slowdown of GDP growth. Because trade dependency of China is high at 60%, the slowdown of the Chinese economy will affect the world economy. China has raised loan interests by 0.27% in October 2005, which will restrain the economy. The growth rate of the world economy will slow down too. The International Monetary Fund forecasted that the growth rate of world economy in 2005 is 4.3%, lower than 5% in 2004. The UN forecasted that the growth rate of world economy is 3.7% in 2004, 3.4% in 2005. The EU estimated 4.5% in 2004, 4.3% in 2005. IEA forecasted that world oil demand will reach 84 million bbl/day in 2005 growing by 1.8 million bbl/day from 2004. World oil supply will reach 85.5 million bbl/day. In the whole, the supply is larger than the demand. While the oil demand in 2004 is 2.5 million bbl/day more than that in 2003, so the growth rate of oil demand in 2005 will slow down (Network Business Department of Chinese International Futures Corporation, 2005).

Under the effect of high oil price, oil-producing countries will increase exploring and mining investment to increase production capacity. Though the supply elasticity of oil is small for a short time, the oil supply will increase gradually with more drilling in a long period. In addition, waking up to that the era of high oil price has come, each country in the world will endeavor to research new energy and technologies, and perform energy-saving methods or use substitute energy, which will increase the oil utility and decrease the growth of oil demand in the mass. Moreover, the long-term oil price will reverse in future if there are no causal incidents seriously affecting the oil supply like those frequently happening in these two years.

4.5.3 Mid-Term and Short-Term Forecast Based on Pattern Matching

4.5.3.1 Common Forecasting Methods of Mid-Term and Short-Term

Mid-term and short-term forecast means the forecasting of oil price less than 1 year in future, i.e., in a period from several weeks to some months. Mid-term and short-term forecast is mainly used by oil companies for a short-term decision. For example, the government buys crude oil in international markets always with a planning period less than 1 year. Mid-term and short-term forecast can be used to optimize the best time to buy. Mid-term and short-term forecasting is more important for enterprise and investors. They frequently perform short-term trade in oil market or related financial markets. Oil fluctuation for a short time will lead to great influence on them. It is firstly considered by deciders how to master the oil price in a short time in future.

Just because the importance of mid-term and short-term forecasts is no less than long-term trend forecast, numerous scholars in the world perform lots of research in the field of mid-term and short-term forecast and present all kinds of effective and useful forecast methods. The common methods of mid-term and short-term forecast include regression analysis, exponential smoothing, ARMA, GARCH, neural networks, grey model and corresponding advanced models. Multivariate models combining above models and other factors important to oil price can be established to forecast.

4.5.3.2 Mid-Term and Short-Term Forecast Based on Pattern Matching

We apply pattern-matching technique to multi-step prediction of crude oil prices and propose a new approach: generalized pattern matching based on genetic algorithm (GPMGA), which can be used to forecast future crude oil price based on historical observations. This approach can detect the most similar pattern in contemporary crude oil prices from the historical data. Based on the similar historical pattern, a multi-step prediction of future crude oil prices can be figured out. In GPMGA modeling process, the traditional pattern matching is not directly employed. Historical data is transformed to larger or smaller scales in the x-axis and the y-axis directions, so that a generalized price pattern to current price movement can be obtained. This treatment overcomes the local deficiency of the traditional pattern modeling and recognition system approach (PMRS), and in addition to this, a matched historical pattern in a larger pattern size can be found. Since the approach takes not only historical similarities but also differences into account, the concept of "generalized pattern matching" is proposed here. It proves a new basis for multi-step prediction by finding more essential similarities through various transformations.

The main point of local approximation is to model the current pattern of a time series by directly matching its current pattern with a past pattern, and then a forecast can be made according to the pattern following the most similar past pattern. PMRS can be considered as a "direct pattern matching" method, since the current pattern and the past pattern are matched without any transformation. Historical data being directly searched to match the past state most similar to the current one implies that current oil prices are expected to change exactly according the historical rules. However, the complexity of oil price movements often makes this kind of direct matching inaccurate. On one hand, historical rules do not always appear in the same way. For example, oil prices were raised by 2 dollars in the past two months, and following oil prices appeared to show a similar movement but with more rapidly rising speed, i.e., it took only one month to rise by 2 dollars. These phenomena may be caused by the differences of the time and market conditions, although the rule behind it may be similar. In other words, historical rules may have only the similarity other than exactly right. On the other hand, small pattern size is usually used for a strong local search in the PMRS method, which can ensure that current pattern and the past pattern match well only in relatively short periods, while seriously deviating in longer pattern sizes. Thus, when performing a multi-step prediction, it is hard to find similar status in the historical time series by using PMRS. Unlike PMRS, which matches the current state and the past one directly, GPM takes the differences between the current state and the past one in addition to their similarities into consideration, and the past one is scaled both in the x-axis and the y-axis directions to match the current one indirectly, so it is suitable to be called "generalized pattern matching". When GPM is adopted, the satisfying past pattern which is most similar to the current one could be found even if a large pattern size is required. Since the similarity between the current and the past patterns in a generalized sense can be kept in a long period,

4.5 Forecast of International Oil Price

a more satisfying result can be obtained when performing a multi-step prediction.

Genetic algorithm, as one of global optimizing methods, simulates the evolving process of the life form in nature. Individuals of one generation exchange information with other individuals through genetic operators such as selection and crossover, so that a new better generation is obtained. Keeping the process iteratively, an optimal solution can be figured out. Genetic algorithm, which can avoid local optimization in the searching process, holds an advantage over the common local search methods. The combination of genetic algorithm and other methods plays a more and more important role in the forecasting field. Here genetic algorithm is employed to search the optimal solution for the above optimization problem, which makes the computation faster and the forecast result more accurate.

4.5.3.3 Empirical Approach of Mid-Term and Short-Term Forecast

Daily Brent and WTI crude oil prices, received from IEA website, are used to make an empirical study. The price unit is dollar/barrel. To deal with a few missing data, a linear interpolation is performed on the original data. In this study, there are 4681 data for Brent daily oil price dating from 1987-05-20 to 2005-07-26. The Brent oil price takes on violent local fluctuation. The fluctuation is kept even in a long period. So do the WTI prices. The WTI data consists of 4933 groups of daily observations from 1986-01-02 to 2005-07-26. To test the forecasting ability of GPMGA, the PMRS and Elman networks are employed as comparative approaches.

For Brent data, the forecasting results of the PMRS method, Elman network and FPMGA methods are illustrated in Fig. 4-44.

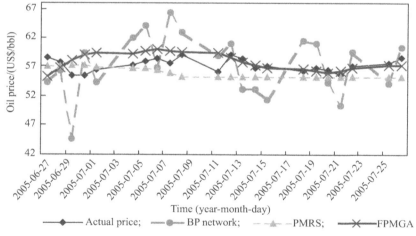

Fig. 4-44 Forecasting results of the three models for Brent data (From 2005-06-27 to 2005-07-26)

Fig. 4-44 shows that the forecasting result of PMRS is basically close to the actual prices in view of the trend, but not in most prices. And the forecasting curve of PMRS is too smooth to exhibit violent fluctuation of oil prices within the short period. On the contrary, the forecasting result of the Elman neural network, which moves downward or upward dramatically, indicates that it is a powerful nonlinear tool. Although rather exact predictions are produced by the Elman network, the forecasting results of the neural network have extreme fluctuations, far more than actual prices. In the obtained results of GPMGA, the predicted values tally with the actual prices not only at quite a few points but also the whole trend. The predicted curve is quite close to the actual curve of Brent prices at almost every point, which shows a better predictive ability than the others. For WTI data, the forecasting results of the three models are shown in Fig. 4-45.

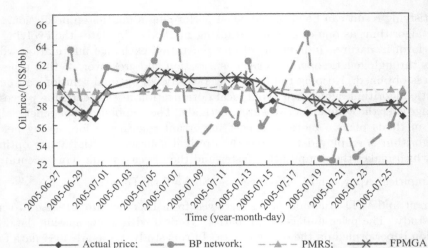

—◆— Actual price; —●— BP network; --▲-- PMRS; —✕— FPMGA

Fig. 4-45 Forecasting results of the three models for WTI data (From 2005-06-27 to 2005-07-26)

Fig. 4-45 shows that the predicted values received from the PMRS model obviously deviate from the actual values in the late period, which is caused by the limitation of the small pattern size determined by PMRS, though the forecasting result in the early period is very close to the actual price. While the Elman network behaves well at only a few points, some of which even overlap the predicted and actual value. However, it behaves poorly at most other points.

Different from the above two models, the fluctuating rules are accurately mastered by GPMGA and a better result is produced notably in the locality but also in the whole prediction period.

The Root Mean Square Error (RMSE) and the Mean Absolute Percentage Error (MAPE) of each model are shown in Table 4-20.

Table 4-20 Predicted errors of three models

Market	Methods	BP neural network	PMRS	GPMGA
Brent	RMSE	4.7635	1.9618	1.7746
	MAPE/%	7.33	2.93	2.43
WTI	RMSE	3.7586	1.5662	1.0909
	MAPE/%	5.50	2.24	1.57

The empirical study, a 21-step prediction of Brent and WTI oil prices (a month roughly), illustrates the predictive ability of the three models. The prediction errors, MAPE, in these models are all below 8%. Among them, the MAPE of the GPMGA model is around 2% both in Brent and WTI, which is much better than the PMRS and Elman network. This result clearly shows the effectiveness of GPMGA in dealing with multi-step predictions of oil prices.

The oil price trend in the early period before 2005-06-24 appears as a rapidly rising trend, even though a decreasing trend right after can be accurately predicted by the GPMGA model, which implies the GPMGA can somewhat more accurately forecast the position of inflection points referring to historical information. Therefore, GPMGA avoids the disadvantage of linear prediction, which tends to tell people that future prices will keep rising from the rising trend of earlier prices. Strong evidence which shows that GPMGA can capture the nonlinear characteristics of price movements in oil time series is obtained from this empirical study.

The fluctuation of oil prices is often related to the macro situation and key events in the world. In general, political events, military conflicts, serious climate abnormalities,

4.5 Forecast of International Oil Price

catastrophes and accidents in important oil-producing areas would lead to sharp changes in oil prices. Therefore, any long-term predictions should be based on them. However, reality is so much more complicated that it is hard to capture the similarity from all events or scenarios that may affect oil prices.

As shown above, the GPMGA model finds the similarity among patterns: starting from Sept. 2003 and Jun. 2005, and gives a somewhat satisfying forecast in Jul. 2005 accordingly. So let's explore the events behind these patterns.

In Sept. 2003, the events behind oil price rapidly rising are: ① The deteriorating situation in Iraq after the war and the problematic postwar reconstruction. There were fears that frustration in the Middle East region might aggravate further leading to a shortage of the oil supply; ② The world economy appears to rebound and recover. With economic recovery in most countries, oil demand increased by 7% compared to before 2003, but oil production only increased by lower than 4% during this period; ③ OPEC adopted the strategy of limiting oil production to maintain prices, cutting oil production even though demand increased; ④ World oil stocks reached their lowest levels in these 10 years. This also included the U.S., whose stocks at the end of 2003 was 12% less than the same period in 2002, amounting to only 0.27 billion barrels, the lowest figure within 20 years.

In June 2005, the influential factors behind rapidly rising oil prices are: ① The rate of recovery in the world economy is still high, over an estimated above 3%, and the sustained strong advancement of the U.S. economy, contributing to greater oil demand; ② The capacity of oil production and refining up to their limits with no easily-attainable ways to greatly increase oil production in order to meet the rising demand; ③ The United States entering into the summer season in which demand will crest, with expectations from the public tending in towards future oil price increases; ④ The persisting chaotic circumstances in the Middle East region and United States uncertainties in Iraq; ⑤ The arrival of the hurricane season in summer and its serious impact on the coastal oil-producing zones such as Mexico bay in the United States, e.g., Hurricane Ivan caused heavy losses last year while hurricanes may cause even worse damages this year; ⑥ The oil stock of United States decreasing from 329 million bbl to 324.9 million bbl in the period of 10, June to 1, July.

4.5.4 Mid-Term and Short-Term Forecast Adjusted by Futures

4.5.4.1 Effects of Futures on Forecast

Futures prices indicate the expectation of people for future spot prices to some degree, so numerous literature endeavors to study whether futures prices can describe future spot price correctly and forecast future spot price correspondingly. Because the importance of oil price in world economy, abnormal oil price changes such as the three oil crises in history seriously harm the economy of each country in the world. Effective forecast methods are required to design strategies for oil change. Therefore, approaches on the relation of oil spots and futures and whether oil futures can improve spot forecast are paid great attention. How to apply futures prices to spot price forecast has been extensively researched. A one-month-ahead oil future is a predictor of short-term forecast of spot price in the future and some forecast models are created.

4.5.4.2 Futures-Weighted Forecast Model based on PMRS

PMRS, a normal pattern matching method, studies the most similar historical pattern to the present pattern in the historical time series, then forecasts according to the historical pattern. However, large bias may appear when PMRS is applied to forecast only using oil spot price. Firstly, futures trade takes a great part in the world oil market. Futures are used by the main oil-import and oil-export countries to avoid risk. Therefore, the change of spot price mainly depends on the change of futures price. If futures price is considered in spot price forecast, the future trend of spot price will be viewed more correctly. Secondly, futures price is an expectation of future spot price and this expectation cannot be neglected in price determination which directly affects the market and the price in the later market. To summarize, it is necessary to combine the instructive effect of futures price and PMRS for improving the forecast. If the futures price in the present pattern is higher than the one in the historical pattern, the oil market will be better and people's expectations are optimistic, so future spot price should be adjusted positively based on the one when futures price is not considered. Contrarily, if the futures price in the present pattern is lower than the one in the historical pattern, the oil market will go worse and people's expectations are pessimistic, so future spot price should be adjusted negatively based on the one when futures price is not considered.

4.5.4.3 Empirical Approach of Forecast Adjusted by Futures

WTI, regular gasoline and heating oil daily spot price and 1-month-ahead daily futures price are used as empirical data (data source: EIA). RMSE and MAPE are used as evaluation indices for forecast result. RMSE indicates average error between forecast price and real price, while MAPE indicates their relative error. Small RMSE and MAPE indicate forecasted oil price is similar to real price and the result is good. We perform the forecast for future one month including 22 data points in our approach. Because the test data includes 100 data points, there are 79 actual forecasts. We take the error standards of these 79 forecasts as final evaluation index. Therefore, the result is tested in a large sample space, which is more precise. At the same time, a naive forecast model and the PMRS are adopted as benchmark and results of different models are compared.

Results of WTI forecast are shown as Table 4-21, where NF means naive forecast model, PMRSF means futures-weighted forecast model based on PMRS.

Table 4-21 Results of different models (WTI)

Model Standard	NF	PMRS	PMRSF
MAPE/%	5.6567	6.0477	5.6326
RMSE	3.3158	3.5688	3.2806

Seen from Table 4-21, for WTI price, MAPE and RMSE of PMRSF are better than the other models, so it gains the best forecast effect. Results of regular gasoline are shown as Table 4-22.

Table 4-22 Results of different models (regular gasoline)

Model Standard	NF	PMRS	PMRSF
MAPE/%	6.6435	6.6021	6.5457
RMSE	10.4	10.315	10.178

Table 4-22 indicates for regular gasoline, MAPE and RMSE of PMRSF are smaller than the other models, so its result is also the best. Results of heating oil are shown as Table 4-23.

In Table 4-23, MAPE and RMSE of PMRSF are also much better than the other two models for heating oil, which indicate a good forecast result.

Table 4-23 Results of different models (heating oil)

Model Standard	NF	PMRS	PMRSF
MAPE/%	6.5065	6.9128	6.3371
RMSE	10.774	11.407	10.557

4.5 Forecast of International Oil Price

In summary, for WTI, regular gasoline and heating oil, PMRSF is better than naive model and PMRS in forecast results. The empirical results indicate the superiority of PMRSF in the short-term forecast of oil price.

4.5.5 Conclusions and Policy Suggestions

4.5.5.1 Revelations from Oil Price Forecast

(1) Oil market is changeable depending on numerous complicated factors.

World oil price mainly depends on oil demand and supply and world economy. Trend of oil price is unstable. In addition, short-term price always deviates from its long-term trend affected by temporary factors such as climate, disaster, political and military action of oil-production countries and market speculation, even in the opposite direction. Basic and non-basic factors interact and make oil price changeable.

(2) Oil price fluctuation affects society in all levels. Especially, high oil price seriously impedes economic development.

With the ration of oil in energy consumption becoming larger and larger, it becomes more and more important. Oil price fluctuation is easier to affect society on all levels, more or less, directly and indirectly. Oil price rising will decrease economic growth, increase inflation, decrease enterprise competitive ability, affect people's lives and threaten financial markets' stability. Especially during the three oil crises, the oil price greatly rising seriously impeded the world economic growth.

(3) Oil price shows similar patterns between history and present because of factors such as seasonality.

The excellent result of mid-and short-term forecast model based on pattern matching indicates the present and the past oil price really take-on similarity and the historical price pattern can be used to forecast the future. Firstly, the oil price fluctuation is periodic. For example, people like to drive for traveling over summer and the United States, the principal oil-consuming country, will reach the oil-using maximum. While in winter, the hot demand for heating oil of the Northern Hemisphere appears. Every year is the same, hence comes the similar oil-consuming pattern, and the oil price also takes on a similar rising and falling pattern. Secondly, a cyclone climate each year induces violent fluctuation in the oil rise. During the period of hurricanes, the oil price takes-on similar change-pattern because of similar factors. Thirdly, peoples' decisions always depend on the analysis of historical data, which leads to the replay of a historical pattern.

(4) In the developing financial market, oil spot price cannot go far from futures price. Futures price indicates the future spot price to a certain extent.

Varieties of financial tools are developing as methods to avoid risk. The international financial market becomes more and more perfect daily. Not only personal investors, but also institutional investors such as corporations and large banks, join financial investment. Large amounts of money emerge, so the futures market even influences (decides) the spot market. Now the pricing of international spot oil chiefly refers to futures price. Therefore, oil spot price is closely tied to the futures price. The futures price indicates the future spot price to a certain extent. The result of the futures-weighted short-term forecast model is superior to other models, which proves this point.

(5) Oil price will gradually revise but still keep high in "future one year".

The world economic growth is estimated to slow down in "future one year" and the oil demand will decrease likewise. What is more, there is enough time for oil-producing countries to enlarge oil investment and production to meet the oil demand increasing quickly. Because the speed of oil demand growth decreases, while oil supply increases continuously, depending on demand and supply relationship, the international oil price will gradually revise in "future one year". If there is no political and military conflict, terrorist events or serious climatic condition, forces driving the oil price to go on rising are small. However, world economic growth will still keep at a high level, although it will slow down a little. A large amount of oil demand is generated with the development of China and India. World oil stock will gradually decrease, the expectation of oil price rising and speculation will, therefore, make oil price rise. All kinds of factors decide when oil price will gradually revise, but it will still keep high in "future one year", in general, fluctuating between 50–60US$/bbl.

4.5.5.2 Countermeasure for High Oil Price

In order to counteract the future effect of high oil price and the price rise which may cause the fourth oil crises, the suggestions below can be considered.

(1) Adjust economic structure and change growth to decrease oil demand.

The two oil crises in 1973 and 1980 led to the serious slowdown of western developed countries. However, the oil price rises in these years affected developed countries only moderately (negatively). Oil price rises by 10US$/bbl will affect the economic growth rate of Japan and EU to fall by 0.4 and 0.5%. For the United States whose oil consumption takes 25% of the world supply and in which 50% of needs depends on imports, its economic growth rate will decrease only 0.3% (Zhang, 2005). The negative effect of oil price rise is smaller and smaller for economies of developed countries, mainly because these countries have gone from industrialized economies to knowledge-based economies. Information and technology have become the strong force of society development. According to statistics, the oil needed by developed countries to generate one unit of GDP, indexed to 1973 to be 100, now has fallen to 40.7. Among the 30 years from 1973 to 2003, the oil and gas needed by the United States to generate 1 dollar GDP falls by 50% (Zhang, 2005). In the same time, the pressure of oil price rise on inflation decreases greatly. However, the approach by Jiao et al. (2005) indicates if oil price rises by 50%, the total GDP in China will decrease by 0.137%, and the price level measured by GDP deflator will rise by 0.675%. So the economy of China is evidently affected by international oil price rise. China is in the industrialized process now and the large oil demand is a normal phenomenon, but we should realize that the industrialized process must quicken, the industry structure must be adjusted, and the growth-form must be transformed in order to relieve the negative effect of oil price rise on the national economy.

(2) Create national strategic oil stock to resist price risk.

Many developed countries such as the United States and Japan have completed a system of strategic oil stock in order to ensure national oil security and resist price risk. As the largest developing country, China has developed quickly and its oil demand has become more and more. Once there are any causal incidents and oil supply is interrupted, disaster will be brought. Therefore, China should gradually form its own strategic oil stock like other developed countries against incidents such as war or significant price rise.

(3) Purchase oil at the right time according to oil price changes.

A large part of oil demand in China depends on imports, and it is estimated to continuously increase in future years. In the past, the Government did not pay attention to what is the best time to purchase oil. Oil was always purchased at a high price, which brought great loss in Chinese foreign exchange and its economy. Oil purchase and imports will be performed under pricing as low as possible, according to forecasts of future oil price (combining all kinds of factors). Additionally, the purchase method should be changed. Not only spot but also futures can be applied in order to limit risk and minimize the cost of oil imports at a high price.

(4) Adopt multi-energy strategy and positively develop new energy.

Coal takes up most of the primary energy consumption in China, and then oil and hydro.

Nuclear energy does not reach 3%, while it is above 15% in developed countries, 40% in France. International nuclear technology is rather mature and China has the base and experience for developing nuclear energy. So it is an option to develop nuclear. At the same time, renewable energy, such as solar, wind, and biology energy should be developed.

(5) Decrease energy consumption and advocate energy saving.

Energy consumption is very large and the energy utility is very low, so the potential of saving energy is immense. The energy consumption of unit production in 2001 is 0.49 kilo oil equivalent, 2.3 times that of the United States, 5.1 times Japan, and 60% higher than newly industrialized countries. In 2003, the crude oil and coal consumption is 7.4% and 31% of the world's, but the corresponding GDP is only 4% of the world's (Zhang, 2005). This energy-utilized mode of high consumption is unsustainable. Therefore, we must adopt energy-saving technology to decrease oil consumption. We should establish energy-saving measures and relative laws and advocate to saving energy, and improve the development of a cyclic economy and energy utility.

4.6 Study of Chinese Oil Pricing Mechanism

There are still some problems with the oil pricing mechanism and domestic oil price, determined by the oil market at present, which cannot completely organize resources rationally or reflect the value of products. In addition, international price risk of crude oil is strengthening the impact on our economy. Both aspects require our government to reform and guide the oil pricing mechanism and the oil market to make the mechanism more rational and complete. We will place emphasis on policies such as oil pricing mechanism and construction of oil enterprises' marketization in this section.

First, we review briefly the historical changes of our oil pricing mechanism from the viewpoint of development. We can see that the reform of our mechanism has been focused on marketization as the long-term goal, but there is a certain disparity between the present and the goal. Next, we discuss problems in oil pricing mechanism's reform and development, briefly, through analyzing the foremost problems that exist in the mechanism.

4.6.1 Overview of Chinese Oil Pricing Mechanism Development

The transition of the price mechanism of countries is basically from relaxing the regulations of oil price to adopting the price mechanism with certain progressive flexibility, and finally realizing the market-based pricing mechanism. With the gradual promotion of an economic system of reform, and foreign currency, and trade system reform, the fast development of the economy and the flourishing economic situation requires the oil price mechanism to keep up with economic development. There is much literature about oil pricing mechanism reform (He, 2004; Gui, 2004; Shi, 2001; Yang, 2002). Since China began to produce crude oil and constitute an oil price in 1955, we have roughly gone through three developments in forming an oil price mechanism, from the mechanism determined completely by government's administration, to one gradually based on oil market pricing, from independent price mechanism to that integrated with the world market. At present our mechanism has been thought of as being basically integrated with the world's oil price. Though problems still exist, marketization degree is improving constantly and the oil price is in more and more close relation with economic and financial activity.

4.6.1.1 Oil Price Mechanism under Traditional Planning Economic System (1955–1981)

With the basic completion of socialist transformation, China entered a highly centralized planned economy era in an all-around way in the early fifties of the last century. In the planned era, the survival of state-owned enterprise that engaged in competitive practices was extremely low, so they needed an energy supply at a low price to survive. Only when the government implements strict control, can the low price of energy supply be maintained. So our country formed strong control of energy and low energy price. During 26 years from 1955 to 1981, as an important strategic material, the oil price's formulation and adjustment were determined by government, which independently fixed prices, without considering changes in the oil price on the world market. Hence, the price established was well below the world oil price. In this planned economy period of centralization, China's oil product and price mainly showed the following traits:

(1) Production and sale of the oil product were run by a state-owned oil company, and trade loss was borne by the state finance.

(2) The country implements overall and highly centralized administratively planned-management to important links, such as oil production, assigning, and selling, etc., and the collection and allocation of funds to oil production, exploring, research expenditure and income from sales are unified by the state.

(3) The disjoint between oil price and product value, production cost and supply and demand of the domestic market made the function fail for controlling resource distribution and price regulating.

So the oil pricing mechanism was in a developmental stage of being under the complete control of the government.

4.6.1.2 Dual Track System Transitioning to Market Price (1981–1998)

Energy control and low-price strategy under a traditional planned economic system still kept its inertia after reform and the opening-up in the market-oriented economy era. Because of insufficient supply in the market and hard-to-get commodity shortage, since 1984, central authorities started the economic system of reform, among which a dual track system was implemented for price system reform.

The dual track system requires implementing two kinds of pricing mechanisms. One is made in unison by the state with the plan-belonging to the monopoly planned pricing. Another is a negotiated price outside the plan. The introduction of a market price is a breakthrough to the single planned price of the past, and recognizes the "law of market". This has far-reaching meaning in both theory and practice. This is a unique initiative along the road to a market economy from a planned economy. From single track, with a planned price, to double track working between planned price and market price simultaneously, has transition characteristics adapted from a planned economy to a planned commodity economy.

The energy price in a planned economy was too low leading to the fact that the energy sector suffered seriously, and whose production was badly insufficient. In order to bring about an advance in the energy sector, the State Council approved the scheme in 1981 that overproduction and oil saving can be sold, according to world oil price; the income of price differential is used in exploration and development by the former oil department. The dual track system became well-known by oil men ever since. In this period, the dual-track system oil market and price showed that:

(1) The price disparity between inside and outside the plan is relatively large, leading to the fact that contradiction and friction between the planned and market behavior were aggravated.

(2) Extensive corruption and publicly Rent Seeking Activities had been caused.

4.6.1.3 Changes to the Market-determined Mechanism of Oil Price after 1998

China became a net importer of crude oil for the first time in 1993, and net importer of the oil product in 1996. The insufficiency of supply had the tendency to increase yearly. In this case, only domestic production ability can not meet economic development's demand. The domestic oil price, which was hard to continue, broke away from world market and operated independently for a long time, which lead to oil price mechanism reform and integration with the world market (since 1998). In June 1998 China extended the great reform to the crude oil and refined oil price forming mechanism, to cooperate with the reforms and reorganization of Petrochina and Sinopec, strengthen the competitive power of our oil enterprise in the world market, adapt to the change of the crude oil price on world oil market, change the mode of the single government price, and begin to implement the way in which domestic crude oil and world crude oil price link together. The domestic crude oil price further draws close to the world oil price. The main content of the reform is, with the foundation that the cost of the factory importing the crude oil and cost of the crude oil transports to the oil plant by land are the same, Petrochina and Sinapec which occupy the absolute monopoly position in the field of petrochemical industry (China National Petroleum Corporation and China Petrochemical Corporation) decide the crude oil price. Settled price of buying and selling is constituted by the crude oil datum price and the agio, in which crude oil base price is determined by the average price of close-quality oil in world market the previous month and the agio is determined through consultation by buying and selling according to the quality price difference of the domestic and foreign oils as well as the market supply and demand and the crude oil shipping and handling cost. The petrol and diesel oil retail sales implement the government guided price, according to import cost from the original State Planning Commission in addition to the domestic reasonable circulation expense to formulate retail standard rate for each place. Petrochina and Sinapec determine the concrete retail price by 5% fluctuation according to this formula (Yi, 2001).

The refined oil pricing mechanism was further reformed in June 2000. It is based on the closing price of similar oil in the future market of Singapore of the previous month, plus premiums with what both sides of supply and demand agreed to confirm the retail prices of domestic refined oil, especially the final petrol and diesel oil. According to the new stipulation, the pricing principle of the refined oil is extremely transparent, and easy to be handled by market speculation. So in October 2001, China carried on the reform to the refined oil pricing mechanism. The refined oil changes, following the method that the Singapore market adjusts the price per month, taking the weighted averages of a basket of prices of New York, Rotterdam and Singapore market as the foundation of the fixed price (price weights kept secret in the three places, adjusting irregularly), the two oil groups have a pricing mechanism with 8% of the floating power. Oil pricing mechanism for this period were for the most part as shown below:

(1) The market price of oil product began to function in the reasonable disposition of oil resources.

(2) Production and sale of oil product were monopolized by several big oil companies, and price failed to reflect changes of domestic market supply and demand.

(3) The base price of refined oil is still controlled by the state, and price change obviously lags behind.

We can see that China's oil price mechanism is striding forward along to marketization, while reviewing the history of changes of the oil price mechanism over nearly half a century. It is only at stage one in the marketization process at present, though the government no longer participates in the formulation of the crude oil price, and partly in refined oil price formulation. The domestic oil price was not determined by supply and demand in the oil market; therefore, the marketization of our oil price mechanism is not over yet. But it has lasted more than six years from the formal reform of the national oil price mechanism of 1998, and in the six years, what is the effect of our reform, what problems still exist, and how to continue the reform in the future? Thinking of these questions can help us to substantiate the direction of the reform.

4.6.2 Problems in Chinese Oil Pricing Mechanism

Following a market-based price-forming mechanism, reform of China's oil pricing will mean that price will be determined by the market, (i.e., both supply and demand), and in the meanwhile oil pricing mechanism will be established after due consideration of the whole economy. In this sense, integration of domestic and world oil price should include several aspects: ① Domestic oil price must reflect the "law of value and rare intensity" of resource; ② Domestic oil price should be favorable to the development of the oil industry to form one's own benign cycle of self-development and accumulation, considering the domestic dependence to the world market at present, at the same time, and price level is basically close to that on the world market; ③ Setting up flexible, open price operation and management mechanism to meet the developing needs of socialist market economy and changing world market (Zhang, 2000) instead of taking oil price in the world market as the domestic price. Considering the principle, the problem exists in the domestic oil price mechanism at present of a weak price function, unreasonable price structure, breaking-off relations between current price and cost, which cannot stimulate enterprises to improve their market efficiency. More specifically, there exist the following problems:

(1) Oil affiliated.

The base price of domestic crude oil is announced by the State Development Planning Commission following the free on-board price of close-quality crude oil on the world market of the previous month, plus additional tariffs. A tangible method is to divide domestic crude oil into light oil, medium oil, medium oil II and heavy oil. The reference oil on the world market is: light oil-Tapiz, medium oil I-Minas, medium oil II-Sinta, and heavy oil-Duri. These kinds of crude oil all have relatively less output and trade, whose market price is easy to be scalped, and the standard is weak.

(2) Standardized operation of both parties.

Two major groups of companies sign a mutual supply agreement every year, according to the crude oil production schedule and resource mutual supply plan. But the current year and quarter agreement lacks enforcement. Because there is no restriction mechanism, sellers already know the trend of the crude oil price. It happens frequently to oversupply when the price is high and there is undersupply, when the price is low, so the buyers are in a totally passive position.

(3) Weak function of the refined oil price.

With reference to the refined oil pricing policy of our country at present, there is none of the parameters to reflect the energy market demand, so the price determined by these parameters only reflects supply and demand of the foreign oil. In fact domestic demand of petrol and diesel oil has a certain uniqueness compared with foreign countries. Specifically, the consumption structure is different; the diesel oil demand is large in the domestic market and the price of petrol and diesel oil that the country confirms is often inconsistent with market demand; and the consumption calendar varies. The world market has its busy season for petrol in summer; and of diesel oil in winter; the domestic market is busy in the second and third quarter for diesel oil. The domestic petrol and diesel oil does not fully reflect the domestic change in demand, which will make the oil price in the busy season unable to improve and to reduce in the slack period, causing prices and supply and demand to swap (Xiao, 2001). This brings about a temporary energy shortage, and threatens the security of the energy strategy.

(4) The price of refined oil obviously lags behind the market.

The refined oil price of our country is decided by taking the weighted average of the refined oil price in the overseas market, i.e., after the rising range of the world refined oil price's weighted average exceeds 8%, our country just carries on the adjustment of corresponding range to the domestic refined oil, and the result is our price, but it obviously lags behind the world market. Oil plant then begins to export large amounts of refined oil, such as diesel oil, to foreign countries to gain profit. In addition, when the crude oil price goes up, the price of refined oil (not adjusted) on the ground, the increasing crude oil price will be turned into production cost and the profit reduced. According to calculations, when crude oil price goes up by 1 US$/ barrel continuously for one year and refined oil price is unchanged, the department of oil and petrochemical industry will pay more than 668 million RMB yuan (Shi, 2003) ,which is very unfavorable to the growth of our oil enterprise.

(5) Risk purchasing mechanism cannot avoid risk.

The Chemical Import and Export Corporation of China, China United Oil Co., Ltd., China International Petrochemical Industry United Co., Ltd. and Ocean Trade Corporation of China received permission in 2002 to become China's first state-owned super enterprises to participate in world oil futures. The releasing of oil futures license meant the formal start of China's oil import-risk purchasing mechanism. However, while being only four major companies at present, China United Oil Co., Ltd. and China International Petrochemical Industry United Co., Ltd., they are, in fact, the subordinate enterprises of Petrochina and Sinopec, which is not a rational relationship. Enterprises receiving licenses should be strictly independent of domestic oil enterprises, thereby reducing risk. Risk-purchasing enterprises utilize various market tools to gain sets of arbitrage, then, after stabilizing oil price fluctuation, they sell oil to enterprises at home at a stabilized price. If we let our (Chinese) oil enterprises, especially large enterprises, participate in the world future markets, the risk of the world markets will fall to consumers. Two major groups can transfer the risk to the oil consumers of domestic market, by raising the price.

(6) Lack of competition in the oil market.

The amount of marketization of the oil consuming market is improving rapidly, but the oil supply is still excessively monopolized. Sinopec, Petrochina and CNOOC nearly dominate the domestic crude oil production; and in crude oil import, only four corporations own exclusive import rights (three described above, plus Sinochem). Other small-scale importers rely on the quotas to survive; but can only share about 20% of quotas. The selling in batches of refined oil is nearly cornered by Sinopec and Petrochina, and in retail, two big companies have monopolized in each area. According to one study, the net profit rate of some foreign oil companies is generally at 4%–7%, and that of CNOOC is up to 30%, Sinopec being about 20% (Li, 2002); far higher than that of any foreign oil company. It means that there are high-monopoly profits in the Chinese oil companies. Obviously, it is impossible for the price formed in such a market to be disposed of effectively, or reflect the rare intensity of the resource, which is unfavorable to the normal functioning of the market mechanism.

4.6.3 Suggestions on Reform of Chinese Oil Pricing Mechanism

4.6.3.1 Problems in Reform of Chinese Oil Pricing Mechanism

The oil price established under a reasonable oil pricing mechanism should meet the following requirements: ① The crude oil price has integral composition; ② Price of crude oil and refined oil should keep have a reasonable relationship with the exchange rate of relevant products; ③ Price of crude oil and refined oil should reflect the market; in addition, the government should enable the oil price to be impartially formed, and should consider development and endurance of the macro economy through macro adjustments and controls (Zhang, 2000).

Crude oil is the same as any other kind of commodity, and whose price structure should include three parts, namely cost, tax and profit. Foundations of price-level should cause enterprises to have the ability to pay various expenses of taxation to the country, according to the regulation; and to re-invest and accumulate wealth after various consumption costs are deducted. Our country has previously joined the World Trade Organization (WTO), and according to its stipulations, our country should unlock the retail market of refined oil, once membership has been for three years, and the wholesale market five years later, the crude oil import rights will be further unlocked. China will cancel the oil import and export quotas by 2004, and the tariffs of different oil products will be further reduced. Therefore, to strengthen the competitive power of oil enterprise in the world market, as soon as possible, and deal with the market-oriented reform, China should guarantee that oil enterprises have reasonable profit and retention to ensure the self-development and sustainability of enterprises.

The price of crude oil and refined oil should have a reasonable exchange rate with relevant products and world prices. Three relations of exchange rate are chiefly reflected in crude oil price: first, the exchange rate between crude oil and refined oil; second, exchange rates between crude oil and other substitute products, such as natural gas and coal, etc.; and third, the exchange rate between domestic crude oil and world crude oil. The relationship of the exchange rate between refined oil is reflected by that between petrol and diesel oil. The crude oil price of our country has changed with the world crude oil price, but the refined oil price has certain disparity and lag-period with the world price of refined oil. This makes price difference between the crude oil and refined oil unreasonable, which has influenced the oil production and supply of our country.

The determination of crude oil and refined oil price should fully reflect the supply-demand relationship of the market. The oil trade regards natural resources as the focus, hence the ability to deal with emergency is relatively poor as little elasticity of supply is unable to increase and change in time with demand, so confirmation of domestic crude oil price should be favorable for the development of the oil industry, increasing supply, and also inhibit the fast growth of consumption of oil and waste. The oil price should be linked closely to the market, reflecting the changes of market supply and demand. There is a need to confirm different consumption prices in different seasons and different consumption areas, and consider its special position in the national economic development at the same time as keeping its price relatively stable.

We should also consider the endurance of the macro economy in the determination of crude oil and refined oil price. The oil industry is a basic industry of the national economy, and the basic energy of the industry, whose price adjustment must have a greater impact on relevant trades. So while confirming and adjusting the crude oil price level, overall planning is necessary. This should fully reflect the pricing principle of the goods, as well as consider the endurance of every department, especially downstream petrochemical industry and processing industry.

In order to ensure that the oil price forms under an oil pricing mechanism and meets these requirements, we should pay attention to the following problems in reform of an oil pricing mechanism:

(1) Establish goal for the price control.

The goal of administering price in developed countries is efficiency and justice. According to their experience and combining the reality of our country, "managing rationally, bearing fairly, regulating demand and giving consideration to the social welfare" should become the main goal of industry's oil-price control.

(2) Role of the government in new energy system.

The government should discuss and enforce the corresponding economic regulating measures at present, and progressively cease administrative management, to stimulate the price and regulate the market and distribute resources rationally. For example, setting up a flexible oil-tax system to realize effective regulation, through a different tax policy and adjustment of taxation method for oil production and supply chain. Only after progressively setting up and improving the economic means of adjusting and controlling the oil market price, could the direct management of the oil price be deregulated. So the role of our government in the new energy system should turn from "price-maker" to progressive "coordinating and supervising", from administrative management to the flexible economic means of utilizing the tax revenue, interest rate, storing etc. to indirect adjustment and control of the domestic oil price, and thereby influencing the pricing policies of oil enterprise.

Besides its changing role in the oil pricing mechanism, government can undertake modifications and adjustment to problems that exist in the current mechanism. The trading of our four kinds of basic crude that is oil affiliated to the world market is small, and easy to be handled by market speculation. Hence we can choose some varieties, and trade with more active trade in the basic crude oil to form more reasonable domestic crude oil pricing. Moreover, both parties' standardized operation and uniting domestic refined oil base price are problems that still exist and refined oil-price lags behind.

(3) Accelerate progressively opening the oil future market to form its own quotation system.

The world oil price is mainly determined on the futures market. China is regarded as one of the main consumers and importers in the world, but has less than 0.1% rights and interests in oil pricing. So we should open a domestic oil future market as soon as possible, and actively participate in the futures of world oil to strive for more influence over pricing. Through the setting up of a domestic oil stock and trade market of the futures, we will achieve the goal of evading the risk, following the supply and demand, adjusting and controlling the market, and guiding oil production, management and consumption.

The common oil pricing method in the world at present is: both sides of supply and demand negotiate directly at the spot price of the crude oil, but the price is not a concrete figure but a particular pricing formula. In this formula, the basic price playing a decisive role is generally the oil futures price of the equal quality. The impact on oil price of the futures market has been larger and larger in the past two years, and people are more and more concentrated on setting up an oil market at home. The impact on the oil market of the futures market is determined by the particular function of the oil futures.

The oil futures function in two respects:

① Price-discovery of oil. Customers of the oil futures market compete on price in the exchange by paying and selling via the broker. Price is generally considered as the leading index of price function and can regulate the supply and demand of various oil products in advance.

② Avoiding risk. The oil and oil-product that the industrial and commercial business circles use, and whose price fluctuates with supply and demand, have an exchange rate and interest rate related to imported and exported oil fluctuating up and down. The fluctuations of the price of oil, exchange rate, and interest rate causes oil production and oil enterprises' profits to fluctuate, which have brought certain risk to their production and business activities. The oil futures market has offered the risk of locking price of raw materials, exchange rates, interest rates change (described above), for oil businesses to make sure to lock the anticipated net profit while receiving the form, and while absorbing internal management and administration.

Setting up a developed futures market in our country cannot merely help to set up the safe and normal domestic oil market, but also improve the ability for enterprises to tackle oil price risk. Our country is a large oil importer-consumer, and if we have the oil futures market and produce the oil pricing system within our country, it will help to fight for oil pricing right and the means of macro adjustments and controls to the market. It will help to set-up a modern oil exchange system to solve the strategic oil reserve problem of our country. Meanwhile, the oil futures have the function of price establishing to reflect the expectancy of supply-demand relationship and price tendency of the future from both sides of the supply and demand to a certain synthetic time and direct production and sale of enterprises. Enterprises can avoid the risk that the price change brings through reversely-operating with the same quantity in the spot market and futures market.

(4) Participate actively in the operation of the world oil market and influence the forming of the world oil price as much as possible.

A problem is, in quite a lot of oil strategic research, the focal point is on how to find the oil. Even the "walking out strategy" is generally for participating in exploring, developing and obtaining "share oil". This research route, following traditional thinking, has many failings, and is unable to meet actual requirements of sustainable development. In our country's new oil strategy, a focal point that should not be neglected is, how to participate in the world market competition omni-directionally and make use of market-oriented means to realize sustainable development and economic security.

The net oil import rose to 69.6 million tons in 2000, but our foreign currency expenditure buying the crude oil only accounted for 5% of the total import value in 2000. Over a longer time in the future, the growth of the gross domestic product will be slightly higher than the growth of the oil import, therefore having the ability to afford the oil imports and the necessary increasing foreign currency expenditure. So just as Dr. Chen Huai of Development Reseach Center of the State Council says, the problem faced in oil import of our country is not that we can not buy or afford the oil but how to buy the oil back favorably and safely. Now we should put more than 70% of the attention in studying the oil market in our country's new oil strategic research. Studying the world market does not study briefly how to bargain in oil purchasing. The competition on the world market is an omni-directional, multi-level competition. We can say, the basic outlet of the oil of our country is to walk out, and the main direction of walking out is to participate in the world market competition in an all-around way. We need to participate in multi-level world market competition such as the stock, futures and property right. We need to influence prices voluntarily through a large amount of repeated business, we need to understand, grasp the competition laws, the price change laws of the world market, and we need the enterprise main body participating in world market competition. To be simple, in our oil strategy, studying the market might be more urgent and important than studying how to find and exploit oil (Chen, 2004). When we are strengthening the cooperation with main oil producing areas, we must make our own oil pricing system, or influences the tendency of the Asian crude oil market price to the minimum limit. Otherwise, even if our country has controlled a large number of oil resources, because the price is controlled by the other countries, our oil security is still unable to be ensured.

An effective method to participate in the operation of world oil market is to set up the risk purchasing mechanism. Only setting up a rational and effective imported oil price forming mechanism whose central part is the risk purchasing mechanism, the domestic oil price may get rid of simply "tracking" the world market.

The basic thinking of the risk purchasing is: Set up one strength which is engaged in risk trade adopting the financial operation mode, really getting involved in the speculative operation, on one hand we can get more world comparable incomes from the price fluctuating of the market, while on the other hand guaranteeing the balanced supply to the domestic demand, enable the import oil price to remain stable within specific limits (Chen, 2001), and reducing influence of the "abnormal" fluctuation in the price of world oil is frequent at the same period. The risk purchasing price can be fixed by application of futures or through the leading pricing way of government on the basis of hearings.

(5) Set up rational oil reserve security system.

Appropriate oil reserve is important for our country. The function of oil reserve lies in increasing the oil supply if necessary and relieving the oil supply and demand contradiction.

4.6 Study of Chinese Oil Pricing Mechanism

When the world oil price rises by a wide margin, oil supply of our country can't increase appropriately accordingly, our economy will suffer serious inflation with quite low economic growth rate, which is an enormous threat to the economic security. So long as our country can control the appropriate oil resources when the world oil price rises by a wide margin to relieve the oil supply and demand contradiction, it can reduce the enormously negative effect on our economy from the oil price rising. It is to be clarified that, through world experience, the oil strategic reserves have never been regarded to stabilize oil price and fluctuation but to ensure oil incessant supply at times of war or natural disaster.

The oil reserve should include two respects: On one hand, reserves of oil, i.e., oil strategic reserves, on the other hand reserves of the oil futures. Reserves cost a lot and have risk of market price, unsuitable to evade risk, and are used mainly for keeping a lookout on oil shortage in times of war or accidental event and restraining the high oil price caused in oil shortage periods. Oil futures reserves are main methods to maintain oil security and reduce the effect on the economy and domestic market from short-term oil price fluctuation.

In addition, we need to link up the oil security with the financial security, setting up an "oil finance" system supported by the banking system to guarantee the healthy and stable development of our economy. This system can link up the oil security with the finance security and become an equilibrium mechanism between oil reserve and foreign exchange reserve. The foreign exchange reserve reached 403,300 million dollars at the end of 2003, but certain risk exists in the world exchange rate which exists too in foreign currency. "Oil finance" system can evade both the oil risk and financial risks through operation of amalgamating oil security with finance security.

Our oil price mechanism at present needs the oil strategic reserves as the material base. Only finishing the national oil strategic reserves as soon as possible, could we strengthen the ability of regulation and control to the oil price. The national oil strategic reserves are one of the effective measures which ensure the oil security too. Our country has already increased oil strategic reserves into the agenda, and has already had some concrete methods. But what are our reasonable strategic storage, the way, and channel to use it, and when can it be put into use, etc? All such questions need to be studied and calculated further.

Before the oil strategic reserves are built up, the rising range of the price of refined oil should be on a basis of average cost profit rate of industry, which is the disparity between the crude oil price and refined oil price in the industry. After the oil strategic reserves are set up, the government should decontrol the price of refined oil completely, only using the oil strategic reserves and economic regulating measure to adjust and control the domestic crude oil price (Shi, 2003).

(6) Set up oil exploration and development fund.

Setting up an oil exploration and development fund is used for replenishing domestic oil exploration enterprises when the world oil price is lower. Because the oil production cost is relatively high, though the oil price mechanism integrated with world price helps our oil enterprises walk out to develop the world oil resources, it may influence the investigation of domestic oil resources and source of funds when the world oil price is low, causing the oil resources reserves to be insufficient and influencing oil safety. Because the oil strategic reserves are only an emergency measure, increasing the proven oil resources reserves could solve the problem of insufficient oil supply fundamentally. The source of the oil exploration and development fund should be mainly drawn a certain proportion from the upward oil price. In legislation, guaranteeing oil exploration should be paid equal attention to with setting up oil strategic reserves.

4.6.3.2 Problems in Evasion of Oil Price Risk

The direct influence on economy of a certain amount of price going up depends on the proportion of oil consumption, accounting for the national income, degree of dependence on imported oil and the ability of end user to reduce consumption or to turn to other energy. Besides these respects, we should also pay attention to the following questions in the evasion of oil price risk and reducing impact on economy from oil price fluctuation:

(1) Adjusting the industrial structure, developing the high-tech industry in a more cost-effective manner, and reducing the proportion in the national economy of high energy consumption.

According to the experience of the developed countries, China should reduce the proportion of the main oil-consumed industry in gross national product, which includes such traditional industries as colored metal smelt, metal forge, steel making, mines, chemical industry and glass making, and develop the low energy consumption industries in a more cost-effective manner such as the high-tech industry and Information & Communication industry, etc.

(2) Reducing the energy intensity and improving energy efficiency.

Sensitiveness of the economy to the fluctuation of oil and energy price is relevant to the proportion in GDP of oil and energy cost (Tatom, 1993). According to Tatom, the impact of high oil price on developing countries like China relying on oil imports is usually greater than for OECD countries. Because developing countries have not only greater dependence on imported oil, but also higher energy intensity (i.e., proportion in GDP of oil and energy cost), and inefficient energy using. Developing countries relying on oil imports use two times larger oil to produce 1 unit of economic output than OECD countries. Though our energy intensity has been reduced to some extent in recent years, there is also disparity compared to that of developed countries. It still has larger potentiality to improve energy efficiency through reducing energy intensity and reduce the impact of oil price fluctuation on the economy.

(3) Improving the production technology of production division.

On one hand, we should closely combine our policy with the national Medium and Long-term Science and Technology Plan, increasing input in research and development in the field of energy, taking key technology as the focal point, mobilize production, teaching, and research to tackle key problems, and accelerating technological progress, and technical innovation of energy industry. On the other hand, we should use many kinds of means such as finance, tax revenue, credit and law to promote industrialization of technical result in energy and offer rewards to enterprises adopting advanced technology. We should improve production in the energy field to make the energy industrial development on a higher technological level. Oil consumption for cars accounts for more than one-third of the total quantity, up to two-third by 2010, so the car fuel efficiency will influence the oil consumption directly. Make every effort to develop through improving economic quality and world competition through accelerating industry's technological progress and changing into increasing intensively from increasing extensively. Change the situation of high investment, low production and get rid of the restriction to economic development of the energy shortage, reduce effect on our economy of the oil price fluctuation and realize the continual, fast and healthy development of the national economy.

(4) Developing and using of the new energy and technology, strengthening the ability to turn to renewable energy.

There is a need to popularize solar energy products in a more cost-effective manner. Our technology has already been mature in this respect, and whose industrialization-scale and application rank first in the world. It is significant for saving regular energy, reducing environmental pollution and fostering new growth engines to use solar energy. Developing nuclear energy and nuclear power is the only clean energy that can offer electricity on a large scale with mature technology, besides water and electricity at present.

(5) Setting up the "oil fund" to support oil enterprises to open up overseas business.

Implement "walking out" strategy actively, evade the risk brought by over concentration of oil imports at present and prevent the effect brought by price fluctuation. Up to now the range of cooperation of our country with overseas oil sources has already been expanded to Russia, Azerbaijan, Kazakhstan of Central Asia, Indonesia, Burma of Southeast Asia, Libya, Iran, Oman and middle and southern American Venezuela and the Middle East, the Sudan of Africa and other places. A lot of collaborative projects between China and foreign countries adopted "share oil", i.e., China participates by shares or investment at a local oil construction project, and obtain certain shares from the output of this project every year. In this case, what our country got is material object, so oil import is unlikely to fluctuate too much by the price. Then sufficient overseas oil output can offset and slow down the impact to economic development of high price of crude oil to a great extent.

(6) Advancing actively the oil marketization reform.

Our oil market should open inward at first before opening to the outside world. On one hand, government should set up the macro adjustments mechanism while decontrolling the oil price to grasp the management and regulation and control right to market; on the other hand, government should break the regional monopoly in domestic oil market and train the market main body actively. Experts think that there are many complicated problems of the oil industry in technology, investment and resources, but most important is lacking a kind of high-efficient, flexible mechanism, which restricted the development of the oil industry. The process of oil marketization lags far behind the changing situation of market economy. Because of monopolization, prices such as oil and gas are controlled for a long time and industry's entry is limited strictly. The oil and petrochemical industry has broken the separation of upstream and downstream, after going through the reorganization in 1998, two large-scale enterprises, Sinopec and Petrochina integrated the upper-stream and downstream set up, added CNOOC and Sinochem who already exist, have formed the competition pattern of oil and petrochemical industry tentatively. But what we need to pay attention to is that, this does not mean it has effective competition in the domestic market. Our country should reduce the control to the oil market access, open the terminal oil selling market and set up the modern oil market. The oil sale belongs to the general competitive business, with lower entry barriers and withdrawal barriers, sufficient market competition, "the Survival of the Fittest" theory to control enterprises to enter and withdraw freely, and realize the high efficiency of the market. Against the background of abundant resources in the world oil market, an open market is an important method that our country has, otherwise, it is unable to guarantee the use of world resources, unable to attract world resources, and unable to improve the security of the oil supply.

(7) Adopting a scrupulous financial and macro economic policy.

In an era of high inflation, high rate of unemployment, exchange depreciation and output-value declining, the economic and energy measure that the government takes will exert an influence for long-term economic development. Though the government is unable to totally dispel the influence brought by the oil price rise, it can use the corresponding financial policy and monetary policy to alleviate these negative effects as much as possible. Improper policy will make the situation worse.

Adopting the overly tightening currency and financial policy may make income and unemployment worse, and adopting expanding currency and financial policy may just delay incomes reducing temporarily, but increase pressure of inflation in the long run, and cause the oil price to rise further.

If the government continues replenishing the oil product to help the poor families and domestic industry, the rise of the oil price may make the financial revenue and expenditure worse. With the rise of the world oil price, the burden of replenishing will aggravate; causing pressure to balance the government's budget and increasing social instability.

The oil crisis will also make companies and consumers lose their confidence, thus they change investment, deposits, expenditure structure, and level. If the government's measure can not be adopted aptly, it will increase these negative consequences during the period. It is unsuitable to adopt the excessive financial policy and monetary policy to try and reduce the impact on the economy from the oil price and fluctuation. The application of these policies must be appropriate, otherwise they will cause greater difficulty instead of reaching the anticipated result.

(8) Implementing the business strategy alliance and expanding overseas development opportunity.

The "business strategy alliance" has already been deemed by numerous entrepreneurs as the most rapid and economic initiative as a type of modern organizational form, and has become an effective format for modern enterprises to improve their world competitiveness. It has been described as "the most important organizational innovation in the end of the 20th century." World major oil companies are strengthening their control and influence to prospect potential resources using scale-capability, strategic alliances and various non-share arrangements.

Basically, the oil industry is located upstream within the whole national economy; its influence on product price in other industries' costs is substantial. Government and enterprise should be shouldering an obligation to its security. It is an irresponsible countermeasure to shift the risk of the world market to downstream through any temporary price mechanism. Tracking simply the oil price of the world market does not accord with the fundamental interests of the national economy, and it can not be regarded as the foremost measure of the oil strategy either. Our government and oil enterprises should take effective measures to improve and drive an oil price mechanism to be actively rationalized, foster a good oil market mechanism and guarantee the energy supply by using effective, rational, sustainable, and stable development of our economy.

4.7 Summary

The oil market in China and the world has been thoroughly researched from six perspectives, i.e., international oil price fluctuation characteristics, interaction between international and domestic oil price, the price change characteristics of crude oil and oil products in China, the impact of international oil price changes on the Chinese economy, international oil price forecast and the Chinese oil pricing mechanism. Consequently, policy suggestions are given based on these studies.

The analysis of international oil price fluctuation is useful to explore the formation mechanism of oil price, to refer to the change in rules over history, master the oil price trend at present and in future, and provide important evidence for the designation of a future oil strategy. Therefore, the wavelet model is adopted to analyze the price level of international oil price in 2005, and the long-term and short-term fluctuation of oil price is discussed. The result indicates that the oil price in 2005, and later, will be at a high level, and take on violent fluctuations. China should pay attention to designing an oil strategy, enlarge oil production, widen oil import methods, and perform an "international cooperation" strategy, so as to eliminate the risk of international price fluctuation.

Allowing for the importance of the oil price, techniques such as long-term trend forecast models based on wavelet analysis, mid- and short-term forecast models based on pattern matching, and futures-weighted multi-step forecast model based on PMRS, are developed and good forecast results are attained using empirical approaches.

As well as analysis of the international oil price, the relationship of domestic and international price is further discussed. Cointegration test, Granger causality, VEC model, and pulse forecast error decomposition models are applied to the study for long-term coin-

4.7 Summary

tegration between domestic and international oil price, the short-term relationship, and asymmetric relationship of their interaction.

The relativity and price ratio analysis of gasoline, diesel and crude oil price, the asymmetric analysis of gasoline and diesel price to crude oil price based on asymmetric error corrected models are included in the studies on domestic oil price. The result indicates the gasoline and diesel price in China takes on distortion, and the distortion of diesel price is more serious than gasoline. And the reflection of gasoline and diesel to crude oil price as material is asymmetric, in which the negative accumulation is larger than positive accumulation.

We also link domestic oil price to macro economy. The effects are calculated by CGE model, including the effect of crude oil price change on the Chinese economy, crude oil production sectors, petroleum sectors and transportation sectors, under high, middle and low technology advances, the ratio change of transportation production and the factor demand and household welfare.

The analysis of domestic oil price and the oil market indicates there are some problems in the oil pricing mechanism in China. It cannot completely reasonably allocate resources and reflect production values. In addition, international crude oil risk already enlarges the impact on the Chinese economy. These two aspects require government to reform the oil pricing mechanism and oil market to perfect the oil pricing mechanism in China.

In summary, the international and domestic problems of oil price are researched systematically and deeply in this chapter, numerous valuable conclusions are obtained, and some ideas and suggestions about oil strategy designation and oil market reform are advanced. Hopefully these would provide certain references for relevant oil policy makers.

CHAPTER 5

Energy, Environment and CO_2 Abatement in China

The Kyoto Protocol came into force on February 16, 2005, which implies that numerous industrialized countries ratified to reduce the amount of six greenhouse gases (carbon dioxide (CO_2), methane (CH_4), nitrous oxide (N_2O), hydro fluorocarbons (HFCs), perfluorocarbons (PFCs) and sulphur hexafluoride (SF_6)) by 5.2% of 1990 levels during the five-year period 2008–2012. This alerts China's government to pay more attention to CO_2 emissions mitigation. As a non-Annex party, China would not be bound in the initial commitment period (2008–2012) to any quantitative restrictions on its greenhouse gas emissions, but China is the second largest CO_2 emitter country next to the United States, so its emissions attract much attention from numerous countries and researchers. The Chinese commitment will be essential in future agreements to reduce greenhouse gases. For China, at least 85% of CO_2 emissions come from fossil fuel use, so energy policies related to CO_2 emissions mitigation become one of the hottest problems in the field of energy strategies and policies.

This chapter focuses on CO_2 emissions from China's fossil fuel use, in order to provide scientific information for the future greenhouse gas emissions mitigation strategies and related energy policies. The problems are as follows:
• Which factors could impact on CO_2 emissions, at different income levels?
• What are the respective roles of rural and urban residents' lifestyle choices on the total energy consumption and the related CO_2 emissions?
• What are the most energy/carbon intensive consumer behaviors?
• The change in China's carbon intensity was at the reverse of those recorded in developed countries during the same period of economic development. Thus, what factors were driving this decline in carbon intensity and could the trend be maintained in future? Moreover, what measures can be adopted to ensure a continual decline in carbon intensity?
• How will China's CO_2 emissions change with the sustainable, rapid economic growth?
• How will the CO_2 emissions change in different regions in China?

5.1 Challenges and Opportunities in Kyoto Era

Global climate change is hitherto the most serious environmental problem and one of the most complex challenges in the 21st century. Climate change is not only the problem of our climate system, but also a problem for the economy-politics-environment, and the core is the economy problem. In order to mitigate effects of global climate change, China should reduce the greenhouse gas emissions and increase the carbon sinks to create CO_2 concentration in the atmosphere at the level before the recent "industry revolution". Greenhouse gases (carbon dioxide (CO_2), methane (CH_4), nitrous oxide (N_2O), hydro fluorocarbons (HFCs), perfluorocarbons (PFCs) and sulphur hexafluoride (SF_6)) are mainly from use of fossil fuel, industrial and agricultural production, land use change and solid waste disposal. Therefore, all the countries are cautious to commit to reducing greenhouse gas emissions because abatement policies will have an impact on economic growth, especially on modernization and sustainable development of developing countries. Climate change will also cause extensive loss, including economic loss, social loss and ecological loss. So, after the Kyoto Protocol went into force, each country, especially developing ones, is faced with new challenges and opportunities.

This section is organized as follows. First, we analyze the relationship between global climate change and reduction in CO_2 emissions. Second, we analyze the relationship between CO_2 emissions and economic development. Finally, we review China's CO_2 emissions and emphasize the possible challenges and opportunities in the Kyoto era.

5.1.1 CO_2 Abatement and Global Climate Change

Fig. 5-1 shows that global climate change was accompanied by the increase of CO_2 concentration in the atmosphere. According to the Third Assessment Report of Intergovernmental Panel on Climate Change (IPCC), the average global temperature has risen $0.6°C \pm 0.2°C$ since 1860, and global mean temperature will be likely to increase by 1.4–5.8°C in 2100. This rate of temperature rise is 2–10 times of the 21st century and not seen on the planet for at least the last 10000 years (The Administrative Center for China's Agenda 21 et al., 2005).

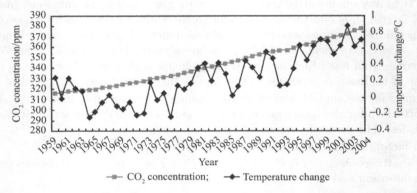

Fig.5-1 Average temperature change and CO_2 concentration in the atmosphere over the period 1959–2004 [CO_2 concentration and global climate change data sourced from CDIAC (2005)]

The global climate warming has attracted much attention, and humans have realized the possible serious consequences brought by global climate warming and recognize that we must adopt some effective measures to solve it. To avoid dangerous climate change, the European Union established a maximum increase in global average temperature of 2°C above pre-industrial levels, which implies that it should reduce greenhouse gases by 15% or even 50% of 1990 levels before 2050. This is roughly equivalent to a stabilization level of 550 ppm CO_2 equivalent or 450 ppm CO_2. The data from CDIAC (2005) shows that CO_2 concentration in the atmosphere in 2004 was 377.43 ppm, and the temperature increased 0.71°C, and global CO_2 emissions in 2002 was 13.4% higher than 1990 emissions. So, each country must adopt more effective measures and technologies to reduce greenhouse gases emissions in order to cause the global climate change not to go higher than 2°C. Based on the emissions of 2002, the global emissions should at least reduce 0.5% every year, but the global greenhouse gas emissions still increase with economic growth. Actually, therefore, the reduction each year is more than 0.5%. The questions, therefore, are how to reduce greenhouse gases emissions, and which countries should be mainly responsible for greenhouse gases emissions reductions, or how to allocate emission reductions in developed and developing countries. Reducing greenhouse gases to mitigate the global climate change is an important task that has a long way to go.

5.1.2 CO_2 Abatement and Economic Growth

The data from CDIAC (2005) show that the United States is the biggest source of CO_2 emissions, and China, India and Japan rank the second, fourth and fifth, respectively.

Figs. 5-2–5-5 show that the change of CO_2 emissions and GDP per capita in the world, China, India, United States and Japan over the period 1960–2002. The data of CO_2 emissions and GDP per capita are sourced from CDIAC (2005) and SIMA (2005) respectively. Fig. 5-2 shows that the changing trend of global CO_2 emissions is similar to the average global GDP per capita. CO_2 emissions and GDP per capita in India increased proportionally (Fig. 5-3). Although China's CO_2 emissions in the year 1997–2000 presented a downward trend, its CO_2 emissions increased in 2001 and 2002, and CO_2 emissions in 2002 are higher than in 1996. At the same time, China's GDP per capita grew continuously, so China's economic growth chiefly drove the CO_2 emissions on the whole. The data of the United States and Japan also show that economic growth essentially causes the increase of CO_2 emissions; CO_2 emissions in the United States and Japan respectively were 21.11% and 12.23% higher than their emissions in 1990. All these illustrate that fossil fuel and cement production will have an important role in economic growth, and CO_2 emissions of both developed countries and developing countries, as well as the whole world, will keep increasing in the future long-term. This is undoubtedly a serious challenge for the realization of committing to the Kyoto Protocol and the establishment of abatement framework in the post-Kyoto era. In view of these, the absolute reduction world-wide is very difficult, because not all countries are willing to limit or restrict the economic and social development for reducing CO_2 emissions, and hope benefits might be gained from other countries' CO_2 emissions reductions.

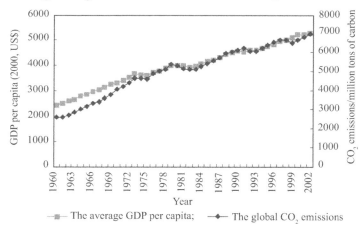

Fig. 5-2 The change of global CO_2 emissions and the average GDP per capita

Therefore, mitigating global climate change not only depends on the reductions of several countries, it should integrate the cooperation between developed and developing countries to improve energy technologies, explore and use renewable energies and carbon capture and storage technologies to protect our environment.

5.1.3 Contemporary Status of Chinese CO_2 Emissions

Fossil fuel use is mainly responsible for greenhouse gases emissions, so China's energy use and fuel mix decide the contemporary status of its greenhouse gas emissions (data from SIMA (2005)).

(1) China's greenhouse gas emissions are large in amount.

The CO_2 emissions of the United States in 2000 were 1528.796 million tons of carbon, which was 22.93% of global emissions. Now China is next to the United States, and CO_2 emissions were responsible for about 11.42% of global emissions, i.e., was 761.586 million tons of carbon. With the development of the economy and increase of energy use, greenhouse gas emissions will increase, and it is estimated that China's CO_2 emissions will be more than the emissions of the U.S.

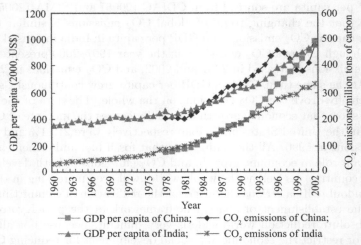

Fig. 5-3 The change of CO_2 emissions and the average GDP per capita in China and India

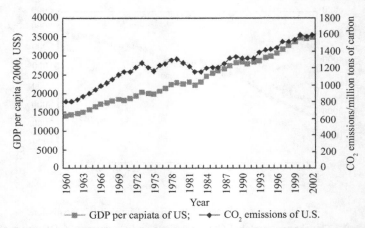

Fig. 5-4 The change of CO_2 emissions and the average GDP per capita in the United States

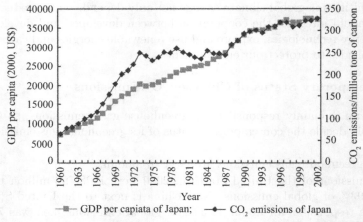

Fig. 5-5 The change of CO_2 emissions and the average GDP per capita in Japan

5.1 Challenges and Opportunities in Kyoto Era

(2) CO_2 emissions intensity is high.

CO_2 emissions intensity is the CO_2 emissions per unit of GDP. Based on the GDP (constant 2000 US$, PPP), China's CO_2 emissions intensity in 2000 was 0.58 kg_{CO_2} / US$, while the average CO_2 emissions intensity of the whole world and of OECD countries was 0.51 kg_{CO_2} / US$, and 0.45 kg_{CO_2} / US$ respectively. This implies that China's CO_2 emissions intensity was 1.14 times and 1.29 times as high as the average global level and OECD level.

(3) CO_2 emissions coefficient is high.

CO_2 emissions coefficient is the CO_2 emissions per unit of energy use. In 2000, China's CO_2 emissions coefficient was 2.45 tons of CO_2/ tons of oil equivalent, and that of OECD and the world was 2.28 tons of CO_2/ tons of oil equivalent and 2.31 tons of CO_2/ tons of oil equivalent.

(4) Energy intensity is higher than that of OECD.

Energy intensity is energy use per unit of GDP. Based on the GDP (constant 2000 US$, PPP), China's energy intensity in 2000 was 0.24 kg oil equivalent/US$, but the average energy intensity of the whole world and of OECD countries was 0.22 kg oil equivalent/US$, and 0.20 kg oil equivalent/US$ respectively.

5.1.4 Challenges and Opportunities Faced by Chinese Energy and Environment

The above analysis shows that mitigating global climate change implies the absolute reduction of CO_2 emissions; but economic development will necessarily cause its increase. As a signatory country of the Kyoto Protocol, the largest developing country and the second largest emitter of CO_2 emissions, China is faced with the contradiction of economic development and mitigation. Now, many kinds of pollution in China are serious, and the share of other greenhouse gases, excluding CO_2 emissions, is the greatest (and relatively high). Moreover, developing the economy, improving residents' living standards, and removing poverty will still be the main tasks over the long-term. By 2020, a "Well-Off Society" will be fully realized, which will cause the increase of energy use and CO_2 emissions. This will have a serious impact on ecology and environment, and increase the pressure to engage in an international pact.

Therefore, in the Kyoto era, China's economic structure, economic growth, and energy use will be faced with serious challenges. But there are also some opportunities due to the CDM projects demanded from developed countries.

5.1.4.1 The Challenges in Chinese Energy and Environment System

(1) The impact on Chinese import/export.

After the Kyoto Protocol went into effect, committed developed countries transferred various energy/carbon intensive industries to China through investment. These industries will improve Chinese economic growth and employment in the short term, but these projects are always large-scale and will require the long-term to remove them, with difficultly over the short-time. These industries use plenty of energy and cause copious pollution. However, it will be a serious challenge for the Chinese economic growth pattern to change from extensive to intensive, and it will have extensive impact on employment and economic growth if they are removed after some years.

In addition, the European Union and other countries may set down policies based on emissions of China's industrial products, which will be a green bulwark. The iron industry and some high-polluting industries may be firstly affected, which will directly affect the Chinese import/export trade.

(2) The impact on energy structure.

The Chinese energy usage largely depends on coal, and 70% is fossil fuel. Based on energy demand forecasting to meet the "well-off society" in 2020, energy use will be about 3.1 billion tons of coal equivalent, and coal use will be about 2.3 billion tons of coal equivalent. Thus, it will make the energy use and energy safety and environmental question even more important. And energy-intensive/carbon-intensive industries will be affected first, and have to pay more money for buying emissions' permits from those countries with lower emissions if China commits to reducing CO_2 emissions in future. Faced with intense resource supply/demand, China can not pollute firstly and protect secondly. Therefore, the Kyoto Protocol will drive the innovation of a fuel structure. The needs of residents, economic structure and industrial structure drove energy use, so it also has an impact on economic structure.

(3) The impact on China's economic structure in the middle- and long-term.

The coal-based fuel mix and industry-based economic structure are mainly responsible for larger CO_2 emissions in China. Second industry is the main user of energy, iron, electricity, chemicals, paper making, cement, and glass industries that are energy-intensive. So the Kyoto Protocol will bring long-term pressure for these industrial sectors. Therefore, the current economic structure and industrial structure have to be adjusted, and technologies be updated to avoid the huge loss if China commits to reducing CO_2 emissions in future.

(4) The impact on China's energy policies.

The Kyoto Protocol and negotiations in the post Kyoto era may have an important impact on China's energy policies, such as:

- The policies on energy efficiency and energy efficiency standards.
- The related policies around CDM projects.
- Renewable energy exploration and use policies.
- The structure adjustment and export/import adjustment policies of energy-intensive industries.
- The exploration and use policies of coal-bed methane.
- The exploration and use policies of fossil fuel.

All adjustments should be based on China's national conditions and international trends.

5.1.4.2 The Opportunities Brought by the *Kyoto Protocol*

The *Kyoto Protocol* has brought many opportunities, which may mitigate the pressure if we can make full use of these opportunities. CDM is a kind of win-win mechanism; it will bring some advanced technologies and capital investment to improve the sustainable development of developing countries, on the other hand, developed countries can decrease the higher abatement cost for reducing CO_2 emissions to stimulate climate change mitigation.

The World Bank estimated that developed countries have to reduce 5.0–5.5 billion tons of CO_2 emissions over the period 2008–2012, and 50% will be realized through the domestic reduction of developed countries, while the rest has to be realized through trade, based on three flexible mechanisms (Sahoo, 2005). Until 2012, there will be about 1–1.5 billion tons of CO_2 emissions reduction through the CDM or JI projects, but up to now, the certified CO_2 emissions reduction was less than 0.3 billion tons of CO_2 emissions. So there will be at least reductions in demand worth US\$ 10 billion, which will induce projects investment worth US\$ 50 billion (Wang, 2005). Therefore, there is much potential to develop the CDM operation. Clearly, China's economy grew rapidly, and energy use obviously increased, the fuel-mix depends on coal, and the average technologies level is relatively low, so the potential to reduce CO_2 emissions is huge. On the other hand, China is the biggest developing country, and the economic potential is enormous and labor supply is sufficient, so the market for CDM projects cooperation is huge, China may, therefore, be the biggest CDM project host country. Moreover, China's government enacted "The Short Rules for Clean Development Mechanism Project Management" on June 30, 2004, which offers an improved policy environment.

Up to December 18, 2005, there were 53 projects registered by EB, and the type of these projects is shown in Fig. 5-6. Most of them are small hydropower projects, and landfill gas projects and small biomass power projects, as well as others. The number of China's CDM projects was three, including a wind power project, a small hydropower project and a landfill gas project (The methane capture projects refer to the methane capture for pig farming).

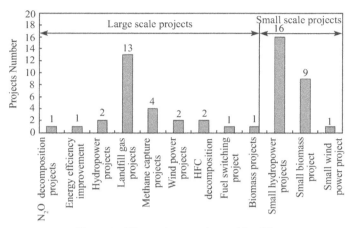

Fig. 5-6 The projects registered by EB

From the above analysis and the actual status of China, the following fields may have the potential for cooperation:

(1) The exploration and use of small hydropower and wind power. China's small hydropower resource below 50MW is about 120GW, and 28.5GW has been explored. The available wind power is estimated to be 253 GW, but its exploration was only 0.14% up to 2000.

(2) The collection and use of landfill gas. The yearly waste is about 140 million tons, and the annual growth is 5%–8%. Now this waste is mainly crammed. With the development of recycling economics, the comprehensive use for waste is an important direction. Whether focused or comprehensive, the application of methane is a valuable opportunity for cooperative CDM projects.

(3) The use of biomass combined heat and power (CHP). In China, biomass is abundant, but the use efficiency is very low and most is used as the primary energy. Biomass can account for 70% of total rural energy use, but most is directly burnt or disposed, and waste is serious and a pollutant to the environment. So China should make full use of biomass to develop CHP, which can improve the rural energy use pattern and increase income.

(4) CH_4 and HFCs technologies, industrial producing process update, especially the update of industrial boilers, and the exploration of coal-bed methane are the other potential CDM cooperative areas.

5.2 Characteristics of Carbon Emissions Trend in China

Carbon intensity is one of the most important indexes in measuring a country's CO_2 emissions. To analyze its change mechanism, therefore, is critical as it provides policy-makers with a clear understanding of the impact of factors that contribute to CO_2 emissions. It can also provide detailed information for future energy strategies and CO_2 emissions reduction policies.

The earliest origin of acknowledging carbon intensity was initiated shortly after the 1973/74 world oil crisis. At that time energy researchers quantified the impact of structure-shift on total industry energy demand, in order to have a better understanding of the mechanism of change in energy use, by adopting Index Decomposition Analysis (IDA) (Ang, et al., 2000). With in-depth research, analysis expanded to investigate the impact on energy intensity, CO_2 emissions, and carbon intensity of structural shifts and final fuel uses. However, most studies of change in carbon intensity focused on research conducted in developed countries rather than in developing ones.

It is argued that in fact it is more important to analyze the carbon intensity change of developing countries, because it would aid in optimizing fuel-mix and economic structure. Moreover, it could provide detailed information on mitigating the growth of energy consumption and the related CO_2 emissions, in order to avoid following the path of "first pollute the environment and then take counter measures".

5.2.1 Trend of Chinese Carbon Emissions Intensity

Fig. 5-7 shows that both primary energy-related carbon intensity and the material production sectors' final energy-related carbon intensity present downward trends, and the difference between them became gradually smaller, especially in 1992 and 1996. The difference between them was 126.60 tC/ million 1990 constant yuan in 1980, 57.87 tC/ million 1990 constant yuan in 1992, 52.31 tC/ million 1990 constant yuan in 1996, and 35.43 tC/ million 1990 constant yuan in 2002. Moreover, the ratio between primary and material production sectors' final carbon intensity presents a downward trend too, which indicates that the efficiency of primary energy usage is increasingly improving, and the energy loss during the process of conversion from primary energy to final energy has decreased. Thus, further improvement of energy efficiency would promote a further decline in carbon intensity.

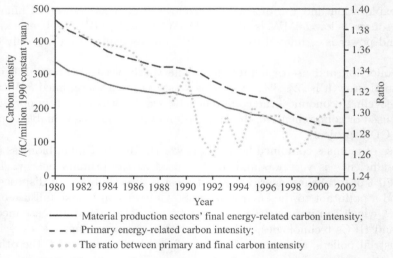

Fig. 5-7 The change in carbon intensity during the period 1980–2002

Fig. 5-8 indicates that final energy-related carbon intensity in the three industry sectors studied here all presented downward trends. Final energy-related carbon intensity in secondary industries declined rapidly, but those in primary and tertiary industries declined slowly. Moreover, carbon intensity in the secondary industry sector was obviously higher than carbon intensity in the primary and some sectors of tertiary industry; further, the change of carbon intensity in primary industries was close to that of tertiary industries. Changes in carbon intensity in secondary industries can be divided into three phases: 1980–1991, 1992–1996, and 1997–2002. Carbon intensity in the secondary industry sector decreased by 34.21% during 1980–1991, 38.05% over 1992–1996, and 44.79% during 1996–2002. The decline in carbon intensity in primary and tertiary industries was slow, and only proceeded rapidly in 1985 because of a sudden decrease in coal consumption in two industries. From these results, we can summarize that the decline of the material production sectors' carbon intensity was mainly attributable to the decline of carbon intensity in the secondary industry sector.

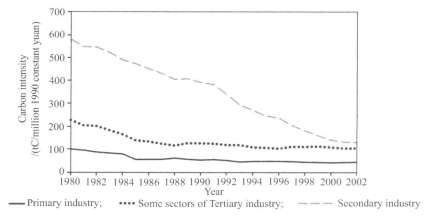

Fig. 5-8 Change in the material production sectors' final energy-related carbon intensity, 1980–2002

In reviewing the above descriptions presented in Figs. 5-7 and 5-8, we find that both primary energy-related carbon intensity and final energy-related carbon intensity in each of the three industry sectors studied have been declining. Thus it is necessary to analyze what driving forces have led to such a decline, whether these forces will maintain this trend, and what measures can be adopted to drive them to decline further.

5.2.2 Comparison of Carbon Emissions Intensity between China and Developed World

As the country with the second largest emission of CO_2, China's energy policies and related CO_2 emissions remain one of the most debated energy problems confronting the world today. Since launching its open-door policy and economic reforms in 1978, China's economy has grown rapidly with its Gross Domestic Product (GDP) increasing at an average annual rate of about 9.57% during 1980–2003. At the same time, carbon intensity has also been declining by 66.94% over this time period.

For the period 1980–2000, the GDP per capita increased from 1067$ to 3425$ (1990 international G-K dollar), which approximately equates to the GDP per capita level of the developed countries in the period 1870–1918. From Fig. 5-9, we can see that the change in China's carbon intensity was the opposite of those recorded in developed countries during the same period of economic development. Thus, the question arises as to which driving forces have led to this trend and whether future increases in carbon intensity can be prevented. However, few quantitative studies have been undertaken regarding this issue.

5.2.3 Methodology and Data

Greening et al. (1998; 1999; 2001) and Greening (2004), respectively, analyzed the changes in aggregate carbon intensity of manufacturing, freight, residential, and personal transportation sectors in ten OECD countries. Consequently, we intend to choose AWD decomposition to analyze the changes of carbon intensity in China, the AWD formulation for decomposing primary energy-related carbon intensity and material production sectors' final energy-related carbon intensity can refer to Davis, et al., 2002; Greening, et al., 1998; Greening, et al., 1999; Greening, et al., 2001; Greening, 2004.

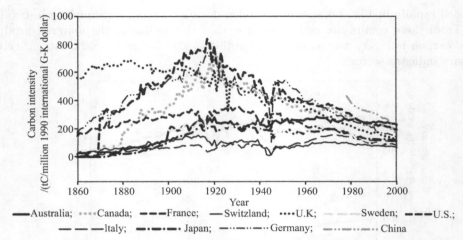

Fig. 5-9 Comparison of carbon intensity between ten developed countries and China

Note: The data of CO_2 emissions are taken from the CDIAC (Carbon Dioxide Information Analysis Center); the real GDP (1990 international G-K dollar) is taken from Maddison, A.

5.2.3.1 AWD Method

(1) The definition of primary energy-related carbon intensity.

Total primary energy-related carbon intensity can be defined as a product of aggregate energy intensity and carbon emissions rate, and the aggregate carbon emissions rate can in turn be expressed as a fuel share weighted sum of fuel emissions rate:

$$G_t \equiv \frac{C_t}{Y_t} = \frac{C_t}{E_t}\frac{E_t}{Y_t} = I_t\frac{C_t}{E_t} = I_t\sum_{j=1}^{n}\frac{C_{jt}}{E_{jt}}\frac{E_{jt}}{E_t} = I_t\sum_{j=1}^{n}e_{jt}R_{jt} \quad (5\text{-}1)$$

$$(1 + \Delta\% G_{\text{tot}}) = (1 + \Delta\% G_{\text{intensity}})(1 + \Delta\% G_{\text{emis}})(1 + \Delta\% G_{\text{fuel}}) \quad (5\text{-}2)$$

(2) The definition of material production sectors' final energy-related carbon intensity.

Total material production sectors' final energy-related carbon intensity may be expressed as the summation over three industries of the product of final energy emission rate, fuel share, energy intensity, and output share.

$$G_t \equiv \frac{C_t}{Y_t} = \sum_{i=1}^{m}\sum_{j=1}^{n}\frac{C_{ijt}}{E_{ijt}}\frac{E_{ijt}}{E_{it}}\frac{E_{it}}{Y_{it}}\frac{Y_{it}}{Y_t} = \sum_{i=1}^{m}\sum_{j=1}^{n}R_{ijt}e_{ijt}I_{it}y_{it} \quad (5\text{-}3)$$

$$(1 + \Delta\% G_{\text{tot}}) = (1 + \Delta\% G_{\text{emis}})(1 + \Delta\% G_{\text{fuel}})(1 + \Delta\% G_{\text{intensity}})(1 + \Delta\% G_{\text{struc}})(1 + D) \quad (5\text{-}4)$$

Table 5-1 provides the definition of variables in AWD formulation.

5.2.3.2 Data Source

The *real* GDP (1990 constant RMB yuan) for the period 1980–2003 is calculated, based on the indices (1978=100) of Gross Domestic Product in the primary industry, secondary industry and tertiary industry. The national GDP is the sum of GDP in the three industries, other than the direct calculation by indices of national GDP, because we find that there is an error between the former and the latter.

5.2 Characteristics of Carbon Emissions Trend in China

Table 5-1 Definitions of variables

Variables	Definitions
i	This refers to three industries; $i=1,2,3$, respectively refers to Primary industry, Secondary industry and Tertiary industry
j	$j=1,2,3$ refers to coal, crude oil, and natural gas among primary energy mix; $j=1,2,3,4$ refers to coal, petroleum products, natural gas and electricity among final energy mix
t	A point in time greater than the base year. In this section t is respectively 1981–2003 and 1981–2002. The base year is the year 1980
E_t	Energy consumption in year t.
E_{ijt}	Energy consumption from the use of final energy type j of industry i in year t
E_{jt}	Energy consumption from the use of final energy type j in year t
e_{ijt}	Energy share from the use of final energy type j in industry i in year t, $e_{ijt}=E_{ijt}/E_{it}$
e_{jt}	Energy share from the use of final energy type j in year t, $e_{jt}=E_{jt}/E_t$
Y_t	The GDP in year t
Y_{it}	The GDP of industry i in year t
y_{it}	The GDP share of industry i in year t, $y_{it}=Y_{it}/Y_t$
I_t	Energy intensity in year t, $I_t=E_t/Y_t$
I_{it}	Energy intensity of industry i in year t, $I_{it}=E_{it}/Y_{it}$
C_t	The energy-related CO_2 emissions in year t
C_{it}	The energy-related CO_2 emissions of industry i in year t
C_{ijt}	The energy-related CO_2 emissions from the use of final energy type j of industry i in year t
C_{jt}	The energy-related CO_2 emissions from the use of final energy type j in year t
c_{it}	The energy-related CO_2 emissions share from the use of industry i in year t, $c_{it}=C_{it}/C_t$
c_{ijt}	The energy-related CO_2 emissions share from the use of final energy type j of industry i in year t, $c_{ijt}=C_{ijt}/C_t$
c_{jt}	The energy-related CO_2 emissions share from the use of final energy type j in year t, $c_{jt}=C_{jt}/C_t$
R_{ijt}	Emissions rate of carbon from the use of final energy type j in industry i in year t, $R_{ijt}=C_{ijt}/E_{ijt}$
R_{jt}	Emissions rate of carbon from the use of energy type j in year t, $R_{jt}=C_{jt}/E_{jt}$
G_t	Carbon intensity in year t, $G_t=C_t/Y_t$
$(1+\Delta\%G_{\text{tot}})_{0t}$	Index of actual changes in carbon intensity between year 0 and year t, where 0 is the year 1980
$(1+\Delta\%G_{\text{emis}})_{0t}$	Index component of estimate of the change in carbon intensity due to a change in energy emissions rate between year 0 and year t
$(1+\Delta\%G_{\text{fuel}})_{0t}$	Index component of estimate of the change in carbon intensity due to a change in energy use share between year 0 and year t
$(1+\Delta\%G_{\text{intensity}})_{0t}$	Index component of estimate of the change in carbon intensity due to a change in energy intensity between year 0 and year t
$(1+\Delta\%G_{\text{struc}})_{0t}$	Index component of estimate of the change in carbon intensity due to a change in industry activity mix between year 0 and year t
D_{0t}	Quotient of actual carbon intensity and estimated carbon intensity, or residual of estimation

Primary energies are coal, crude oil, natural gas, and hydro-power. The share of hydro-power is relatively small: only 7.8% in 2002. Moreover, there are no CO_2 emissions for the use of hydro-power. So we do not consider hydro-power, and just analyze the coal, crude oil, and natural gas. Primary energy consumption is taken from the State Statistical Bureau (2004).

The final energies considered are coal, petroleum products, natural gas and electricity, in which coal includes raw coal, cleaned coal, other washed coal, and briquettes; petroleum products include crude oil, gasoline, kerosene, diesel oil, fuel oil, LPG, refinery gas and other petroleum products. The consumption of coal, petroleum, natural gas and electricity by different industries are also taken from the State Statistical Bureau (1987; 1990; 1992; 1998; 2001; 2004). The CO_2 emissions for the period 1980–2003 are calculated in terms of the energy consumption and the carbon emission coefficients provided by the Energy Research Institute in 1991 (Zhang, 2000).

Because the 1980–2003 period analyzed is a relatively short term, we assume that the carbon emission coefficients of coal, oil and natural gas are constant. In fact, these coefficients have changed over time because of a change in grade of fuels; these changes are so small that they can be negligible when we study the macro changes in carbon intensity. The carbon emissions coefficient of electricity, however, is available given the change in fuel mix used in the generation of electricity and the technological improvements in generation. So, the impact of emission rate on carbon intensity is restricted to the impact of change in electricity emission rate.

5.2.4 Empirical Results and Discussion

(1) The primary energy-related carbon intensity analysis.

Fig. 5-10 indicates that the change in primary energy-related carbon intensity was mainly due to the change in primary energy intensity that took place during the 1980–2003 period.

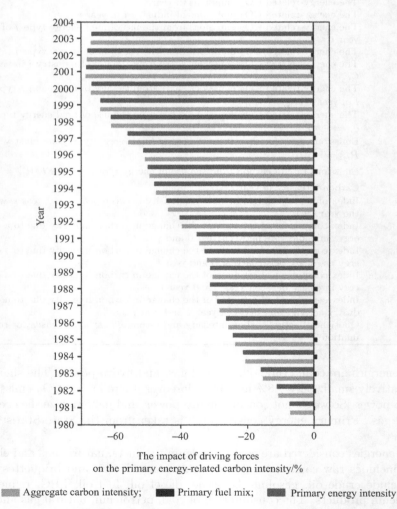

Fig. 5-10 The impact of different driving forces on primary energy-related carbon intensity over the period 1980–2003 in China

5.2 Characteristics of Carbon Emissions Trend in China

Our results show that more than 99% of decline in primary energy-related carbon intensity for the period 1980–2003 was attributable to real energy intensity declines.

The primary fuel mix partly offsets the decline of the related carbon intensity that occurred during the years 1980–1998, and increases the decline in carbon intensity over 1999–2003 because the share of coal in the primary energy mix was lower than that in 1980. The impact of the primary fuel mix on the related carbon intensity was small because the change in the primary fuel mix was not obvious during 1980–2003. In 1980, the share of coal in total primary energy consumption was 75.21%, the share of oil was 21.56% and the share of natural gas was 3.23%, whereas the share of coal dropped to 72.46%, the share of oil rose to 24.51% and the share of natural gas decreased slightly to 3.02%.

Therefore, there is potential for continued decline of primary energy-related carbon intensity in the future. Clearly this needs to be accomplished with further decrease of primary energy intensity and adjusting the share of coal in the primary fuel mix. If we only decrease of primary energy intensity and do not adjust the primary fuel mix, the impact of the primary fuel mix on related carbon intensity will partly offset the driving force of primary energy intensity for related carbon intensity. Moreover, there will be opportunity to adjust the primary energy mix in China in terms of the transition to energy mixes, like those found in developed countries. In brief, it will be possible for the primary energy-related carbon intensity to continue to decline in China over the long term.

(2) The material production sectors' final energy-related carbon intensity analysis.

As the basis of our examination of primary energy-related carbon intensity, we analyze the impact of structure, final energy intensity, emission rates for generation and final fuel mix on the change of the material production sectors' final energy-related carbon intensity. The final energy consumption of material production sectors was approximately 62% of total final energy consumption, and the related CO_2 emissions were 64% of total fossil fuel-related CO_2 emissions during 1980–2002.

Fig. 5-11 illustrates that the decline in final energy intensity was driving the decrease in the related carbon intensity. Emission rates for generation were also driving the decline in carbon intensity and, as the impact increased, the technologies used to improve the generation sector enhanced the efficiency of fuel for generation. Moreover, the structure drove the decline of carbon intensity over the years 1980–1986, and partly offset the decline over the years 1987–2002. Notably, three industrial sectors became carbon-intensive over the period 1987–2002, by paying great attention exclusively to the development of the secondary industry sector. Its share was higher than it had been in 1980, and increased to more than 50% in the years 2000–2002. The change in final energy mix partly offset the decline in carbon intensity which shows that the final energy mix became carbon-intensive.

Compared with the final energy-related carbon intensity in 1980, the final energy-related carbon intensity in 1986 decreased 30.09% due to the decline in final energy intensity and electricity emissions rate, and the development of the structure. The actual decline was 27.14% due to the offset impact of final fuel mix; from the period 1988 to 2002, the decline of final energy intensity exceeded the decline of carbon intensity over the period 1988–2002. The final energy-related carbon intensity in 2002 decreased 84.08% because energy intensity decreased by 75.68% and electricity emissions rate decreased by 8.40%. But the actual decline was only 68.09%, because the final fuel mix effect is an increase of 18.60% and structure effect is an increase of 23.23%.

So, the decline of the material production sectors' final energy-related carbon intensity was mainly due to the decline of final energy intensity. But the additional decline of carbon intensity cannot depend solely on decline in energy intensity, the change of structure and final fuel mix. Likewise, the change of products-production and final energy use in secondary industries will further increase the decline of carbon intensity.

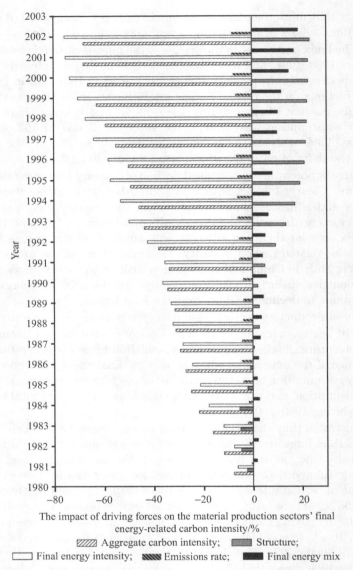

Fig. 5-11 The impact of different driving forces on the material production sectors' final energy-related carbon intensity during 1980–2002 in China

5.2.5 Conclusions

Through a quantitative analysis of driving forces behind the decline in primary energy-related carbon intensity, and the material production sectors' final energy-related carbon intensity, we have found that energy intensity was mainly driven by the change of primary energy-related carbon intensity and material sectors' final energy-related carbon intensity. For changes in primary energy-related carbon intensity, our results show that more than 99% of decline in primary energy-related carbon intensity for the period 1980–2003 was attributable to real energy intensity decline. For the change of material production sectors' final energy-related carbon intensity, the decline of final energy intensity exceeds the decline of carbon intensity over the period 1988–2002.

These results of primary energy-related issues only indicate that the structure change of primary energy-use can further increase the decline of carbon intensity, but it can not tell us how to adjust the primary energy mix.

The analysis for material production sectors' final energy-related carbon intensity, however, may give us this answer. The results show that structure partly offset the decline over the period 1987–2002, and the change in final energy mix also partly offset the decline in carbon intensity over the period 1981–2002. This suggests that final energy use in three industries was carbon-intensive, and the development of these three industries became carbon intensive. We can deduce that the offset impact of primary energy use on the decline of carbon intensity was mainly caused by meeting the need of three industries for final energy use and the development of the secondary industry sector.

Therefore, the change of primary energy use structure should focus on the three industries' final energy use; especially the secondary industry sector. The change of final energy use in the secondary industry has two considerations:

(1) The adjustment of products' production. Government can encourage residents' consumption needs toward less energy/carbon-intensity to drive the development of non-energy/carbon-intensive sectors. Also, government should decrease the export of energy/carbon-intensive products, and even transfer the production of some energy/carbon-intensive products.

(2) Enhancement of energy efficiency and decrease energy consumption per unit product through technology improvement. National and regional governments should give assistance on tax and loans.

Additionally, the electricity emission rate also drove the decline of material production sector's final energy-related carbon intensity. Therefore there is still potential for improvement. The decrease of coal consumption for electricity generation, the shutdown of small power generation plants, and the use for hydro and nuclear can greatly decrease the electricity emission rate.

In our opinion, carbon intensity has declined extensively in recent years. However, we argue that there is still a great deal of opportunity for further decline of carbon intensity, with China's continual economic growth. Through our analysis, we can confirm that the study for further decline of carbon intensity should mainly focus on the decline of energy intensity, especially the change of energy and carbon intensity to analyze how secondary industries become carbon-intensive which is very important for the change of carbon intensity in the future.

5.3 Impact of Population, Economic Growth and Technology on CO_2 Emissions

Global warming is occurring, and scientists have reached a consensus that the massive accumulation of greenhouse gases is the main reason for the recent rise in global temperature (IPCC, 1995). Thus every country shares the responsibility to curb the rapid growth of greenhouse gases emissions in order to mitigate global climate change worldwide. To do that efficiently, research concerning which factors have an impact on CO_2 emissions and the extent of their impact has been of great importance, since these impact factors will directly influence the constitution of CO_2 abatement measures, policies and strategies for each country.

5.3.1 Impact Factors of CO_2 Emissions

CO_2 emissions are determined by technology levels, affluence, energy structure, economic structure, population constitution and so forth, but these impact factors play different roles in explaining the growth of CO_2 emissions. Some traditional studies typically claim that increasing energy consumption is the main cause of growing CO_2 emissions, without considering the impact of population and technology on these emissions (Shi, 2003). On the other hand, some studies claim that population, economy and technologies are all key factors determining CO_2 emissions (Engelman, 1994; Cole, et al., 1997; Meyerson, 1998; Schmalensee, et al., 1998; Ye, 1998), and further claim that their impact on CO_2 emissions is heterogeneous across different countries (Shi, 2003). In that case, the question arises: what are the relations among different economies, population, technology levels and CO_2 emissions in each country?

To solve this question, many studies and discussions have already taken place. Dietz and Rosa (1997), Shi (2003) and York et al. (2003) analyze the relationship between population and CO_2 emissions by applying the STIRPAT model. Doing so, Dietz and Rosa (1997) and York et al. (2003) have found that the elasticity of emissions with respect to population change is nearly 1, but Shi (2003) finds that the elasticity is between 1.41 and 1.65. However, these findings focus on the whole of CO_2 emissions and the population of many countries and lack of analysis for a particular country. So conclusions lack advice for CO_2 emissions abatement strategies of each country.

In view of these conclusions, from a historical perspective, we employ the STIRPAT model using the Partial Least Squares regression to analyze the impact factors of global CO_2 emissions, the total CO_2 emissions of China and countries at the high, upper-middle, lower-middle, and low income levels, using data from the period 1975–2000. This study finds that in this period, the greatest impact on global CO_2 emissions is the growth of the global economy. Globally, the proportion of population aged 15–64 has the least impact. The impact of population on total CO_2 emissions at the high and low income levels is relatively greater, and in these countries the impact of the percentage of the population aged 15–64 on CO_2 emissions is very large, indicating that human behavior is very important for environmental change. In the low income level countries, the impact of GDP per capita on total CO_2 emissions is vast. At the lower-middle income level, the impact of energy efficiency is very large, but the impact of GDP and energy efficiency on total CO_2 emissions at the high income level are relatively great compared with those of other income levels, which implies that the high income countries may further optimize their economic structures and strengthen energy saving. Altogether, these findings underscore that the impacts of population, affluence and technology on total CO_2 emissions are different at different levels of development. Thus policy-makers should consider these issues fully when constructing long-term strategies for CO_2 abatement.

5.3.2 STIRPAT Model

Ehrlich and Holdren (1971; 1972) were the first to use IPAT to describe how our growing population contributes to our environment, both positively and negatively. This took the form of an equation combining environmental impact (I) with population size (P), affluence (A, per capita consumption or production), and the level of environmentally damaging technology (T, impact per unit of consumption or production), known as $I=PAT$. This equation is a widely recognized formula for analyzing the impact of the population on environment (Harrison, Pearce, 2000), and is still used for analyzing the driving forces of environmental change (York, et al., 2002).

There is some controversy about $I = PAT$. We get the proportionate impact of environmental change by changing one factor and simultaneously holding other factors constant. To overcome the limitation of these models, York, Dietz and Rosa (2003b) reformulate IPAT into a stochastic model, naming it STIRPAT (for Stochastic Impacts by Regression on Population, Affluence, and Technology), in order to analyze the non-proportionate impact of

5.3 Impact of Population, Economic Growth and Technology on CO_2 Emissions

population on environment. The specification of the STIRPAT model is:

$$I_i = aP_i^b A_i^c T_i^d e_i \tag{5-5}$$

The model keeps the multiplicative logic of the equation $I = PAT$, treating population (P), affluence (A) and technology (T) as the determinants of environment change (I). After taking logarithms, the model takes the following form:

$$\ln I_{it} = a + b(\ln P_{it}) + c(\ln A_{it}) + d(\ln T_{it}) + e_i \tag{5-6}$$

The subscript i denotes that these quantities (I, P, A and T) vary across observational units; t denotes the year; b, c, and d are the exponents of P, A, and T, respectively; e is the error term, and a is the constant. Eq. (5-6) presents the linear relationship between population, affluence and technology.

The determinants P and A (Dietz, Rosa, 1994) as well as T (York, Dietz, Rosa, 2003a) are decomposed. We revised Eq. (5-6) by incorporating the percentage of the population living in the urban areas (urbanization), and the percentage of population aged 15–64, giving us Eq. (5-7). Here U refers to urbanization, and L refers to the percent of population aged 15–64.

$$\ln I_t = a + b_1(\ln P_t) + b_2(\ln U_t) + b_3(\ln L_t) + c(\ln A_t) + d(\ln T_t) + e_i \tag{5-7}$$

5.3.3 Trends of Population, Economic Growth, Technology and CO_2 Emissions at Different Income Levels

5.3.3.1 Data Set

In term of STIRPAT model, we analyze the impacts of population, GDP per capita and GDP per unit of energy use on CO_2 emissions using the data over the year 1975–2000.

Affluence is measured by GDP per capita (1995 constant US$). As there is no clear consensus on valid technology indicator (York, et al., 2003), Shi (2003) used two variables denoting technology: manufacturing output as a percentage of GDP and services output as a percentage of GDP; York et al. (2003) used industries output as a percentage of GDP; York et al. (2003) used a percentage non-service GDP of GDP. Technology in this section is measured by energy intensity, that is energy use per 1995 constant PPP$ GDP (kg of oil equivalent per 1995 constant US$). The less the energy intensity is, the higher the efficiency is of economy activities, and the less the CO_2 emissions are. Population is decomposed to two variables: the percentage of population aged 15–64 and the proportion of the population living in urban areas. Ordinarily, the higher the percentages of population between the ages of 15–64 and of urbanization are, the more energy consumption there is, but, at the same time, the higher the awareness is of environmental protection and technology. Table 5-2 shows the definitions of variables used in this section.

Table 5-2 Definitions of variables used in this section

Variable	Definition	Unit of measurement
CO_2 emissions	Emissions from industrial process stemming from the burning of fossil fuels andthe manufacture of cement	kt
GDP per capita	Gross domestic product divided by mid year population	1995 Constant US$
Population	Mid-year population	Number
Technology	Energy use per 1995 constant PPP$ GDP	kg of oil equivalent per 1995 Constant PPP$
Urbanization	Proportion of population living urban areas	Percent
Population aged 15–64	Ratio of population aged 15–64 over the total population	Percent

The data used in this study are all from the Statistical Information Management and Analysis (SIMA) database of the World Bank (1997). In conformance with other studies, we express the unit of CO_2 emissions as the unit of carbon. The conversion ratio is a unit of carbon to 3.664 units of CO_2 emissions (Engleman, 1998).

In order to contrast the effects of population, affluence and technology on CO_2 emissions at different developmental levels, this study analyzes all the countries in the world and all the countries at the high, upper-middle, lower-middle and low income levels, and analyzes the historical characteristics of the relations between total CO_2 emissions and the impact factors between the years 1975 and 2000. The World Bank's definitions of the income levels are: low income countries are those with a GNP per capita of 765 US$ or less in 1995, lower-middle income countries are those with a GNP per capita between 766 US$ and 3035 US$, upper-middle income countries are those with a GNP per capita between 3036 US$ and 9385 US$, and the high income countries are those with a GNP per capita above 9386 US$ (World Bank, 1997). The CO_2 emissions of population in different income levels are their respective aggregates of all countries in different income levels. For example, CO_2 emissions in the high income level are the aggregate CO_2 emissions of all high income countries. GDP per capita in different income levels are their average GDP per capita, and urbanization, population aged 15–64 in different income levels are their average urbanization and, population aged 15–64. These data are directly from SIMA database of the World Bank.

5.3.3.2 Descriptive Analysis of Variables

Fig. 5-12 shows the CO_2 emissions at each income level plus China. From 1975 to 2000, the global CO_2 emissions, the CO_2 emissions of China, and the total CO_2 emissions of countries at the various income levels all present upward trends despite some dips. The global CO_2 emissions has an overall upward trend and grew by 42.9%. Total CO_2 emissions of countries at the high income level also have an upward trend and grow 28.8% despite a dip in 1980. The total CO_2 emissions of countries at the upper-middle level grew by 38.0% between 1975 and 1987 and decreased by 19.9% from 1988 to 2000, thus growing by 65% in the aggregate. The total CO_2 emissions of countries at the lower-middle income level had been growing and grew by 58.6%, but between 1989 and 2000, emissions decreased by 8.7%, so on the whole, emissions grew by 44.8%. For the low income countries, emissions grew by 162%. For China, emissions grew 126% during 1975–1996, then decreased 16.52% during 1996–2000, and on the whole grew by 88.94%.

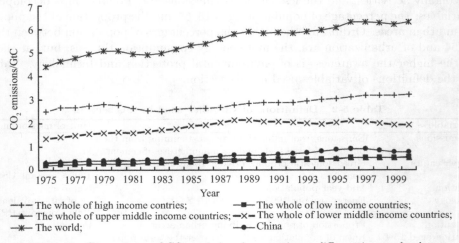

—+— The whole of high income contries;   —■— The whole of low income countries;
—▲— The whole of upper middle income countries; —×— The whole of lower middle income countries;
—*— The world;   —●— China

Fig. 5-12 Comparisons of CO_2 emissions in countries at different income levels

5.3 Impact of Population, Economic Growth and Technology on CO_2 Emissions

Fig. 5-13 shows that GDP per capita has an upward trend overall during the 1975–2000 period. Real GDP per capita grew 42.5% at the global level, 75.2% at the high income level, 30.0% at the upper-middle income level, 65.2% at the lower-middle level, 67.6% at the low income level and 400% in China.

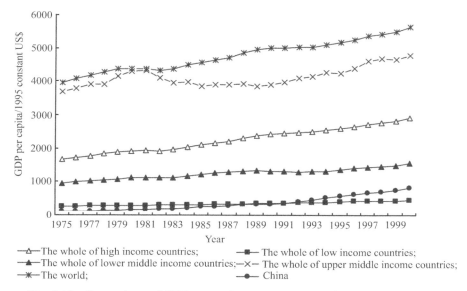

Fig. 5-13 Comparisons of GDP per capita in countries at different income levels

Fig. 5-14 shows upward trends for the 1975–2000 period. The global population increased sharply and grew by 46.0%. For the countries at the high income level, population had increased slowly and grew by 19.5%, and its growing rate only was 0.72%. The population of the upper-middle income countries had been increasing slowly and grew by 49.1%. The population grew by 43.1% at the lower-middle income level, 76.6% at the low income level and 28.44% in China.

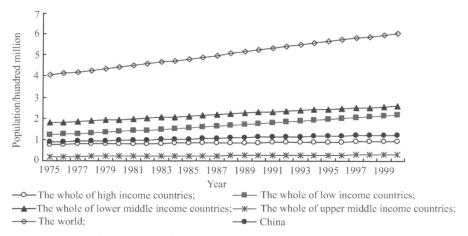

Fig. 5-14 Comparisons of population in countries at different income levels

Fig. 5-15 compares the energy intensity in 1975–2000 at the various development levels. Apart from the upper-middle income level, energy efficiency had improved worldwide, in China and at the other income levels. Energy use per 1995 constant PPP$ GDP had increased by 33.0% worldwide, 40.9% at the high income level, 49.4% at the lower-middle income level, 16.3% at the low income level and 233% in China, but increased 25.8% at the upper-middle income level.

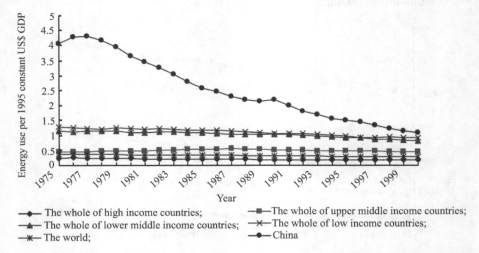

Fig. 5-15 Comparisons of energy use per 1995 constant PPP$ GDP in countries at different income levels

From these statistical data, we find that the change of CO_2 emissions, population, GDP per capita and energy use per 1995 constant PPP$ GDP worldwide appear to be coincident, and population growth outpaces growth of emissions. The changes in GDP per capita and energy use per 1995 constant PPP$ GDP outpace the changes in emissions at the high income level. The growth of emissions outpaces that of population at the upper-middle income level, but energy use per 1995 constant PPP$ GDP had a downward trend. The change of emissions, population, GDP per capita and energy use per 1995 constant PPP$ GDP at the lower-middle income level are also coincident, and, more importantly, the growth of GDP per capita outpaces that of emissions. For countries at the low income level, the change in emissions outpaces changes in population, GDP per capita and energy use per 1995 constant PPP$ GDP. For China, the changes in GDP per capita and energy use per 1995 constant PPP$ GDP significantly outpace the growth of emissions. Obviously, the relationship between the changes in these factors and the growth of emissions differs at different development levels. In that case, we must ask how these factors interrelate such that they have an impact on the growth of CO_2 emissions on the Earth.

5.3.4 Results Analysis and Discussions

We treat CO_2 emissions as the dependent variable and establish the STIRPAT model.

Table 5-3 shows that the correlation coefficients are higher among China's CO_2 emissions, GDP per capita, energy intensity, and population. The STIRPAT model using the OLS (Ordinary Least Squares) regression can not reflect the real relationship among CO_2 emissions and impact factors because standard error of parameter estimation about regression coefficients enlarges, the confidence interval widens, the stability of parameter estimation declines, and furthermore the test statistic of regression coefficient is not significant or the parameter estimation is not correct (Yi, 2002). Therefore, we use the PLS (Partial Least Squares) estimation method in order to establish the STIRPAT model for population, affluence, technology and CO_2 emissions, thus avoiding the multi-collinearity among variables. The results, as calculated by SAS V8 Software, are shown in Table 5-4.

5.3 Impact of Population, Economic Growth and Technology on CO_2 Emissions

Table 5-3 Correlations among CO_2 emissions, population, GDP per capita and energy intensity in China

Variable	CO_2 emissions	GDP per capita	Energy intensity	Population
CO_2 emissions	1.000 000			
GDP per capita	0.967 077	1.000 000		
Population	0.977 957	0.993 532	1.000 000	
Energy intensity	−0.953 012	−0.997 810	−0.989 028	1.000 000

Table 5-4 PLS regression of CO_2 emissions: the effects of population, affluence and technology: 1975–2000

Variable	High income	Upper-middle income	Lower-middle income	Low income	World	China
C (constant item)	4.47	4.83	2.865	2.69	1.66	−9.59
GDP per capita	0.57	0.33	0.44	0.26	0.68	0.06
Population	0.54	0.21	0.28	0.33	0.30	0.37
Energy intensity	0.20	0.13	0.71	0.22	0.59	0.32
Urbanization	0.57	0.23	0.23	0.33	0.24	0.14
Percentage of population aged 15–64	−0.70	0.17	0.57	0.23	0.34	0.73
Index of models Number of latent variables	5	4	2	3	2	4
Minimum Root Mean PRESS (predicted residual sum of squares)	0.1082	0.22	0.20	0.0716	0.14291	0.140723
Probability	1	0.32	0.35	1	0.44	1
PLS components explaining the percent of independent variables	100.00	99.9986	99.1574	99.9013	99.2672	99.9952
PLS components explaining the percent of dependent variables	99.2487	96.6894	96.9409	99.6651	98.5485	98.9294
R^2	0.94	0.94	0.95	0.93	0.95	0.93

The regression coefficient of Table 5-4 shows that during the past twenty years, the effect of the percentage of population aged 15–64 on CO_2 emissions in high income level is negative, which has some similarity with the Boserupian view that the impact of population on the state of the environment is likely either a non-relationship between two variables, or even a negative elasticity. However, the effects of population, affluence and technology on emissions accord with the Malthusian view that a larger population could result in increased demand for energy for power, industry, and transportation, hence increasing fossil fuel emissions.

In the PLS regression approach, VIP (Variable Importance in Projection) reflects better the explanatory potential of each independent variable for each dependent variable. VIP shows the importance of every independent variable when explaining the dependent variable. If a predictor has a relatively small VIP value (Wold (1995) considers less than 0.8 to be "small"), then it is a prime candidate for deletion. The equation is:

$$\text{VIP}_j = \sqrt{\frac{p}{Rd(Y;t_1,\cdots,t_m)} \sum_{h=1}^{m} Rd(Y;t_h) w_{hj}^2} \tag{5-8}$$

Where: VIP_j is the VIP of x_j; p is the number of independent variables; $Rd(Y;t_1,\cdots,t_m) = \sum_{h=1}^{m} Rd(Y;t_h)$ is the accumulative explanatory capability; $t_1,\cdots,t_m$ are components extracted in the variable X; w_{hj} is the j of w_h, which was measured by the marginal contributions of x_j for constitution t_h, and for any $h = 1, 2, \cdots, m$, $\sum_{j}^{p} w_{hj}^2 = w_h' w_h = 1$.

Figs. 5-16–5-22 show the VIP of each factor. The VIP of population, the percentage aged 15–64, GDP per capita, energy intensity and urbanization are all more than 0.8, so each independent variable plays an important role in explaining the growth of CO_2 emissions.

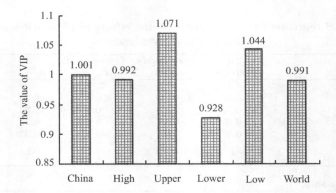

Fig. 5-16 The VIP of population

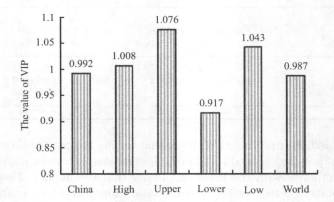

Fig. 5-17 The VIP of urbanization

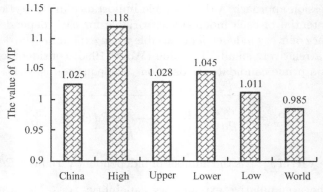

Fig. 5-18 The VIP of population aged 15–64

5.3 Impact of Population, Economic Growth and Technology on CO_2 Emissions

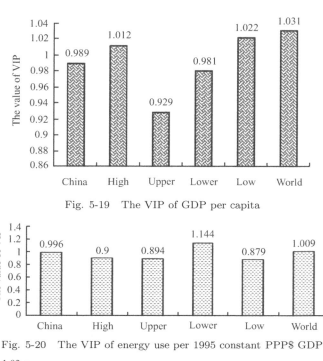

Fig. 5-19 The VIP of GDP per capita

Fig. 5-20 The VIP of energy use per 1995 constant PPP\$ GDP

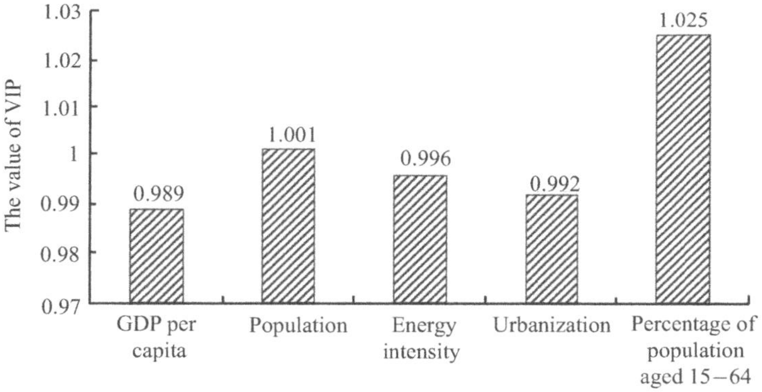

Fig. 5-21 The VIP of population, GDP per capita, energy intensity, urbanization and the percentage of population aged 15–64 in China

Above results and analysis indicate the following relations:

(1) During the period 1975–2000, the impact of population on emissions is the greatest at the upper-middle income level, followed by the low income level, and is the least at the lower-middle income level. Thus, for emissions abatement, it is important for the upper-middle income countries to mitigate the growth of population and optimize people's production and living fashions.

(2) The effect of urbanization on emissions is similar to that of population, which is to say the greatest at the upper-middle income level followed by the low income level, and the least in the lower-middle income level. This shows that higher urbanization further increases energy consumption per capita and increases CO_2 emissions because energy efficiency, energy saving technology and awareness of environmental protection are still low. In our analysis, with the countries at the upper-middle income level, urbanization should implement energy-using fashions and advocate energy-saving by the population living in urban areas during the process of urbanization.

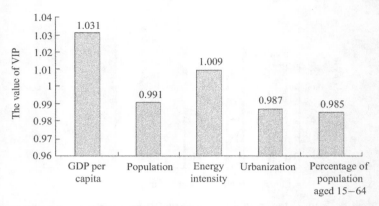

Fig. 5-22 The VIP of the global population, GDP per capita, energy intensity, urbanization and the percentage of population aged 15–64 world-wide

(3) The effect on emissions of the percentage of population aged 15–64 is the greatest because most of the VIP are more than 1, and its effect is greatest, and, more importantly, negative, at the high income level, while it is still great but positive at the lower-middle income level. Because of the high awareness of environmental protection, energy efficiency and energy-saving technology, the labor force at the high income level is expected to reduce the use of cars and use products that reduce environmental damage; in this case, the increase of labor force favors the abatement of CO_2 emissions. This illustrates that the role of population on CO_2 emissions is not only constrained by impersonal conditions, but also affected by subjective awareness; that is, the behavior of people can greatly affect CO_2 emissions. Thus the existence of the B in the equation $I = PABT$ is important.

(4) The role of GDP per capita is greatest at the worldwide level followed by the low income level, and is the least at the upper-middle income level. Except for the role of GDP per capita at the high income level, the others present a downward trend with the increase of economic development level, which shows us that CO_2 emissions per unit GDP in the developed countries is lower than those in the developing countries. This is in agreement with the fact that developed countries have some advantage in energy efficiency, economic structure, energy-consuming structure, energy technology and so forth. For the countries at the high income level, due to their economic structure, product structure and consumption behavior, higher GDP per capita can induce more energy consumption and more CO_2 emissions.

(5) The effect of energy intensity is the greatest for the lower-middle income countries, and is similar for other income levels. Because of the high investment, maintenance costs and long R&D cycles of technologies, the improvement of energy efficiency is relatively slow. Thus its impact on CO_2 emissions is relatively small. For the countries at the high and upper-middle income levels, the effects of reducing CO_2 emissions through improving energy efficiency is not obvious because of the difficulty in developing new technologies on the basis of relatively reasonable energy-consuming and economic structures. For the low income countries, because of high technology costs, their economic and energy-consuming structures and technological barrier, it will take a long time to improve energy efficiency. However, for the lower-middle income countries, the effect of abating CO_2 emissions will be obvious through decreasing energy intensity due to the optimization of economic and energy structures, increased energy consumption, and relatively low energy efficiency and energy technology use.

(6) For China, the percentage of population aged 15–64 has the greatest impact on CO_2 emissions, followed by population, energy intensity, urbanization, and GDP per capita. This shows that the labor force's production and living fashions have the greatest effect on emis-

sions. As a lower-middle income country, China could reduce emissions greatly by decreasing intensity. In addition, energy efficiency and energy-saving technology are relatively low and public awareness of environmental protection is not high, so higher urbanization essentially increases CO_2 emissions. Thereafter, the optimal strategy at present for reducing China's CO_2 emissions is: controlling the growth of population, changing the labor force's production and living styles, and improving energy efficiency.

(7) At the global level, GDP per capita has the greatest impact on emissions, followed by energy intensity, and the impact of the percentage of population aged 15 to 64 is the least. Thus it seems that global CO_2 emissions reductions depend on the improvement of technologies, because of the continued growth of economic development, population, and urbanization worldwide, especially at the lower-middle income level. Moreover, the greenhouse effect is a global phenomenon, but the costs of technologies for reducing CO_2 emissions need to be taken on by each country, and, importantly, the growth of GDP will increase CO_2 emissions so that almost all governments think that the emissions reduction works against economic growth. Thus they are not willing to develop or even use new technologies to reduce CO_2 emissions. Therefore, the reduction of global emissions is still a long-term process.

5.3.5 Conclusions

Through quantitative analysis of population, GDP per capita and energy efficiency in countries at differing income levels, we find:

(1) The effect of population on the CO_2 emissions is great, especially the percentage of population aged 15 to 64, and its effect in the high income countries is negative. However, the fact that the effect is positive at the other income levels supports the Malthusian perspective, which holds that population growth increases CO_2 emissions. This indicates that awareness of environment protection and effect in improving the environment quality varies at the different income levels, technological levels and economic conditions. When income per capita reaches the higher level, people resort to science and technology to optimize the energy consumption structure and reduce CO_2 emissions, thus improving the self-existing environment. With increases in the percentage of population aged 15 to 64, CO_2 emissions increasingly reduce. Contrarily, when the income per capita is relatively low, CO_2 emissions increase following an increase in the percentage of population aged 15 to 64, as tested by the results of the model. This illustrates the importance of the B in the equation $I = PABT$. Thereafter, policy-makers should take heed of the role of production and consumption behaviors on CO_2 emissions reduction at the various income levels when constituting long-term strategies for CO_2 emissions reduction. Therefore, the impact of human behavior on CO_2 emissions should be treated as the next object of analysis, through which we can determine which are the most energy/CO_2-intensive behavior activities, what is the role of behavior choice on the economy, energy consumption and CO_2 emissions, and how much of total energy consumption and CO_2 emissions is a product of behavior activities.

(2) The impact of the growth of GDP per capita on CO_2 emissions presents an approximately descending trend with the increase of economic development level.

(3) The effect of the enhancement of energy intensity on CO_2 emissions abatement is restricted by the level of economic development and energy-consuming structures. For the high, low and upper-middle income countries, the effect is small, but it is great for the lower income countries.

(4) The effect of urbanization on CO_2 emissions reduction is also restricted by the level of economic development, energy structure, and energy consumption per capita, the gap between the rural and the urban and so forth.

The PLS regression results of the STIRPAT model fully illustrate that the impacts of population, economy and technology on CO_2 emissions are different in the countries at different development levels. The main policy suggestions for CO_2 emissions abatement in China are as follows.

First, China as the second largest producer of the amount of the emissions in the world and a lower-middle income country, it is very important to develop the economy in comparison to CO_2 emissions reduction. During the process of becoming an upper-middle income country, it may optimize the economic and product structures, otherwise energy efficiency would be difficult to improve, and might present a downward trend such as that of the current upper-middle income countries.

Second, because China is in the process of becoming a high income country while having the greatest population in the world, it is important for the government to influence the behaviors of people, because currently more and more people with high incomes and educations are likely to have bigger houses, smaller household type, cars, more and bigger appliances, more entertainment opportunities, fashions and personal travel and so forth. These activities inevitably induce more energy consumption and CO_2 emissions, and so they also create a serious challenge for energy safety and CO_2 emissions mitigations.

5.4 Impact of Lifestyle on Energy Use and CO_2 Emissions

Energy use and the related CO_2 emissions are influenced by technology efficiency, and also by personal lifestyles. In the late 1980s, some energy researchers began to pay attention to the impact of consumers' behaviors on energy use, and introduced the concept of lifestyle into the study of personal energy consumption; thereby trying to find the impact of personal lifestyle on the total energy consumption and environmental change. Schipper and Bartlett (1989) concluded that: "about 45%–55% of total energy use is influenced by consumers' activities for personal transportation, personal services, and homes", they thought that: "significant changes in energy demand will be driven by the mix of personal activities and their locations ... besides energy prices and incomes". Lenzen (1998) analyzed the impact of consumers' activities on energy consumption and greenhouse gases emissions in Australia using the input-output model. Weber and Perrels (2000) made a quantitative analysis of the impact of some lifestyle factors on the 1990s and 2010s energy demand, and the related emission in West Germany, France and Netherlands. Pachauri and Spreng (2002) analyzed the direct and indirect energy demand of Indian households using input-output tables. Reinders et al. (2003) evaluated the average energy requirement of households in eleven European Union member countries, based on household data of expenditure. Bin and Dowlatabdi (2005) studied the relationship between consumer activities and energy use and the related CO_2 emissions using the Consumer Lifestyle Approach (CLA). Rees (1995), Daly (1996) and Duchin (1998) argued that: "most environmental degradation can be traced to the behavior of consumers either directly, through activities like the disposal of garbage or the use of cars, or indirectly through the production activities undertaken to satisfy them".

China is a typical binary economy and socially diverse country; there is a significant difference in many aspects between urban and rural regions that has existed since 1949. Therefore it is critical to analyze the impact on the energy use and the related emissions of lifestyles of urban residents and rural residents respectively. Presently, it is widely recognized that China is a "transitional economy", whose transition will bring deep change for its people, employment patterns, living standards, consumer behaviors and similar sociological indicators. Whilst socio-economic development will take place, it is likely that resource issues and environmental problems will manifest themselves.

Firstly, changes are due to urbanization. There is great difference in energy consumption between rural and urban populations. Rapid urbanization will bring more and more rural residents to become urban residents, which will change their energy consuming behavior and increase the demand for electricity, oil and natural gas. Secondly, the difference of lifestyle between rural residents and urban residents is becoming gradually smaller. In rural regions, more and more commercial energy, such as coal, oil, electricity, and natural gas are used, and gradually substituted for the use of non-commercial energy, such as straw and firewood. Meantime, the income increases of rural residents will lead to greater commodities' demand, and increases in their living expenditures for durable goods, such as refrigerators, computers, motorcycles and similar fast-moving-consumer goods (FMCG). All of which will lead to greater demand for electricity. On the other hand, more and more rural residents will purchase and use efficient machines driven by electricity and oil for agricultural production (to reduce the labor). So the total development and modernization of rural regions will improve the living standards of rural residents, and will also lead to a high energy requirement. Thirdly, the living consumption structure of urban residents will change increasingly. The first evidence of improvement of living consumption structure has focused on the home appliances during the period 1980–1995; and interestingly, refrigerators, color TVs and washing machines were the most fashionable goods. After 1995, the living consumption structure was changed again, and the consumption of housing, cars, communication and electronic products, culture and education, and holiday and travel became new "hot" consumption increases. Therefore, the change of urban and rural living consumption structure could have a significant impact on the future energy use and related CO_2 emissions.

It is clearly evident that the study of the impact of residents' lifestyle on energy use and related CO_2 emissions in China is extremely important. Based on a Consumer Lifestyle Approach (CLA) used by Bin and Dowlatabadi (2005), we analyze the following questions: How do we study the impact of lifestyle changes of China's residents on energy consumption and the related CO_2 emissions? What are the respective roles of rural and urban residents' lifestyle choice in the total energy consumption and the related CO_2 emissions? What are the most energy/CO_2-intensive consumer behaviors?

5.4.1 Relationship of Lifestyle, Energy Use and CO_2 Emissions

Bin and Dowlatabadi (2005) illustrated the relationships among different lifestyles, energy consumption and the related CO_2 emissions. They argued that consumers might use energy directly, whilst on the other hand, consumers need to buy and use a range of products in order to meet their basic needs for commodities such as: clothing, food, housing and traveling. The production and processing of these commodities would obviously lead to extensive energy consumption. Therefore, the impact of consumers' lifestyle on the energy consumption and the related CO_2 emissions can be divided into direct impact and indirect impact. Because China is a transitional economy, we discuss the direct and indirect impact of urban and rural residents' lifestyle on the energy use and CO_2 emissions. Details of categorization for residents' living behaviors are depicted in Table 5-5. When discussing the impact of rural

Table 5-5 Residents' living behaviors categorization

Residents' living behaviors categorization	Urban residents' living behaviors categorization	Rural residents' living behaviors categorization
Direct influences	Home energy use including lighting, appliances, cooking, space heating and water heating; personal travel	Home energy use including lighting, appliances, cooking, space heating and water heating; personal travel
Indirect influences	Food; clothing; residence; household facilities, and services; medicine and medical services; transport and communication services; education, cultural and recreation services; miscellaneous commodities and services	Food; clothing; household facilities, and services; medicine and medical services; transport and communication services; education, cultural and recreation services; miscellaneous commodities and services

residents' lifestyle on energy use and CO_2 emissions, we do not consider the impact of the residence. The rationale for this decision lays in the fact that the end-use energy shown in the *Construction in Energy Statistics Yearbook* did not contain energy use of rural construction. And there is almost no heating and gas supply in China's rural regions. Moreover, there is also a significant difference in water supply between urban and rural regions.

5.4.2 Data and CLA Method

5.4.2.1 Consumer Lifestyle Approach (CLA)

Drawing upon the work of Bin and Dowlatabadi (2005), the term consumers refers to the entity that purchases and uses products and services for the purpose of individual or household consumption. Lifestyle is a way of living that influences and is reflected by one's consumption behavior. Lifestyle is influenced by many factors and also leads to consequences. The CLA (Bin, Dowlatabadi, 2005) contains the following factors:

(1) External environment variables, such as cultural background, social consumption attitude and technology development, which form external context to a consumer's decision-making process.

(2) Individual determinants, such as attitudes, personal preference and consumption motives, which are personal psychological variables influencing a consumer's decision making.

(3) Household characteristics, such as household size, income and location, and housing area, which form household context for a consumer's decision making.

(4) Consumer choice, such as purchases and use of service and equipment.

(5) Consequences, such as energy use and the related environmental changes, which are the results of consumer behavior.

The role of the external environment in the lifestyle is the most significant, and it directly influences the other four aspects. Lifestyles in different countries, and even different regions in a country, are different due to the alternative external environment. It was therefore deemed appropriate to adopt the CLA to analyze the impact of lifestyles of urban and rural residents on the energy use and the related CO_2 emissions in China.

5.4.2.2 Data

Home energy, final energy use for heating and electricity generation, and final energy consumption of different sectors are all taken from *China Energy Statistics Yearbook* 1997–1999, 2000–2002. The consumption expenditure of urban and rural families is taken from the *China Statistics Yearbook* (2000; 2001; 2002; 2003). The gross outputs of different sectors are taken from *China Industry Statistics Yearbook* (2001; 2002; 2003). The population of urban and rural regions is taken from *China Statistics Yearbook* (2000; 2001; 2002; 2003), *China Population Statistics Yearbook* (2003), *China Rural Statistics Yearbook* (2003).

5.4.2.3 Calculations

(1) Home energy use.

Home energy including lighting, appliances, cooking, space heating and water heating that is direct living energy use, and can be taken from *Energy Statistics Yearbook*.

The calculation of the related CO_2 emissions of home energy is presented below:

$$End\_use\_CO_2 = \sum_m (\text{Fuel}_m \times CO_2 \text{ Coefficient}_m) \tag{5-9}$$

Where: Fuel_m refers to the fuel m, m kinds of fuels respectively refer to coal, petroleum, natural gas, electricity and heat; CO_2 Coefficient$_m$ is the carbon coefficient of fuel m, which is from Energy Research Institute in 1991 (Zhang, 2000).

5.4 Impact of Lifestyle on Energy Use and CO₂ Emissions

(2) Personal travel.

The vehicles for personal travel include cars, motorcycles, buses, planes and trains. In view of the availability of data, we only consider the energy use of domestic cars and motorcycles. The energy use and the related environmental impacts are relatively smaller when residents choose public transport. Moreover the acquisition of these data is difficult. But domestic cars and motorcycles are convenient for travel; and the impact of using them on energy consumption and the environmental changes are greater. The number of domestic cars and motorcycles is taken from *China Statistics Yearbook* 2000, 2001, 2002, 2003; and the number of kilometers of travel per year for motorcycles and their respective gasoline consumption per 100 kilometer is taken from the China Fuel Efficiency Background report of the Development Research Center of the state council (DRC) of the People's Republic of China et al. (2003). The average travel (kilometer) of domestic cars per year is taken from the results of Yao, Jiang (2003).

Based on the number of home cars and motorcycles, the number of families, the number of kilometers traveling per year for home cars and motorcycles, and gasoline consumption per 100 km of home cars and motorcycles, we can get the personal travel energy use.

(3) Indirect energy use.

We use an average annual resident's consuming expenditure to calculate indirect energy use by urban and rural residents' behaviors. These consumer expenditures includes residents' expenditures on food; clothing; residence; household facilities, and services medicines and health care; transport and communication; education, cultural and recreational services; miscellaneous commodities and services.

The indirect energy use calculated in this section does not contain the energy use incurred through the preparation of imported products and services because we assume that the energy intensity and carbon intensity of imported products and service are the same as those of domestic products and services. So the results calculated may be different from the actual energy consumption due to imports. But compared with the number of imported products and services purchased or used by China's residents, the domestic products and services are the main source of consumption. Hence, the results reflect the impact of lifestyle on energy use and CO₂ emissions, and can also identify energy/carbon-intensive behaviors.

$$\text{Energy}_u = \sum_i (EI_i \times X_i) \times UP$$
$$\text{Energy}_r = \sum_i (EI_i \times X_i) \times RP \quad (5\text{-}10)$$

$$CO_{2_u} = \sum_i (CI_i \times X_i) \times UP$$
$$CO_{2_r} = \sum_i (CI_i \times X_i) \times RP \quad (5\text{-}11)$$

Where: Energy_u refers to the indirect energy use of urban residents; $Energy_r$ refers to indirect energy use of rural residents. i refers to the kind of consumption i; X_i refers to the expenditure of the consumption i; UP refers to the number of urban residents; RP refers to the number of rural residents; EI_i refers to the energy intensity of the sectors whose products or service correspond to the consumption i; CI_i refers to the carbon intensity of the sectors whose products or service correspond to the consumption i.

5.4.3 Direct and Indirect Impact of Lifestyle on CO₂ Emissions

Following the previously discussed (above) methodology, we calculate the direct and indirect impact of urban and rural residents' lifestyle on energy use and the related CO₂ emissions in the year 1999–2002.

5.4.3.1 Direct Impact of Residents' Lifestyle on Energy Use

(1) Direct energy use of urban residents.

Fig. 5-23 shows that the share of coal in the urban residents' home energy use presented a declining trend, and the share of coal was the greatest in the period 1999–2001. But the share of petroleum was the largest in 2002. The share of electricity, and heat and natural gas was rising, which shows that final home energy of urban residents was in transition. Currently, the number, and kinds, of home appliances of urban residential families are rising, and home appliances not only include common appliances such as washing machines, color TV, and refrigerators, but also air conditioning, computers, microwave ovens, etc. Interestingly, the number of air conditioners in 2002 was twice as many as that of 1999, which means that the requirement of urban residential families for electricity will become greater; that inturn could make the threat of electricity supply shortages more serious.

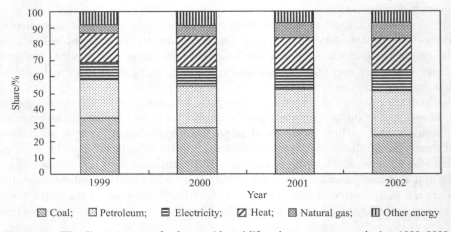

Fig. 5-23 The direct impact of urban residents' lifestyle on energy use during 1999–2002

(2) Direct energy use of rural residents.

The home energy use of rural residents was approximately 3.4% of the total final energy consumption in 1999–2002. Fig. 5-24 shows that the share of coal was gradually decreasing whilst accordingly the share of oil and electricity were rising (but the share of coal was still more than 60%). Heating and cooking in most rural regions are usually dependent on the direct use of coal whose efficiency is very low, with a thermal efficiency of around 30% for coal briquettes that are used for cooking (Wang, et al., 2003). Thus, this will lead to more coal waste. The challenge therefore is to improve energy efficiency of heating and cooking in rural regions. This should become one of the important research issues for energy conservation.

Although the population of rural regions was 1.7 times more than that of urban regions in the period 1999–2002, the home energy use of urban residents was more than that of rural residents (Fig. 5-25). This is because the kinds, and number, of home appliances in urban families is far more than those of rural families. Also, rural residents use plenty of crop residues and firedamp as well as the common commercial energies. With the improvement of urbanization and living conditions of rural residents, the direct home energy use will continuously rise.

5.4 Impact of Lifestyle on Energy Use and CO$_2$ Emissions

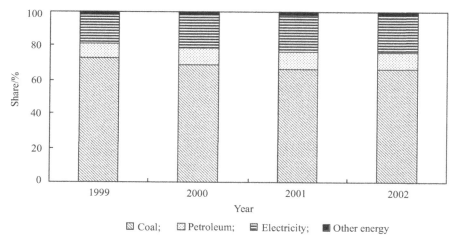

Fig. 5-24 The direct impact of rural residents' lifestyle on energy use during 1999–2002

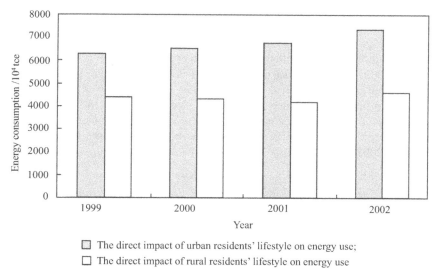

Fig. 5-25 The comparison of direct impacts of rural and urban residents' lifestyle on energy use in the period 1999–2002

5.4.3.2 Indirect Energy use of Residents' Lifestyle

The indirect energy use by urban residents' lifestyle was 14.10% of total energy consumption in 2002. The most energy-intensive urban residents' behavior was in the residences during the period 1999–2002. In this period indirect energy use was 8.91% of the total energy consumption; the next was food whose indirect energy use was 2.30%, education, cultural and recreation services had indirect energy use of 1.65%, clothing had indirect energy use of 0.38%, medicine and medical services had indirect energy use of 0.33%; transport and communication services had indirect energy use of 0.159%; miscellaneous commodities and services had indirect energy use of 0.155%, and the least was household facilities and services whose indirect energy use was 0.12% in 2002 (Fig. 5-26).

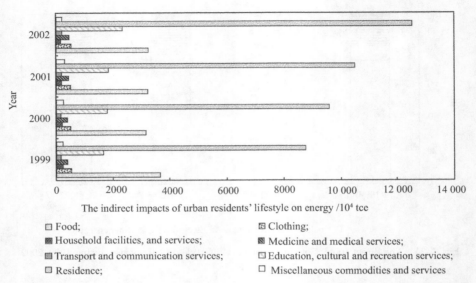

Fig. 5-26 The indirect impact of urban residents' lifestyle on energy use in the period 1999–2002

The indirect energy use by rural residents' lifestyle was 2.33% of total energy consumption in 2002. Fig. 5-27 shows that the most energy-intensive rural residents' behavior was food in the period 1999–2002, whose indirect energy use was 1.34% of the total energy consumption, the next was education and cultural whose indirect energy use was 0.60%, then medical insurance whose indirect energy use was 0.12%; clothing whose indirect energy use was 0.11%, others whose indirect energy use was 0.07%; and transportation and communication whose indirect energy use was 0.05%. The least was house appliances whose indirect energy use was 0.04% in 2002.

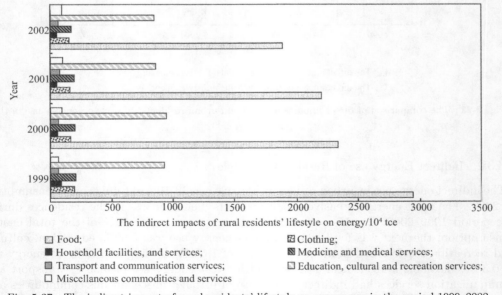

Fig. 5-27 The indirect impact of rural residents' lifestyle on energy use in the period 1999–2002

In the period 1999–2002, the energy use caused by the urban residents' residence, education, cultural and recreation services, medicine and medical services, and transport and communication services was rising. But the energy use caused by food, clothing, household facilities and services, and miscellaneous commodities and services was declining (Fig. 5-26). Although the expenditure on food for urban residents is the greatest expenditure, the total indirect energy use caused by urban residents' living behaviors presents an upward trend (Fig. 5-28). This is because the growth of expenditures of residence, education, cultural and recreation services, medicine and medical services, and transport and communication services was more rapid than that of 1999 whose respective growth was 37.53%, 59.12%, 75.12%, 101.59%; and with energy intensities being greater. However, total indirect energy use caused by rural residents' behavior was decreasing (Fig. 5-27) in the period 1999–2002. Although the expenditure of food, clothing, household facilities and services, medicine and medical services; transport and communication services; education, cultural and recreation services, miscellaneous commodities and services was rising, energy use caused by these behaviors was declining due to the reduction of energy intensity and the lower growth of these expenditures. This can tell us that blind stimulation for residents' consumption may lead to more requirements for energy. For example, if the expenditure increases 100/year per capita for urban residents, the causal indirect energy use will increase 200 million tce and the related CO_2 emissions will increase 141.8 million tons of carbon which is 10.15% of total energy use caused by urban residents' lifestyle, and 9.5% of the related CO_2 emissions.

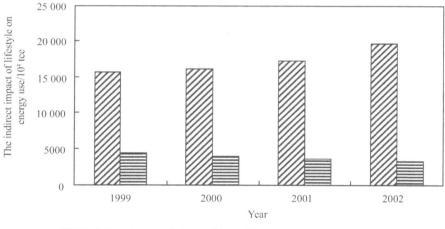

☒ The indirect impact of urban residents' lifestyle on energy use;
☲ The indirect impact of rural residents' lifestyle on energy use

Fig. 5-28 The comparison of indirect impacts of urban and rural residents' lifestyle on energy use in the period 1999–2002

The energy intensity of residences is high because the production of electricity and heat needs plenty of coal; and there is much coal waste during production, conversion and distribution. So here the potential for conservation of energy is the greatest. We can consider how to use home appliances which have high conservation energy standards, while pursuing ideas on how to enhance the energy efficiency of these related sectors. The energy intensity of education, cultural and recreation services is also relatively high due to the high energy intensity of paper production. So we can reduce paper waste and recycle paper to reduce energy waste. Moreover, the energy use caused by the lifestyle of education, cultural and recreation services was increasing, which can show that the consumption level of urban residents is developing toward a higher education, culture and recreation level. We suggest that the issue is to fully understand how to educate urban residents to rationally use cultural products and services, whilst decreasing energy intensity of paper production using new technologies. Such issues should attract greater attention because the growth of expenditure for education, cultural and recreational services is rapid within the context of enhancing education and living conditions.

5.4.3.3 Direct Energy Use by Personal Travel

From Fig. 5-29, we can deduce that the energy use by rural residents for personal travel was twice that conducted by urban residents. The reasons are as follows:

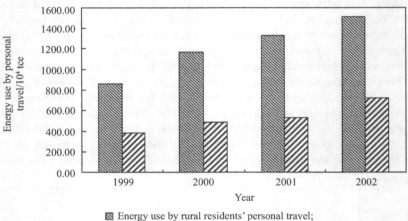

Fig. 5-29 The energy use by urban and rural residents' personal travel in the period 1999–2002

(1) The population of rural regions is more than that of urban regions, and rural families are 2.18 times larger than urban families. Also, motorcycle popularity in rural regions is higher than that of urban regions; specifically: in rural regions it is approximately 22.80% and 19.14% in urban regions.

(2) With the development of the economy and the increase of income per capita, the price of motorcycles has become lower, so purchasing and using motorcycles is no longer a luxurious consumption for rural residents. They have substituted for the bicycle as the primary means of transport.

(3) Bigger cities often control the number of motorcycles in use. On the other hand, the popularity of domestic cars is very low (only 0.58%) due to their high price and maintenance costs.

Consequently, it is very important to decrease gasoline use per 100 kilometers for motorcycles.

5.4.3.4 The Impact of Residents' Lifestyles on the Energy Use and the Related CO_2 Emissions

(1) The impact of rural and urban residents' lifestyle choices on the total energy consumption and CO_2 emissions.

Fig. 5-30 shows that approximately 26% of total final energy consumption and 30% of CO_2 emissions were caused by residents' lifestyles and the economic activities to support these demands in the period 1999–2002. From 1999 to 2002, the impact of residents' lifestyle on energy use is respectively 25.69%, 26.04%, 26.20% and 26.43%, which underestimates the impact of residents' lifestyle on energy consumption and CO_2 emissions, because we do not consider the indirect energy consumption and CO_2 emissions caused by rural residents' residence. For example, the direct electricity consumption of rural residents in 2002 is 83.248 billion kWh, if the average electricity price is RMB 0.5 yuan/kWh, then the expenditure of

5.4 Impact of Lifestyle on Energy Use and CO_2 Emissions

rural residents on electricity use in 2002 is RMB 41.624 billion yuan (the expenditure of rural residents on residence is RMB 234.85 billion yuan), then the indirect energy use is 36.29 million tce, which is 2.58% of total final energy consumption in 2002. If the government, in an unforeseen way, stimulates residents' behaviors whose energy intensities are relatively high, and does not rationally harmonize the relationships among all kinds of consumption behaviors, then it will make future energy supply security more serious.

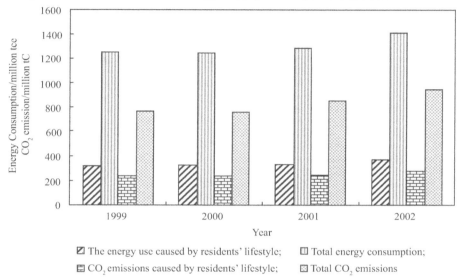

Fig. 5-30 The comparisons between energy use caused by residents' lifestyle and total energy consumption, between CO_2 emissions caused by residents' lifestyle and total CO_2 emissions in the period 1999–2002

(2) The impact of urban and rural residents' lifestyle on the energy use and the related CO_2 emissions.

Focusing upon 2002, we analyze in detail the different impact of urban and rural residents' lifestyle on the energy use and the related CO_2 emissions. For the urban residents (Table 5-6), we found that the indirect impact of lifestyle on energy use and the related CO_2 emissions was higher than the direct impact; and the indirect energy use was 2.44 times that of direct energy use; and indirect CO_2 emissions was 2.78 times that of direct CO_2 emissions caused by the lifestyle of residents. However, for rural residents, the direct impact of lifestyle on energy use and the related CO_2 emissions was higher than the indirect impact; and the direct energy use was 1.86 times that of indirect energy use.

According to Fig. 5-31, for urban residents, the most energy-intensive residents' behavior was those in residence during 2002; the impact on energy use was 45.1% of total energy use caused by urban residents' lifestyles. The next was direct home energy use, food, education, cultural and recreation services, whose respective impacts on energy use were 26.43%, 11.66%, 8.37%. The impact of residents' behavior on energy use was 91.56% of total energy use caused by lifestyle. Meantime, these four behaviors were the most carbon-intensive of residents' behaviors; whose respective impact on CO_2 emissions were 43.82%, 24.47%, 12.85%, 9.74% of CO_2 emissions caused by urban residents' lifestyles; which in total was 90.88%.

Fig. 5-32 shows rural residents. The most energy-intensive residents' behavior was direct home energy (in 2002). The impact on energy use was 48.94% of the total energy use caused by rural residents' lifestyle; with the next being food, education, cultural and recreation services, and personal travel; whose respective impact on energy use was 20.07%, 8.99%,

16.13%. The impact of the residents' behavior on energy use was 94.13% of total energy use caused by lifestyles. Meanwhile, these four behaviors were the most carbon-intensive residents' behaviors; and whose respective impact on CO_2 emissions were 51.97%, 20.51%, 9.70%, 11.17% of CO_2 emissions caused by rural residents' lifestyle; and which in total was 93.35%.

Table 5-6 The energy use and the related CO_2 emissions caused by residents' lifestyle in 2002

Residents' lifestyles		The impact of urban residents' lifestyle on energy use and the related CO_2 emissions		The impact of rural residents' lifestyle on energy use and the related CO_2 emissions	
		Energy use /10^4 tce	CO_2 emissions /10^4 tC	Energy use /10^4 tce	CO_2 emissions /10^4 tC
The direct impact of lifestyle on energy use		7340.78	4944.66	4597.12	3828.50
Energy use by personal travel	Energy use by motorcycles	638.77	346.85	1515.40	822.86
	Energy use by domestic cars	87.08	47.28		
The indirect impact of lifestyle on energy use and the related CO_2 emissions	Residence	12528.0	8854.90		
	Food	3240	2597.20	1885.1	1511.40
	Education, cultural and recreation services	2325.1	1968.10	844.46	714.80
	Medicine and medical services	464.30	394.98	174.85	148.74
	Clothing	541.76	505.56	149.35	139.99
	Transport and communication services	224.44	206.84	71.80	66.17
	Household facilities, and services	168.82	155.55	54.38	50.10
	Miscellaneous commodities and services	218.11	185.66	100.06	85.17
	Total indirect impact	19710.23	14868.78	3280.00	2716.38
	Total impact of lifestyle	27776.86	2020758	9392.52	7367.73

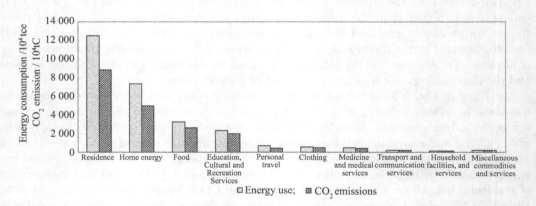

Fig. 5-31 The energy use and the related CO_2 emissions caused by urban residents' lifestyles in 2002

5.4.4 Conclusions

Based on CLA, we analyzed the direct and indirect impact of residents' lifestyles on energy use and the related CO_2 emissions. One could argue that a study of the impact of lifestyles on energy use and the related CO_2 emissions in developed countries is less important than a study of developing countries, as there is a need to understand and to mitigate for growth of energy consumption and global climate changes.

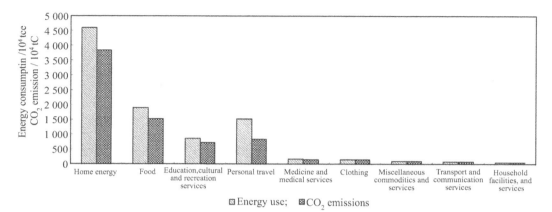

Fig. 5-32 The energy use and the related CO_2 emissions caused by rural residents' lifestyle in 2002

The results quantify conclusively that residents' lifestyles can have an important and significant impact on energy use and the related CO_2 emissions. Meanwhile, we also suggest that one of the most efficient measures for energy conservation is the change of lifestyles, that is the transition from luxurious consumption (household area of residents in high-income level is too great) to frugal consumption (reduction of use of air conditioning as soon as possible); domestic private cars from high gasoline consumption and emission to low gasoline consumption and emission; home appliances from high power to energy conservation and environmental protection. All these are not easily carried out, however. This study provides decision makers with extensive valuable information for establishing energy policies and strategies.

From the above analysis, we know that there are large amounts of indirect energy use by residents' consumption behavior in the total energy usage. A "stimulation consumption" policy is always an important tool within the government's macro economic control. Therefore, we would argue that while sustainable consumption behavior should be encouraged, there is a need for research to identify what the sustainable consumption mode should be.

5.4.5 Policy Implications

These results identify that the direct and indirect impact of urban and rural residents' lifestyle on energy use are different. So it is necessary to institute policies to influence urban and rural residents' lifestyle.

5.4.5.1 The Policy Suggestions for Urban Residents' Lifestyles

For the urban residents, we know that the indirect impact of lifestyle on energy use and the related CO_2 emissions was higher than the direct impact. So, focusing on the energy/carbon-intensive behavior, we argue that: for consumption, government should peremptorily constrain some lifestyles into a rational range; in so doing, it should enhance the energy conservation level for production.

(1) Instituting the standard of a housing area per capita. In order to reduce living energy use per capita for heating, cooling, lighting and electricity, we suggest that government should regulate a higher price for the part that is in excess of the stipulated standard of the housing area.

(2) Instituting the standard of temperature difference between outdoor and indoor. That is to say, that the temperature indoors during winter should not be too high or too low in summer.

(3) Instituting the energy conservation standards of home appliances. Government should encourage energy conservation technology through the use of policies and tax on products. Also decrease energy consumption of products during their production and use.

(4) Developing full-scale production of "green foods", encouraging consumption of primary food; decreasing the use of fertilizers and pesticides, and reducing processed food production. Additionally, inducting residents to a healthy mode of eating; that will also improve their health.

5.4.5.2 The Policy Suggestions for Rural Residents' Lifestyles

For the rural residents, we know that the direct impact of lifestyle on energy use and the related CO_2 emissions was higher than the indirect impact, so our policy suggestions focus on technology progress for improving daily living standards to reduce the direct use of coal.

(1) Enhancing the thermal efficiency of home coal stoves in rural families.

(2) Popularizing the use of liquid gas in the rural intensive population regions and substituting the direct burning of coal for cooking.

(3) Designing heating appliances in rural regions; and introducing economical and convenient warming facilities to reduce the low efficiency use for coal, and enhancing the heating capability of rural housing to prevent heat loss.

(4) Improving the transportation conditions of rural regions to decrease gasoline use per 100 kilometers for motorcycles.

5.5 Forecasting China's CO_2 Emissions in 2020

5.5.1 Energy Consumption and CO_2 Emissions

CO_2 emissions from fossil fuel combustion is the major source of greenhouse gas in China. Global warming, caused by increasing emissions of CO_2 and other greenhouse gases as a result of human activities, is one of the major threats now confronting the environment. CO_2 accounts for the largest share of total greenhouse gases, and its impact on the environment is also the greatest. If anthropogenic CO_2 emissions are allowed to increase without limits, the greenhouse effect will further destroy the environment for humans and all other living beings, threatening the existence of humankind. In order to control the continuous global warming and protect the living environment, the Kyoto Protocol to the United Nations Framework Convention on Climate Changes, signed in Kyoto, Japan in 1997, sets detailed emissions mitigation commitments for the 38 major industrialized countries. The Kyoto Protocol came into effect on Feb. 16, 2005.

The appearance of the Kyoto Protocol brought the research of CO_2 emissions to the forefront. Silberglitt et al. (2003) discussed the U.S. energy requirements and related CO_2 emissions for the year 2020 under different scenarios; Savabi and Stockle (2001) investigated the impact of increased atmospheric CO_2 and temperature on water balance, crop production, plant growth, and soil erosion; Roca and Alcantara (2001) considered the case of Spain in the period 1972–1997 as an example to analyze the role of energy intensity and the relationship between CO_2 emissions and primary energy in order to explain the evolution of CO_2 emissions by unit of real GDP. Gielen and Moriguchi (2002) discussed the STEAP model (Steel Environmental Assessment Program) and apply it for analysis of the interaction of CO_2 taxes, technological change, market structure and trade effects for iron and steel. Clinch et al. (2001) described the development of a computer model to enable a bottom-up assessment of the technical potential for energy saving in the domestic sector

5.5 Forecasting China's CO_2 Emissions in 2020

using Ireland's dwelling stock as a case study. The national savings in energy costs, CO_2 and other environmental emissions, as well as the capital costs resulting from the implementation of various energy-saving retrofit measures across the dwelling stock, are predicted; Hsu and Chen (2004) proposed a multi-objective mix integer model to investigate strategies to reduce CO_2 emissions from the power sector of Taiwan; Sun (1999) analyzed the nature of CO_2 emissions Kuznets curve and indicated that the real situation of energy use should be considered when the environmental policy is made.

Although the protocol did not set an explicit CO_2 reduction obligation for China and other developing countries, these nations still face great pressure from the environment. In 2003, CO_2 emissions caused by fuel combustion in China were about 0.849 billion tons of carbon (tC), accounting for 13.1% of the world's total, second only to the United States, the largest CO_2 emitter worldwide (IEA, 2003). The first commitment period of Kyoto Protocol ends in year 2012. And the second commitment period will start soon. By that time, could China still hold the advantage of a low per capita emission? How will the total emission level in China be? These will decide the pressure China is going to undertake in future mitigation negotiations.

Therefore, it is necessary to perform scenario analysis for China's future CO_2 emissions under different development paths of major driving forces.

5.5.2 Forecasting CO_2 Emissions Based on Energy Consumption

The object of this study is to forecast CO_2 emissions from fossil fuel combustion, which is decided by consumption, structure and potential emission coefficients of fossil fuels. Scenarios of future fossil fuel consumption and related structure are obtained with input-output model (for detail analyzing process of energy requirement scenarios please see 3.2 in chapter 3).

Given fossil fuel consumption, the CO_2 emissions are computed with the approach recommended by IPCC (Economia & Energia, 2001). The calculation steps are as follows:

(1) Introduce a conversion coefficient δ to change the unit of energy consumption from "oil equivalent" to "10^6kJ".

$$\delta = 41.868 \times 10^6 \text{kJ/toe} \qquad (5\text{-}12)$$

(2) Obtain carbon content multiplied by potential carbon emissions factor matrix (see Table 5-7).

(3) Correct non-oxidized carbon with the fraction of oxidized carbon (see Table 5-7).

Table 5-7 Potential carbon emission factors of leading fossil fuels & their fraction of oxidized carbon

Fuel	Potential carbon emissions factor[1]/(kgC/10^6kJ)	Fraction of oxidized carbon[2]
Coal	24.78	0.98
Crude oil	21.47	0.99
Natural gas	15.30	0.995

Source: [1] Workgroup 3 of the National Coordination Committee on Climate Change (Xue, 1998).
 [2] IPCC (PRC Ministry of Science and Technology Economy & Energy—NGO, 2001).

Summarizing the above analysis, let M^K be the CO_2 emissions in region K, then

$$M^K = \delta \times Q \times O \times D^K \qquad (5\text{-}13)$$

Where: Q is a diagonal matrix with h dimensions, whose element q_j is the carbon emission factor of fossil fuel j (unit: kgC/10^6kJ); O is a diagonal matrix with h dimensions, whose element o_j is the fraction of oxidized carbon of fuel j; D^K is the energy consumption of region K.

5.5.3 Forecasting CO_2 Emissions under Different Growth Paths

5.5.3.1 Model Size

Corresponding with the forecasting of energy requirements, eight economic regions are considered in this study, i.e., Northeast (NE), Beijing-Tianjin (BT), Northern Coastal (NC), Eastern Coastal (EC), Southern Coastal (SC), Central (C), Northwest (NW) and Southwest (SW), with four sectors taken into account for each region, i.e., agriculture, manufacturing, construction and services (including freight transport, communication, commerce, catering and non-material production services).

5.5.3.2 Scenario Setting

In line with the forecasting of energy requirements, five scenarios are set in this study as shown in Table 5-8.

Table 5-8 Scenario description

Scenarios	Scenario Description
Scenario **BAU** (Business-As-Usual)	Assume that in the coming years up to 2020, China's economy could maintain its current growth rate and realize a relatively high growth rate; *per capita* income achieves the preset 'well-off', 'developed' society objectives; population and urbanization rate grow at a medium speed; technology advance at a medium rate and achieves the National Energy-Saving Layout in each analysis year
Scenario **L** (Low economic growth)	Assume that various challenges and risks constrain economic development and urbanization advancement
Scenario **H** (High economic growth)	Assume a higher economy growth rate on the base of scenario BAU
Scenario **HP** (High economic growth + High Population)	Assume a higher population growth rate on the base of scenario H
Scenario **HT** (High economic growth + High Technology)	Assume technology achieves greater improvement on the base of scenario H

5.5.3.3 Results

In this section, the analysis of results for CO_2 emissions for year 2010 and 2020 are discussed.

(1) CO_2 emissions.

CO_2 emissions for year 2010 and 2020 are presented in Figs. 5-33 and 5-34, respectively. Results show that, for year 2010, total national CO_2 emissions will be 1.471–1.591 billion tC, regional requirements will be 0.244–0.260 billion tC for Northeast, 0.046–0.050 billion tC for Beijing-Tianjin, 0.225–0.244 billion tC for Northern Coastal, 0.196–0.210 billion tC for Eastern Coastal, 0.083–0.097 billion tC for Southern Coastal, 0.382–0.419 billion tC for Central, 0.147–0.159 billion tC for Northwest, 0.139–0.152 billion tC for Southwest; for year 2020, total national CO_2 emissions will be 1.653–2.255 billion tC, regional requirements will be 0.262–0.358 billion tC for Northeast, 0.056–0.077 billion tC for Beijing-Tianjin, 0.269–0.368 billion tC for Northern Coastal, 0.236–0.324 billion tC for Eastern Coastal, 0.107–0.146 billion tC for Southern Coastal, 0.404–0.548 billion tC for Central, 0.167–0.228 billion tC for Northwest, 0.152–0.206 billion tC for Southwest.

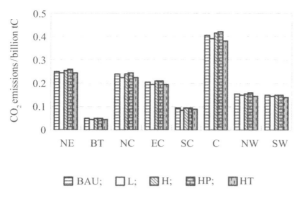

Fig. 5-33 Regional CO$_2$ emissions for year 2010

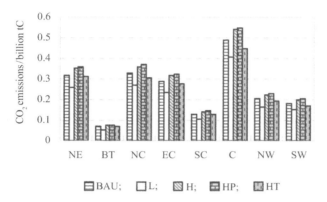

Fig. 5-34 Regional CO$_2$ emissions for year 2020

In both analysis years, all regions reach their highest CO$_2$ emissions in scenario HP. In all scenarios the highest emissions appear in region Central, followed by Northeast; the lowest emissions appear in region Beijing-Tianjin, followed by Southern Coastal.

During the period 1997–2020, in all scenarios, the highest annual growth rate of CO$_2$ emissions appears in Eastern Coastal (3.62%–5.05%), closely followed by Southern Coastal (3.53%–4.94%); the lowest annual growth rate appears in Southwest (1.88%–3.23%).

(2) Per capita CO$_2$ emissions.

Fig. 5-35 compares per capita CO$_2$ emissions in each scenario with the world average in 2004. Per capita emissions will be 1.061–1.135 tC/capita and 1.115–1.494 tC/capita for year 2010 and 2020 respectively. It can be seen from Fig. 5-35 that in year 2010, per capita emissions in scenario BAU, H and HP are close to the world average in 2004; in year 2020, per capita emissions in all scenarios are higher than the world average in 2004 except those in scenario L. If the commitments of the Kyoto Protocol are effectively carried out, the time when China's per capita emissions reach or surpass the world average is possible to be earlier than year 2020.

Figs. 5-36 and 5-37 display regional *per capita* CO$_2$ emissions for year 2010 and 2020 respectively. In all scenarios, in both analysis years, *per capita* CO$_2$ emissions in region Southern Coastal and Southwest are much lower than the national average; and those in region Central are slightly lower than the national average; *per capita* emissions in the other five regions are higher than the national average. In 2010, in all scenarios, the highest per capita emissions correspond to Northeast, followed by Beijing-Tianjin; in 2020 in all scenarios, the highest *per capita* emissions correspond to Beijing-Tianjin, closely followed by Northeast.

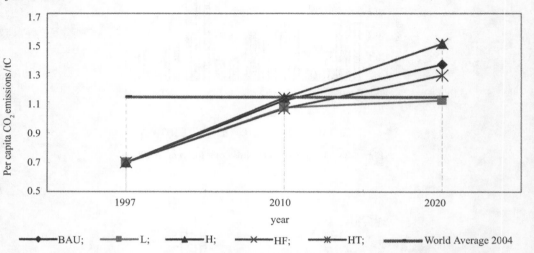

Fig. 5-35　China's per capita CO_2 emissions in 2010 and 2020 vs. world average level in 2004

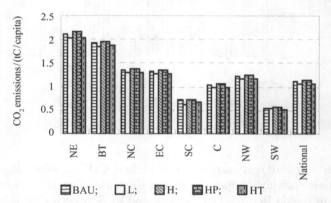

Fig. 5-36　Regional CO_2 emissions per capita for year 2010

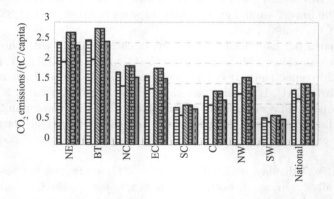

Fig.5-37　Regional CO_2 emissions per capita for year 2020

5.5 Forecasting China's CO_2 Emissions in 2020

(3) CO_2 driven vs. CO_2 emitted.

The results presented in Figs. 5-33 and 5-34 are CO_2 emissions generated by each region, which is driven directly by final demand in the region itself and also indirectly by demands from other regions.

Figs. 5-38 and 5-39 display the relative error between CO_2 emissions driven by one region and those emitted by that region. Here, the emission driven by one region refers to the emissions produced either inside the region or in other regions in order to satisfy the final demands in that region. That is, it excludes the CO_2 emissions attributed to the region's exports, while it includes the emissions taking place in other regions, but resulting from the satisfaction of imports of this region. A positive relative error implies that emissions driven by a region are larger than those it generates, and *vice versa*.

It can be seen from Figs. 5-38 and 5-39 that in both analysis years, the relative errors clearly exist in most regions. The regions, emissions driven by which are higher than those it generates, include region Beijing-Tianjin, Eastern Coastal, Southern Coastal, and the two western regions. Of the five regions, in 2010 the largest positive relative error corresponds to Southern Coastal (49.62%–56.84%); whilst in 2020 the largest positive relative error corresponds to Beijing-Tianjin (34.23%–34.89%).

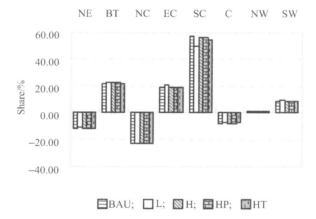

Fig. 5-38 Relative errors between CO_2 emissions driven by one region and those generated by that region for year 2010

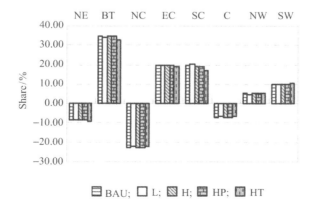

Fig. 5-39 Relative errors between CO_2 emissions driven by one region and those generated by that region for year 2020

Among all the regions with negative relative errors, Northern Coastal has the lowest errors. Emissions driven by this region are 23.04%–23.42% lower than those it generates in year 2010, and 22.32%–22.57% lower than those it generates in year 2020.

The above results show that there existing obvious relative errors between emissions driven by one region and those emitted by that region, which implies that when performing environmental policy reform different identifications of a region's emission obligation may incur evidently different impacts on that region.

For example, on the basis of scenario BAU, suppose a carbon tax of RMB 50 yuan/tC is going to be imposed in year 2020, net tax payments for each region and the total interregional transfer under different taxing schemes are illustrated in Table 5-9. Here Emitter-pay refers to the principle that a region's emission responsibility is identified as those produced by that region; Driver-pay refers to the principle that a region's emission responsibility is identified as those driven by that region; net tax payment of a region equals its total carbon tax payments minus its total production tax subsidy or lump-sum transfer. A positive net tax payment implies that the region benefits from the policy, and *vice versa*. In order to maintain revenue neutrality, all the revenues from carbon tax are reimbursed to production sectors or households. Here the indicator total interregional transfer is used to assess the political feasibility of each taxing scheme. It is calculated by summing up all the positive regional net tax payments.

Table 5-9 Regional net tax payment and the total interregional transfer under different taxing schemes

Reimbursement scheme of carbon tax revenue	Identification of emission responsibility	Regional net tax payment billion RMB yuan								Total interregion at Transfer billion RMB yuan
		NE	BT	NC	EC	SC	C	NW	SW	
Reduce	Emitter-pay	7.2	−1.4	3.1	−9.1	−9.4	5.5	5.0	−0.8	20.8
Production tax	Driver-pay	5.9	−0.3	−0.6	−6.3	−8.1	3.8	5.5	0.05	15.3
Lump-sum	Emitter-pay	7.3	1.6	3.8	2.9	−3.4	−3.5	1.0	−9.6	16.5
Transfer to households	Driver-pay	6.0	2.8	0.1	5.7	−2.2	−5.2	1.5	−8.7	16.1

If all the carbon tax revenues are used to reduce production tax, assuming a uniform production tax subsidy rate amongst all regions, under the Emitter-pay principle the beneficial regions would include region Beijing-Tianjin, Eastern Coastal, Southern Coastal and Southwest. Under the Driver-pay principle, the net loss of Northeast and Central are respectively about 19% and 31% smaller than those under the Emitter-pay principle; the net benefits of Beijing-Tianjin, Eastern Coastal, and Southern Coastal are respectively about 82%, 31%, and 14% lower than those under the Emitter-pay principle; Northern Coastal changes from a net payer under the Emitter-pay principle to a net receiver, while Southwest changes from a net receiver under the Emitter-pay principle to a net payer; the net losses of Northwest are about 11% higher than those under the Emitter-pay principle in all scenarios. As for the total interregional transfer, under the Driver-pay principle the total interregional transfers are about 27% lower than that under Emitter-pay principle, implying a much higher political feasibility at the national level. However, under the Driver-pay principle the total net losses of the two western regions are about 35% higher than those under the Emitter-pay principle, which may lead to a conflict with the West Development Policy being implemented in China.

If all the carbon tax revenues are used as a lump-sum transfer to household, dividing among regions according to population, under both principles the beneficial regions will all be Southern Coastal, Central and Southwest. The total interregional transfers under the

5.5 Forecasting China's CO_2 Emissions in 2020

two principles are also similar. However, the impacts on regional net tax payments by the two principles also obviously differ from each other. For example, under the Emitter-pay principle, the net losses of Northwest are about 35% lower than those under the Driver-pay principle; while the net benefits of Southwest are about 10% higher than those under the Driver-pay principle. In this sense the Emitter-pay principle may be more favorable to the West Development Policy.

From the above example and discussion it can be seen that, different principles of responsibility identification will have obvious different impacts on most regions. The effects of a given reimbursement scheme would be evidently affected by the identification of regional emission responsibility. The principles of emission responsibility identification would be one of the most important determinants of whether the policy goal of a certain taxing scheme could be realized. Therefore, in the future when performing environmental policy reform, careful decisions should be made regarding which kind of emissions a region is supposed to be responsible for.

(4) CO_2 emissions intensities.

Figs. 5-40 and 5-41 illustrate the results of CO_2 emissions intensities for year 2010 and 2020, respectively. It can be seen that, under the assumption of uniform rates of technology improvement for all regions, up to year 2020 there will be distinct differences among regional emission intensities. In all scenarios in the two analysis years, the regions whose emission intensities are higher than the corresponding national average include: Northeast, Northern Coastal, Central and Northwest. The highest emission intensities correspond to Northeast, closely followed by Northwest. In all scenarios in the two analysis years, the regions whose emission intensities are lower than the corresponding national averages include Beijing-Tianjin, Eastern Coastal and Southern Coastal. Both the lowest energy and emission intensities correspond to Southern Coastal.

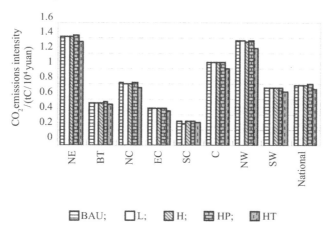

Fig. 5-40 Regional CO_2 emissions intensity for year 2010

5.5.4 Policy Implications

Summarizing the above analysis results of the model, the following policy recommendations are proposed:

(1) Policies promoting changes in energy structure should be devised to control the quickly increasing CO_2 emissions. Changes in energy structure should be promoted in Eastern Coastal.

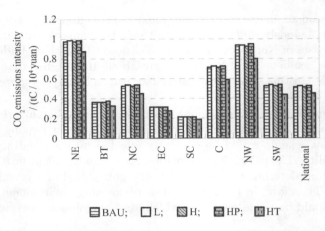

Fig. 5-41 Regional CO_2 emissions intensity for year 2020

The above forecast results show that even in scenario HT, where technology advances the fastest, CO_2 emissions increase rapidly. Moreover, the increase of per capita emissions will quite possibly reach or exceed the worldwide average within 20 years. The results also show that as time goes on further enhancements in energy-saving technology will become more and more difficult. Therefore, it is desirable to establish energy and environmental policies in favor of cleaner energies as early as possible to accelerate the changes in energy structure and to support sustainable economic development.

The results show that the ratio of CO_2 emissions in Eastern Coastal, when compared with the national total, is 13.19%–13.30% in 2010, and rises to 14.28%–14.62% in 2020. In all scenarios, the highest annual growth rate of regional CO_2 emissions during 1997–2020 corresponds to Eastern Coastal. In both analysis years, the *per capita* CO_2 emissions in this region are higher than the national level. Thus this region should also be regarded as one of the key areas for energy saving and CO_2 mitigation. The model results also show that, in both analysis years, in all scenarios, the energy intensity of this region is second-lowest among all regions (see 3.1.5), and thereby lower than the corresponding national average-level. That is, in both analysis years the energy end-use efficiency of this region could surpass the national average goal, where it is set up to year 2020 the national average energy cost per unit output for all the main industry products would reach, or be close to, world advanced-level. Therefore, in the near future it would leave relatively limited potential for this region to accelerate its energy end-use efficiency beyond that in its BAU case. Making use of its economic and technical advantages, with new technologies such as energy substitution or renewable energy, changes in energy structure should be promoted in Eastern Coastal to accelerate its energy saving and CO_2 mitigation.

(2) Emission responsibilities of each region should be carefully identified.

Model results show that there exists obvious relative errors between emissions driven by one region and those emitted by that region. Such relative errors would make the identification of emission responsibility a very important determinant of whether the policy goal of certain environmental policy reform could be realized. As shown by the simple analysis of carbon tax in Sec. 4.2, the regional effects of the tax revenue reimbursement scheme would be significantly affected by the principles of emission responsibility identification. Therefore, in the future when performing environmental policy reform, carefully considered decisions should be made about which kind of emission should be taken as the one that a region is supposed to be responsible for.

(3) The potential for CO_2 mitigation in China is limited in the next 20 years, and thus

decision making should be based on this key point.

From today until the 2020, China's GDP is expected to maintain a high growth rate. While the special coal-dominated energy structure is not expected to change much in the near future, and because the efficiency of coal is much lower than that of oil and natural gas in most cases (He et al., 2002), the decline of energy intensity will likely be limited over the next 15 years. At the same time, the high cardinal number of coal, to a great extent, will likely limit the decrease of CO_2 emissions by total fossil fuel consumption, which can still cover the majority of total primary energy consumption. Consequently CO_2 intensity will not likely drop considerably faster in the near future.

5.6 Summary

This chapter analyzes the relationships between global climate change and CO_2 emissions reduction, economic development and CO_2 emissions, and the current status of China's CO_2 emissions. We consider that the main question for China's CO_2 emissions is the contradiction between economic development and CO_2 emissions reduction. The economic structure, economic growth pattern and fuel mix will be faced with serious challenges.

This chapter analyzes the change of China's carbon intensity, and our results show that:

(1) The change in primary energy-related carbon intensity was largely due to the change in primary energy intensity that took place during the 1980–2003 period, and more than 99% of decline in primary energy-related carbon intensity for the period 1980–2003 was attributable to real energy intensity decline.

(2) The structure shift increased carbon intensity over the year 1987–2002, and final fuel mix also made final carbon intensity increase over the period 1980–2002.

(3) The change of primary energy use structure should focus on the material production sectors' final energy use and structure among sub-sectors, especially the secondary industry

Based on STIRPAT model, we quantified the impact of population, affluence and technology on the total CO_2 emissions. We found that:

(1)The impact of population on CO_2 emissions is great, especially the proportion of the population between ages 15 and 64. It has a negative impact on the total CO_2 emissions of countries at the high income level, but the impact is positive at other income levels.

(2) The effect of the decline of energy intensity on CO_2 emissions abatement is restricted by the level of economic development and energy-consuming structures. For the high, low and upper-middle income countries, the effect is small, but it is great for the lower income countries.

(3)The impact of the growth of GDP per capita on CO_2 emissions presents an approximately descending trend with the increase of economic development levels.

(4) Among impact factors affecting CO_2 emissions, the impact of GDP per capita is greatest, the next is energy intensity and population, and the role of the share of fossil fuel in the primary energy and the carbon coefficient of fossil fuel is rather small.

Over the year 1999–2002, approximately 26% of total energy consumption and 30% of CO_2 emissions are a consequence of residents' lifestyles, and the economic activities to support these demands. If the government, in an unforeseen way, stimulates residents' behavior of those whose energy intensities are relatively high, and does not rationally harmonize the relationships among all kinds of consumption behaviors then it will make future security of energy supply more serious.

(1) In 2002, for urban residents, residence; home energy use; food; and education, cultural and recreation services are the most energy-intensive and carbon-emission-intensive activities, and the impact of these behaviors on energy use was 91.56% of total energy use caused by lifestyles.

(2) For rural residents, home energy use; food; education, and cultural recreation services; and personal travel are the most energy-intensive and carbon-emission-intensive activities. The impact of these behaviors on energy use was 94.13% of total energy use caused by lifestyles.

Therefore, we would argue that while sustainable consumption behavior should be encouraged, there is a need for research to identify what should the sustainable consumption mode be.

Based on qualitative and quantitative analysis, we developed a multi-regional input-output model and applied it to obtain CO_2 emissions scenarios for different regions for year 2010 and 2020. Results show that:

(1) For year 2010, total national CO_2 emissions will be 1.471–1.591 billion tC, regional requirements will be 0.244–0.260 billion tC for Northeast, 0.046–0.050 billion tC for Beijing-Tianjin, 0.225–0.244 billion tC for Northern Coastal, 0.196–0.210 billion tC for Eastern Coastal, 0.083–0.097 billion tC for Southern Coastal, 0.382–0.419 billion tC for Central, 0.147–0.159 billion tC for Northwest, 0.139–0.152 billion tC for Southwest; for year 2020, total national CO_2 emissions will be 1.653–2.255 billion tC, regional requirements will be 0.262–0.358 billion tC for Northeast, 0.056–0.077 billion tC for Beijing-Tianjin, 0.269–0.368 billion tC for Northern Coastal, 0.236–0.324 billion tC for Eastern Coastal, 0.107–0.146 billion tC for Southern Coastal, 0.404–0.548 billion tC for Central, 0.167–0.228 billion tC for Northwest, 0.152–0.206 billion tC for Southwest.

(2) Per capita emissions will be 1.061–1.135 tC/capita and 1.115–1.494 tC/capita for year 2010 and 2020 respectively. In 2010, in all scenarios, the highest per capita emissions correspond to Northeast; in 2020 in all scenarios, the highest per capita emissions correspond to Beijing-Tianjin. If the commitments of the Kyoto Protocol are effectively carried out, the time when China's per capita emissions reach or surpass the world average is possible to be earlier than year 2020.

(3) There exists obvious relative errors between emissions driven by one region and those emitted by that region, therefore different principles of responsibility identification will have obvious different impacts on most regions. The principles of emission responsibility identification would be one of the most important determinants of whether the policy goal of a certain taxing scheme could be realized.

(4) Under the assumption of uniform rates of technology improvement for all regions, up to year 2020 there will be distinct differences among regional emission intensities.

Summarizing the above model results, the following policy recommendations are proposed.

(1) Policies promoting changes in energy structure should be devised to control the quickly increasing CO_2 emissions. Changes in energy structure should be promoted in Eastern Coastal.

(2) Emission responsibilities of each region should be carefully identified.

(3) The potential for CO_2 mitigation in China is limited in the next 20 years, and thus decision making should be based on this key point.

CHAPTER 6

Strategic Petroleum Reserves and National Energy Security

As the second largest oil consumer, China's oil import dependency was more than 40% in 2004. The international oil price has risen continually over the last years, but China, as one of the main oil import countries, has no international right to speak on oil prices, and Chinese oil imports are traded in cash rather than in the futures market. In the complex international energy geopolitics context, China's oil import mainly depends on the unstable Middle East. Although China has began to establish a national Strategic Petroleum Reserve (SPR), it will take some time to make SPR one of the effective measures to ensure national energy security.

Therefore, China's energy security problems are not ones to be optimistic about.
• What hidden troubles exist in China's energy security?
• Can 90 days' SPR level, required by IEA, be suitable for China's economic development and national energy security?
• What is the optimal SPR level in 2010 and 2020?
• How does China's oil import diversification index change?
• How to decrease China's oil import risk index compared with those of the main oil importers?

Focusing on these questions, we established some models based on quantitative and qualitative methods, and analyzed comprehensively the above questions, and finally present some suggestions to ensure China's national energy security.

6.1 China's Energy Security

National energy security is one of the categories of modern society. Since the definition of energy security was presented, the concept has experienced a process of continuous development and enrichment. At this stage energy security has become a focus of universal concern; particularly in the period of high oil prices, the energy crisis have evolved as the train of the economic crisis. The relationship between energy and economics has emerged with the process of industrialization and has become more prominent and strengthened. Only after the process of industrialization, does energy becomes a factor of production, just like capital and labor, which can impact on a country's economy.

6.1.1 Definition and Connotation of Energy Security

In the 1970s, energy security first drew people's attention to supply security (economic security). Many industrialized countries have suffered great economic losses after the first global oil crisis. Therefore, some developed countries which had suffered from the energy crisis set up the International Energy Agency (IEA) in 1974, and formally proposed the national energy security concept, the core of which is focusing on the stability of oil supplies and prices. After the 1980s, as the global warming increased and atmospheric quality of the environment rapidly fell, people gradually reached a consensus which was about environmental protection and sustainable development issues. In 1997 the Kyoto Protocol was signed, and marked the beginning that the world was to redefine the concept of energy security. That is, in the country's energy development strategy, increasing the use of energy should not lead to any major threat to human being's survival and to the ecological environmental development. Up to now, the concept of energy security has included two aspects: economic security and ecological environmental safety.

National energy security definitions can be summarized as follows: The energy economic security (supply safety) and the energy ecological environmental safety (using safely). The former refers to the stability of normal demand of energy supply which meets the country's survival and development; the latter refers to the fact that energy consumption and usage should not lead to any major threat to the ecological environment of human's own survival and development. It is generally believed that, the energy supply security includes two implications: first, the energy is sustained and no serious shortage, by the standard of IEA, the supply shortage gap is less than 7% of the previous year's import volume. Second, there are no unbearably sustained high oil prices.

The connotations of the national energy security change greatly within different social development stages. In the period of industrialization, social development is the main theme. All the energy consumption and usage are used to meet the demand of rapid economic and social development. Energy supply is a strong constraint to economic development. Furthermore, because of technical and economic constraints, alternative energy technologies have not yet achieved a breakthrough, and the strategic energy reserves and other issues have not been effectively resolved. The country and industrial production sector's capacity to respond to the energy emergency is very limited; therefore, in this period, the connotation of energy security is mainly to protect the security of energy supply.

In the industrialized mature period, because the productivity has achieved a very high level, people's material life is more affluent, and energy efficiency has made great progress compared with the early stage of industrialization. The energy consumption structure of the whole society and industrial sectors is mainly of high-quality energy, and clean energy and renewable energy accounted for a large proportion of energy consumption structure. The alternative technologies of primary energy use have made some progress. The country has a better strategic energy reserve system and a scale of reserves which can deal with some unexpected events. Industrial production, household consumption and other sectors have some ability to adapt and control the short-term of energy supply shortage. Environment protection becomes the new focus of people's concern. Therefore, in this period, the content of energy security includes not only protecting the security of energy supply, but also not leading to any major threat to human being's survival and development of the environment.

The connotations of the national energy security also changes a lot within different international situation developments. During war, energy, especially the strategic value of oil is mainly reflected in its uniqueness. During the World War II the energy supply situation in the war became a crucial factor. Several world oil crises after 1970s were due to war in the Middle East or some local armed conflicts, which caused world oil supply disruption or a shortage, resulting in the fact that international oil market supply could not satisfy the market demand, oil prices have soared sharply, national energy security lapsed, and inflation worsened. Therefore, over a long period, energy security in wartime has become the focus of concern of many nations' energy security strategy. In a state of war, the energy supply problem is different from the general supply security issues. It not only involves security of energy supply, it is more related to the war's progress and nation's security. During wartime, energy security facilities, such as the oil reserve facilities, domestic oil production base, the oil port, transnational oil pipeline and so on, are most vulnerable to attack and destruction; therefore, in wartime, the main content of national energy security is the energy supply security, that is, energy production, petrochemical, energy reserves and transport and other issues of security.

In the periods of peaceful development, with the growth of the world economy and the expansion of energy use, the strategic value of oil has been gradually transformed into the economic value of the basic energy. At the present stage, with the process of economic globalization accelerated, oil's attributes as an important strategic commodity become even more evident. In light of the global view, the economic risks such as market, prices and other risks, is replacing the threat of war, embargoes and other security risks to some extent, and becoming the main risk factors for many countries' energy security. The oil supply is still affected by international political events, however, as a special commodity, and its supply and demand still mainly depends on the market mechanism. Therefore, in periods of peaceful development, the main connotations of national energy security are economic security and energy use security, that is, the main risk factor is the market and price risk rather than a supply disruption risk. It might be argued, therefore, that the use of fossil energy and nuclear energy should not lead to any major threat to human being's survival and change to the environment.

6.1.2 Contemporary Status of China's Energy Security

Due to historical reasons, China has carried out a self-sustainable energy development strategy. Given that China has abundant coal reserves, it has become one of the few countries in the world that uses coal as a basic energy, straying far from the mainstream of the world's trend, that is the energy structure in which oil and gas are the main fuels. However, with China's energy consumption structure focusing on progressive optimization, the ratio between petroleum and natural gas in primary energy consumption has gradually increased. In 2002, China became the world's second largest oil consumer, and China's economic development increasingly relied on oil. Constrained by the national oil production capacity, in 2004 the oil import dependence was more than 40%, so China's energy security became more prominent.

(1) The percentage of high efficiency and clean energy in the primary energy consumption is relatively low.

Low efficiency and high pollution energy (coal, coke) account for a sizeable proportion of China's primary energy consumption structure. Oil, natural gas and other clean and efficient energy account for a very small percentage. China Statistical Yearbook 2005 (State Statistical Bureau, 2005a) shows that, in 2004, oil accounts for about 22.7% of the total energy consumption, natural gas only accounts for 2.6% of the total energy consumption, coal accounted for 67.7% in primary energy consumption, and a lot of the coal was directly used for combustion, about 45% was used in industrial boilers, furnaces, cooking and heating; only 35% coal was used for power generation or co-generation. Therefore, there is a large space of the optimization of the primary energy consumption structure.

(2) Oil import channels and means are single.

Against a background of a complicated international energy geopolitical stage, the developed countries have adopted diversification of oil imports strategy to reduce their risk of oil imports and protect the oil supply security. Although the Middle East region is rich in oil resources, many of its petroleum resources' exploration and development are controlled by Western multinational oil companies. Therefore, developed countries' oil imports include a large proportion of overseas shared oil. China's investment in the overseas oil and gas resources development and utilization is still limited. The overseas shared oil in China's oil imports only takes up a small proportion. Moreover, China's crude oil imports come from about 30 countries, while those of the United States imports from more than 60 countries. China's oil imports diversification strategy still is heavy and has a long way to go.

(3) Strategic petroleum reserves have just been started.

Strategic petroleum reserves are the energy reserve for the national strategic needs. The purpose is to decrease the impact on economic development brought about by a shortage of oil supply, thereby reduce the state's macro-economic losses caused by a sudden rise in oil prices. With China's sustained and rapid economic development, in 1993, China became a net importer of oil from an oil exporting country; and in 1996 turned into a net importer of crude oil. According to the forecast, in 2010, China's demand for oil will be about 350 million tons, the domestic output is about 180 million tons, and the imports dependence will be more than 48%. Before 1993, some energy experts suggested China should build national strategic petroleum reserves. However, at that time, China's oil import dependence was still low, and the establishment of national strategic petroleum reserves would cost huge amounts of money, so the strategic petroleum reserve has not been a priority. As China's oil import dependence grows rapidly and the international crude oil prices keep soaring, at the beginning of this century, China's oil supply security issues became prominent. Therefore, establishing a national strategic petroleum reserve has been raised as an important agenda. In 2004, China formally launched its strategic petroleum reserve projects. The first project includes four reserve bases, in which the largest is in Zhenhai, Zhejiang Province, which was completed at the end of 2006, and the others were completed in 2008.

(4) The efficiency of energy use is low and the intensity of carbon emissions is high.

The average efficiency of coal use is very low. Only very small quantities of coal have been turned into electricity. Most parts of coal are used for direct combustion by end-users, which will not only cause a lot of energy waste, but also cause great environmental pollution. According to the energy efficiency evaluation and calculation method proposed by the United Nations Economic Commission for Europe (ECE), in 1992, China's efficiency of the energy exploitation was 32%, the middle efficiency was 70% and the end-use efficiency was 41%. In 1980, 1992 and 1995, China's energy efficiency was 25.86%, 28.7%, 34.31%, respectively. While the world Organization for Economy Cooperation and Development (OECD) countries' energy efficiency was 32% in the 1970s, and 41% in the early 1990s (Wei et al., 2003). This shows that there is still a big gap between China's energy efficiency and developed countries'. China's energy efficiency is only equivalent to that of the OECD countries in late 1970s. Carbon emission intensity is the carbon dioxide (CO_2) emissions per unit gross domestic product (GDP). Because China's energy consumption structure is coal-based, and the energy use efficiency is low, China's carbon emissions intensity is high. According to the purchasing power parity terms in 2000, China's CO_2 emissions intensity was 0.58 kg CO_2/US\$, was 1.14 times than that of the world's average level, and was 1.29 times than that of the OECD countries. The emission intensity of the world average level and the OECD countries was 0.51kg CO_2/US\$ and 0.45kg CO_2/US\$ respectively (Wei et al., 2003).

6.1.3 Hidden Troubles in China's Energy Security

As the impact on China's economy caused by the three large-scale oil crises in history was very small, the security of China's energy supply has not been given sufficient attention. However, with rapid growth of China's extensive economy, oil demand is expanding rapidly; moreover, due to domestic oil production constraints, China has to depend heavily on foreign oil to meet its domestic economic development needs. The major sources of China's oil imports are the Middle East and the West African region, and the oil transportation mainly depend on foreign tankers. China's overseas share oil is limited. Besides, China's oil trade is all spot transactions rather than futures, so as the world's second largest oil consumer, China's oil supply security faces many hidden troubles:

(1) Oil imports excessively depend on the Middle East, and the oil transports mainly depend on foreign oil tankers, so the imports risk is getting greater.

In 2004, China imported 120 million tons of crude oil, mainly from the Middle East (where there is great turmoil). As China's oil imports are mainly concentrated in the Middle East, North Africa, and Southeast Asia, the main channel of China's oil imports is maritime tanker transport. Some 90% is transported by foreign fleet companies and only 10% by our domestic fleet, which focuses on Southeast Asia. Moreover, 80% of the imports have to get across the insecure Malacca Strait. Therefore, China's oil imports are faced with many risks: such as, price risk, diversification risk, supply risk amongst a range of additional risks.

(2) Domestic emergency oil added production capacity and reserves are very limited.

In 2004, China's output of crude oil was about 174 million tons, as the oil prices had been in high priced, domestic oilfields such as Daqing, Shengli and other oil production enterprises were at full production capacity. However, some major foreign oil production enterprises have surplus production capacity in preparation of an emergency event and to alleviate any supply crisis. Due to China's limited surplus production capacity, it is difficult for domestic enterprises to increase production and ease a crisis when the supply shortages appear. At present, China's strategic petroleum reserve projects have not yet been completed, so without available short-term national strategic petroleum reserves, it is hard to imagine how we can deal with the shortage of any oil supply, if say, hurricane "Katrina" appeared in China.

(3) It is obviously unreasonable that oil imports only use the spot purchase, and refined oil exports lack macro-control.

At present, China continues to import crude oil, subject to the national allocated planned indicators each month, and China is the only country that negotiates oil imports in cash rather than in futures. So there is some "out of the ordinary" trading phenomenon: such as, buying when the oil prices are high, rather than when the prices are low, and the higher the oil prices are, the more oil is imported. China's refined oil exports lack any national macro-control; therefore it is difficult to protect oil supply security. In August, 2005, Guangzhou, Shenzhen and other places which suffered "oil shortage" had first-hand knowledge of the consequences of an "oil crisis". However, after this bitter experience, when we were rethinking about this "oil shortage", we were surprised to find out that at the same time when the domestic oil supply was suffering an unprecedented tension, the exports of refined oil increased substantially in first half year of 2005. The data from China customs statistics show that in the first six months of 2005, the refined oil exports had reached 7.59 million tons, an increase of 48.6% compared with that in the same period in 2004, while the petroleum products imports decreased by 21.1% than the same period in 2004. It is indeed puzzling and difficult to understand, when the domestic oil were in shortage, and even in crucial moments of the "oil shortage", China's refined oil exports had increased dramatically, while the domestic refined oil production had grown slowly. In the first half year of 2005, China's gasoline and diesel production only increased by 2.4% and 9.2% compared with the same period in 2004, the output of jet fuel decreased by 9% compared with the same period in 2004. However, in the same period, three oil exports had increased as high as 31.6%, 21.8% and 130%. Therefore, the nation's weak macro-control on petroleum products market further aggravated the "oil shortage".

(4) Lack of the warning mechanism of national oil supply security.

The warning standard of oil supply security formulated by IEA to its member countries is: "If the oil supply shortage is 7% more than the previous year's imports, then enter the warning state." China has not established the perfect oil market monitoring mechanism, the warning system of oil provision, emergency measures, and other mechanisms, so the "oil shortage" was sustained for a long time in Guangzhou, Shenzhen and other places in August 2005. Therefore, there is a heavy burden and a long way to go for China to achieve the oil supply security target.

6.2 Study on Optimal Scale of Chinese Strategic Petroleum Reserves

6.2.1 Contemporary Status of Chinese Strategic Petroleum Reserves

With continuous development of China's economy, the oil demand is increasing rapidly.

According to the 2005 China's Oil and Gas Industry Annual Report, it is estimated that China's oil demand in 2010 will be approximately 350 million tons, while domestic production will be only 180 million tons. This implies that China will import 170 million tons, giving an import dependence of nearly 50%. However, most of the oil imports came from the unstable Middle East, 80% of which have to get across the perilous Malacca Strait.

China's oil imports are negotiated in cash (other than futures); so there are some strange trading phenomena: such as buying when the price is high rather when the price is low; and when the oil price is high, more oil is imported.

Therefore, China's oil supply security has many risks that can directly influence economic security and social stability. The SPR is an emergency petroleum store maintained by the government to decrease economic and social reparation resulting from a shortage. It is important to establish the SPR for China and so help deal with petroleum use crisis and ensure energy and economic security. Hence the Chinese government began to prepare for the establishment of SPR in March 2004. The first project included four stockpile bases, and the largest was completed in December 2006, which reserve is 5.2 million cubic meters of oil. Others were completed at the end of 2008. If China had its own SPR, a petroleum crisis would not have occurred in China's Guangzhou and Shenzhen in August 2005. This event caused significant negative effects on economic development and social stability.

It is important to distinguish between strategic reserve and operational and speculative reserves; although strategic, operational and speculative reserves constitute physically the same inventory. The former constitutes that part of the petroleum inventory that will not be used unless an emergency has occurred, and is deployed mainly in order to provide lead-time between the occurrence of the emergency and measures required to resolve it. It also provides a bargaining position or prevents hostile actions, and thus ensures short-term protection and short-term demand (Samouilidis, Berahas, 1982).

Industrialized countries suffered greatly during the two oil crises in the 1970s. The United States Department of Energy (1988) estimated that the economic loss from the two oil shocks of the 1970s were 1.2 trillion US$ (evidently estimated in 1987 or 1988 dollars). Post 1973–1974, some industrialized countries set up the IEA (International Energy Agency), which required that each member be committed to maintaining a reserve that is equivalent to 90 days of net oil imports. In order to decrease the negative effects of oil shortages, a strategic petroleum reserve is essential for dampening shocks to the economic system and providing an interval during which efforts can be made to mediate disputes. Thus, the larger the SPR, the less the economy is affected in a negative way. Thus the fundamental two questions are: How many days of net oil imports will be in China's SPR? Will an SPR of 90 days of net oil imports be suitable for China?

In fact, security of energy supply is a relative and uncertain problem. The degree of oil supply security is decided upon not only by the level of SPR, but also by the possibility of short-term oil supply disruption and the possible loss to the economy. Currently many countries would like to establish their SPRs as large as possible to prevent any oil supply disruption through minimizing the total cost and making energy demand and economy cost an equilibrium state. Clearly there exists a trade-off between the stockpile and the vulnerability to supply disruptions. The larger the inventories held, the lower the vulnerability, but at the same time the higher the inventory holding costs. Therefore, the loss of GDP may be different for each of the countries. Even if oil supply disruption and the dependence on oil imports could be the same, the SPR in different countries could be different. Some industrialized countries or regions have their own optimal SPR to meet their needs (see Table 6-1). Table 6-1 shows that the stockpiles of the United States and Japan are larger than those of other countries and regions. Chinese Taipei is the smallest. Then for China, what level of SPR can be optimal?

6.2 Study on Optimal Scale of Chinese Strategic Petroleum Reserves

Table 6-1 The SPR of some industrialized countries or regions in 1999 (Paik, et al., 1999)

Countries/regions	SPR controlled by government million barrels
United States	563
Japan	315
South Korea	43
Chinese Taipei	12
Europe	325
Total	1258

6.2.2 Optimal Petroleum Reserves Model Based on Decision Tree

6.2.2.1 Model and Methodology

An alternative way to structure a decision problem pictorially is using a decision tree. A decision tree depicts chronologically the sequence of actions and outcomes as they unfold. For a terminal decision problem based on prior information, the first fork (square node) corresponds to the action chosen by the decision maker, and the second fork (round node) corresponds to the event. The numbers at the end of these terminal branches are the corresponding payoffs (or losses) (Ravindran et al., 1987).

The decision tree has been widely used as a modeling approach. However, very little is known from the literature on how the decision tree performs in predicting an optimal stockpile reserve. The decision tree is a non-parametric modeling approach, which recursively splits the multidimensional space defined by the independent variables into zones that are as homogenous as possible in terms of the response of the dependent variables (Vayssieres, et al., 2000). This relates the decision variables and parameters with selected possible events; as illustrated in Fig. 6-1.

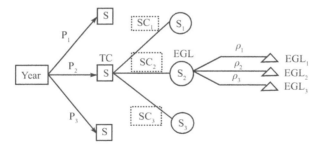

Fig. 6-1 The decision tree model structure

Each decision concerning the stockpile of different periods (for simplicity, in Fig. 6-1 only three alternatives are presented) is associated with several oil prices and loss of GDP caused by shortages due to oil supply emergencies (again only three are presented). Any branch of this decision tree model expresses a scenario to be evaluated by the decision maker.

The sudden rising of oil prices creates three types of economic loss to the economy for oil importing countries: ① transfer of wealth from oil import countries to oil export countries; ② loss of the potential GDP; ③ macroeconomic adjustment loss. The transfer of wealth is a transfer payment, and is equal to the quantity of oil import multiplied by the difference between the monopoly price and the competitive market price of oil. The GDP loss considered in this chapter refers to the sum of the previously discussed three losses.

Because decision-makers do not know exactly what oil supply disruption will occur and what oil prices will be in the future, they can assume or reckon the possibility of possible disruptions and prices based on experience and the history of disruptions, then make decisions. The optimal SPR is a decision with risk. There are many risks for decision-makers to consider prior to establishing SPR: the GDP loss will be too large if long-term oil supply disruption occurs and the size of SPR is too small in the future; the expensive reserve investment and maintenance cost for establishing SPR will be wasted if oil supply disruption will not occur or short-term disruption will occur in the future. Therefore, this chapter analyzes the optimal SPR within scenarios of different oil prices using decision tree analysis.

The decision tree is the figure form of expected value (EV) method. The decision dots are presented in small panes; the arcs linked with decision dots refer to possible SPR decisions, that are 30, 60, and 90 days of net oil imports. The remaining ends of arcs are condition dots presented in circles; the arcs linked with condition dots are possibility arcs above which are shown possible oil supply disruptions. The other ends are result dots presented by small triangles; there is a figure behind each which refers to the cost of every decision under different conditions (Figs. 6-2–6-5).

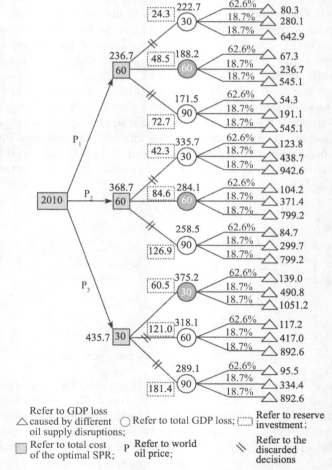

Fig. 6-2 The decision tree of SPR analysis for China in 2010 scenario I
(Note: The other units are always 100 million dollars, besides possibilities)

When the possibility of oil supply disruption occurring is ρ, then the expected cost of every decision is $E(\alpha, \rho)$ and total expected cost of national SPR is $E(\alpha)$. Because the possible condition is discrete, the possibility of condition i is ρ_i. Then

$$E(\alpha) = \sum_{i \in S_2} \rho_i P(\alpha, i) \quad \forall \alpha \in S_1 \tag{6-1}$$

6.2 Study on Optimal Scale of Chinese Strategic Petroleum Reserves

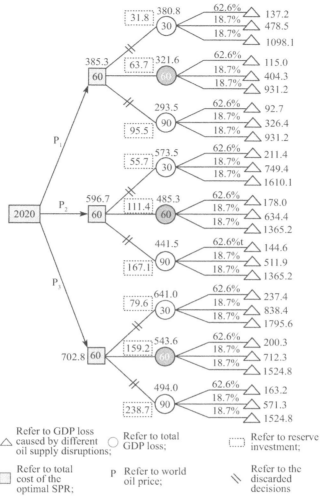

Fig. 6-3 The decision tree of SPR analysis for China in 2020 in scenario I

Where: S_1 is the set of every decisions; S_2 is the set of different conditions.
When minimizing total cost, then decision α^* is optimal.

$$\min_{\alpha \in S_1} E(\alpha) = E(\alpha^*) \tag{6-2}$$

6.2.2.2 The Assumptions

The empirical analysis in this chapter is based on the following assumptions:

(1) China's SPR base is being established and has no stockpile. So we do not know how much the exact maintenance and investment costs are. We assume that the proportion among oil import cost, investment cost and maintenance cost in China's SPR investment is 75.4%, 22.9%, 1.7%, respectively, which refers to the experience of the United States.

Greene et al. (1998) posited that although the SPR can be effective against a short-lived, random shock, it is relatively ineffectual against a strategic shock of two years or more. It is quite clear that SPR alone can not resolve a major supply emergency. So the SPR can not by itself provide long-term security. But Samouilidis and Berahas (1982) believed that the amount of the SPR held is a function not only of the holding costs and petroleum prices, but also of the displaced shortage costs (either monetary or social) and the other damages incurred by an energy emergency. So we consider both opinions.

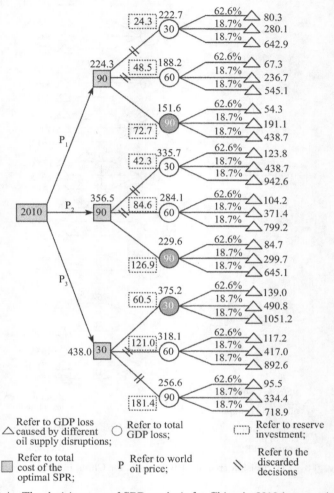

Fig. 6-4 The decision tree of SPR analysis for China in 2010 in scenario I

(2) We assume that the impact of the SPR of 60 and 90 days of net oil import on the decreasing GDP loss are the same if long-term oil supply disruption occurs.

(3) We assume that the impact of different SPR on the decreasing GDP loss is different whatever oil supply disruption occurs.

(4) Based on some forecasts for China's future oil demand and production, we assumed that the China's net import of crude oil is 130 million tons in 2005, 150 million tons in 2010, and 200 million tons in 2020, respectively.

6.2.2.3 The Source of Data

The world oil prices, China's oil consumption and oil production for each year are taken from *BP Statistical Review of World Energy* 2004 and *China Statistics Yearbook* 2003; the GDP of China is taken from *China Statistics Yearbook* 2003. The GDP and the oil demand of China in 2005, 2010 and 2020 are taken from Chen (2003). The historical information of oil supply disruption is taken from EIA (Energy Information Agency) of the United States Department of Energy.

6.2 Study on Optimal Scale of Chinese Strategic Petroleum Reserves

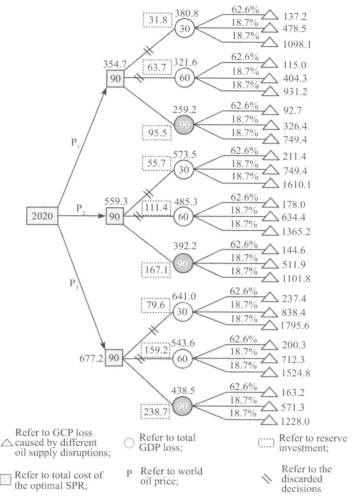

Fig. 6-5 The decision tree of SPR analysis for China in 2020 in scenario II

6.2.3 Discussion on Different Reserve Scales

The Chinese government began to establish national SPR in 2004. The biggest reserve will be finished in December 2006. So it is not possible to own national SPR now. The optimal SPR of 2005 calculated in this chapter is theoretical (and is not discussed). The optimal SPR under different assumptions and oil prices in 2005, 2010, and 2020 can be found in Table 6-2.

Table 6-2 The optimal SPR in different assumptions and oil prices in 2005, 2010, and 2020

Year	20US$/bbl		35US$/bbl		50US$/bbl	
	Scenario I	Scenario II	Scenario I	Scenario II	Scenario I	Scenario II
2005	60 days	60 days	30 days	30 days	30 days	30 days
2010	60 days	90 days	60 days	90 days	30 days	30 days
2020	60 days	90 days	60 days	90 days	60 days	90 days

6.2.3.1 Scenario I Analysis

For scenario I, the assumption that the impacts of the SPR of 60 and 90 days of net oil import on the decreasing GDP loss are the same if long-term oil supply disruption occurs.

The results in 2010 show that:

(1) The optimal SPR is 60 days of China's net oil imports when oil price is US$20/bbl and the economy growth is around 7%.

(2) The optimal SPR is also 60 days of China's net oil imports when oil price is 35US$/bbl and the economy growth is around 7%.

(3) The optimal SPR is 30 days of China's net oil imports when oil price is 50US$/bbl and the economy growth is around 7%.

When oil price is 20US$/bbl and 35US$/bbl, the reserve cost per barrel is relatively low, the contribution of SPR of 60 days for decreasing the GDP loss caused by oil supply disruption is great compared with the SPR of 30 days, so the total cost for SPR of 60 days is less than that of 30 days. Because we assume that the contribution of SPR of 60 days and 90 days to decreasing the GDP loss caused by long-term oil supply disruption are the same, the reserve investment of 90 days is far more than that of 60 days. Therefore, the expected total cost of SPR of 60 days is the least.

When oil price is 50US$/bbl, the reserve cost per barrel is relatively high, and the GDP loss caused by oil supply disruption is less than the increased reserve investment for 60 and 90 days of SPR, although the contribution of SPR of 60 days to decreasing the GDP loss caused by oil supply disruption is great compared with the SPR of 30 days, so the total cost for establishing SPR of 60 and 90 days is larger than that of 30 days.

Therefore, the optimal SPR in 2010 for China is 60 days of net oil import if the fluctuation of world oil price is 35US$/bbl or less; the optimal SPR for China should be 30 days of net oil import if the fluctuation of world oil price is 50US$/bbl or higher (Fig. 6-2).

The results in 2020 show that the optimal SPR for China is 60 days of net oil import, when oil prices are 20US$/bbl, 35US$/bbl, or 50US$/bbl and the economic growth is around 7% (Fig. 6-3). It is estimated that China's oil demand will reach 450 million tons in 2020; and the domestic production will be 200 million tons. So the dependence on oil imports will be more than 55%, and the impact of the fluctuation of world oil prices on economic development will be greater than now; which makes the establishment of a larger SPR necessary. Theoretically, the greater the size of SPR in 2020 is, the less the GDP loss is. But the possibility of long-term oil supply disruption occurring is only 18.7% and the contribution of SPR of 60 days and 90 days to decreasing the GDP loss caused by long-term oil supply disruption is the same.

The possibility of short-term oil supply disruption occurring is 62.6% and the reserve investment of 90 days is more than that of 60 days, although the contribution of SPR of 90 days to decreasing the GDP loss caused by oil supply disruption is greater than that of 60 days. So the total cost for establishing SPR of 90 days of net oil import is higher than that of 60 days. Therefore, the optimal SPR for China will be 60 days of net oil import in 2020 whatever the world oil price is.

6.2.3.2 Scenario II Analysis

For scenario II, the assumption that the impact of different SPR on the decreasing GDP loss is different when various oil supply disruptions occur. The results in 2010 show that (Fig. 6-4):

(1) The optimal SPR is 90 days of China's net oil imports when oil price is 20US$/bbl and the economy growth is around 7%.

(2) The optimal SPR is also 90 days of China's net oil imports when oil price is 35US$/bbl and the economy growth is around 7%.

(3) The optimal SPR is 30 days of China's net oil imports when oil price is 50US$/bbl and the economy growth is around 7%.

We assume that the contribution of alternative SPR to the decreasing GDP loss is different when different oil supply disruptions occur. When oil price is 20US$/bbl and 35US$/bbl, the reserve cost per barrel is relatively low, the contribution of SPR of 90 days to decreasing the GDP loss caused by oil supply disruption is greater compared with the SPR of 60 and 30 days. So the total cost for SPR of 90 days is less than that of 60 and 30 days. Therefore, the expected least total cost of SPR is 90 days.

When oil price is 50US$/bbl, the reserve cost per barrel is relatively high, and the GDP loss caused by oil supply disruption is less than the increased reserve investment of 60 and 90 days SPR, although the contribution of SPR of 60 days and 90 days to decreasing the GDP loss caused by oil supply disruption is greater compared with the SPR of 30 days. So the total cost for establishing SPR of 60 and 90 days is larger than that of 30 days. Therefore, the optimal SPR in 2010 for China is 90 days, net oil import if the fluctuation of world oil price is 35US$/bbl or less; the optimal SPR for China should be 30 days' net oil import if the fluctuation of world oil price is 50US$/bbl or higher.

The results in 2020 show that the optimal SPR for China is 90 days of net oil imports, when oil prices are 20US$/bbl, 35US$/bbl, or 50US$/bbl and the economy growth is around 7% (Fig. 6-5).

Because the oil demand of China will be greater, the dependence on oil imports and the share of oil in the primary fuel mix will increase, and the impact of the fluctuation of world oil prices on economic development will be greater than now. This supports the argument for establishing a larger size of SPR. Therefore, the optimal SPR for China will be 90 days of net oil imports in 2020, whatever the world oil price is.

6.2.4 Results

The security value of strategic petroleum reserve is a very complex problem. In order to estimate the optimal SPR level, all factors related to SPR must be quantified, but this is difficult to accomplish, so we make a lot of assumptions. Although the approach omits many features of the problem, it considers the most important decision variables that greatly influence the optimal stockpile. Thus, it gives the decision maker basic information such as storage and shortage costs.

For the same period, the optimal SPR of the lower oil price scenario is always larger than or equal to that of the higher oil prices scenario; which is in completely accord with the rule "buy low and sell high". The optimal SPR will increase from 30 days to 90 days of net oil imports with greater requirements for energy and high dependence on oil imports, which is necessary for economic development and energy security.

Our results show that the optimal SPR level is sometimes not the largest strategic stockpile. When oil prices are low, 60 days and 90 days of net oil imports are suitable for China's economic development and security in 2010 and 2020. When oil prices are high, the optimal SPR should be 30 and 60 days of net oil import for 2010 and 2020. Therefore, the oil prices can influence the optimal SPR.

6.3 Risk Assessment of Oil Imports and Countermeasures

With the changes of economic and military strength of countries all over the world, new geopolitics for world energy have formed gradually. As the second oil consumption country in the world, following the United States, whether China could play a major role and important part in numerous and complicated energy geopolitics, and whether China could gain safe and sufficient oil resources in new energy reallocation, will be the key to influence Chinese future oil imports and energy supply risks.

6.3.1 Overview of World Oil Trade

The world oil resources are distributed extremely unbalanced, that is, the abundant oil resources of countries require relatively small oil, whereas the reserves of countries with larger quantity of consumption are leanness, even almost none, thus the fact that the production and consumption separated seriously caused world oil trade to be highly centralized and monopolized, which is shown as Figs. 6-6 and 6-7. Oil export center mainly in the Middle East, Russia, west African, south and middle America, Southeast Asia and north Africa, accounting for 81.7% of the total world exports, among which the exports of the Middle East and Russia are the most, which reached 1.294 billion tons in 2004, accounting for 54.2% of the world oil exports. And oil consumption countries center mainly in United States, China, Japan, Europe and other Asian-Pacific area countries, the oil imports of these countries and regions were 2.146 billion tons in 2004, accounting for 90.1% of the world oil imports. However, the imports of the United States, Japan and China were up to 1.065 billion tons, which was 44.7% of the total world oil imports. So the world oil trade is actually that between several main export areas and a few bigger importing countries.

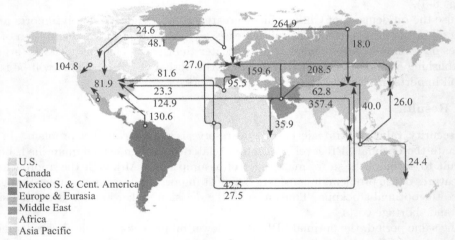

Fig. 6-6 World oil trade in 2004
(Note: Data from BP 2005)

At the end of Iraq war in 2003, the United States already controlled the abundant oil resources in Iraq basically, which helped to reserve sufficient oil for its future energy supply. The formally contracted "Taishet-Nakhodka oil pipeline" between Japan and Russia indicated the success of Japan to transit its oil import to newly born oil-exporting country from the traditional sources, and Russia will offer 50 million tons of crude oil to Japan every year, which is a successful step taken by Japanese government to realize the diversification of oil import too. European Union is also progressively strengthening the energy cooperation with Russia and the Caspian area, whose oil import from Russia has already exceeded two times that of the Middle East area, thus European Union is gradually getting rid of the oil imports risk. But there is still not a significant reform up until now in oil imports of China, that is, the import from most areas is increasing, and the share from the Middle East area is still up to 40%. We are in a severe import situation.

The overall oil imports of four countries and regions, that is, United States, the European Union, Japan and China, accounts for 70.3% of general quantity of world imports, so the changes of import sources and quantity in these countries and regions influence directly the equilibrium and stability of supply and demand of world oil. In fact, the oil import risk

6.3 Risk Assessment of Oil Imports and Countermeasures

is affected by many kinds of factors, such as import dependence, size of national strategic petroleum reserves, international energy geopolitics relationship and transportation of imports, etc. Compared to the United States, European Union and Japan, there are more risks in China's oil imports. Its dependence increased quickly, already up to 41.3%, and it is estimated that would be up to 60% in 2020. The national strategic petroleum reserves construction would not be started formally until 2004, and there are four reserve bases during the first stage of plan, among which the largest Zhenhai base is expected to be completed at the end of 2006 and the others in 2008. The total storage of 2010 will reach up to 30 days net import quantity. About 80% of China's oil imports has to get across the Malacca Strait, additionally, 90% is transported by foreign fleets, so there are a lot of risky factors in China's oil imports.

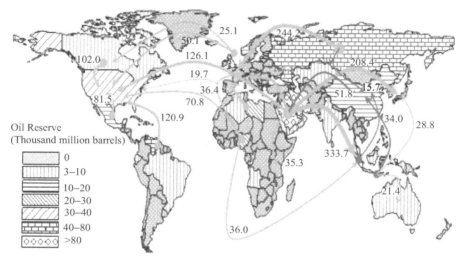

Fig. 6-7 World oil reserves and import in 2003 (oil import: million tons)
(Note: Data from BP, 2004)

Petroleum is playing an important role in China's energy consumption and economic development, and with the high-speed development of China's economy, our oil gap is greater, with the import dependence rising to 41.3% in 2004 from 2.3% in 1994, having increased nearly 20 times, which is shown in Fig. 6-8. China's oil import dependence increased relatively slowly in 1993—1999, while since 2000, it soared rapidly, going up by 2.5 times in only 5 years. The high dependence increased China's oil import risk greatly, and then influenced energy security and national security. Thus, how to assess and analyze the import risk objectively, and which effective measures are to be taken to reduce the risk, have become the hot issue that domestic and international scholars pay close attention to.

Because of the turbulent political environment and intense international relations in some areas, the crude oil price has remained at a high level since 2003, soaring even more in the second half of 2004, and continually hitting new peaks. Therefore, China will undoubtedly suffer enormous economic and foreign currency losses, for its crude oil imports in cash rather than in futures. So it is generally agreed to realize the sources diversification of China's oil imports and to set up an oil futures market. To general merchandise, since its risk coefficient is the same, the smaller the diversification index is, the more scattered the risk is, and the lower the risk index is, but because of the particularity of oil import, the risk coefficient of every oil supplier is not the same, and the diversification index is not in direct scale with its import risk. The index would not be reduced with the increase of quantity of import sources, so the question is mainly how to realize the diversification of oil imports to reduce

China's imports risk effectively.

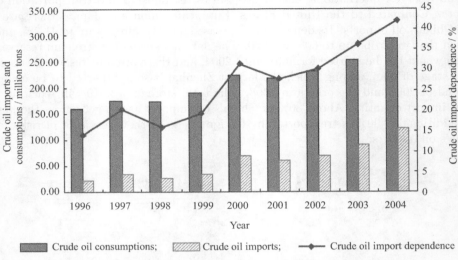

Fig.6-8 The crude oil import dependence of China over 1996–2004
(Note: Data from *Yearbook of China's Foreign Economics Relations* and *Trade/China Commerce Yearbook*)

6.3.2 Risk Evaluation of Crude Oil Import–Based on HHA Method

The Hirschman-Herfindahl Agiobenebo (HHA) method which is improved on the foundation of classical theory HHI (Hirschman-Herfindahl Index) used for measuring the concentration degree of product, and now is extensively applied to evaluate such fields as the market share of product, the diversification of national import and export, the market share of some merchandise, and is suitable for measuring the dependability of supply and marketing (Neff, 1997; Agiobenebo, 2000). Here is the definition formula of the present diversification index:

$$\text{HHA} = \sqrt{H}/100 = \frac{1}{100} H^{1/2} = \sqrt{\sum_{i}^{n} S_i^2} \qquad (6\text{-}3)$$

Where: S_i is the share of crude oil import from supply country i.

We adopt HHA method to calculate the diversification index of oil import of several main importers (United States, European Union, Japan and China) in the year 2001–2003. HHA method generally supposes the risk of every supplier is the same, which is not suitable to oil import, because the risk of every country importing oil from different areas is different, even if it is the same area to different importers. So, on the premise of considering coefficient of the risk, we can get the risk index of a certain country using the HHA approach, as the Eq. (6-4) shows:

$$R = \sqrt{\sum_{i}^{n} w_i^2 S_i^2} \qquad (6\text{-}4)$$

Where: R is the risk index of oil import; w_i is the risk weight coefficient of the area i.

6.3.3 Weight Coefficient of Oil Import Risk–Based on AHP Method

Because of the intricate world energy geopolitics, the difference of oil import methods (pipeline, shipping, railway, etc.), transportation route and fleet ownership (domestic oil fleet, foreign oil fleet), the import risks of the same area are different in different countries. During the first oil crisis in the 1970s, the Arab countries did not reduce the oil exports to Japan, because of the great transition of Japanese foreign policy (declaring to support Palestine people). So the first world oil crisis has less influence on Japan's economy than that of the western developed countries.

Secondly, European Union is the biggest overseas developer of the Middle East; United States controls most oil of the Middle East; Japan's overseas oil exploitation is far higher than China. Additionally, China and Japan oil imports from the Middle East have to get across Malacca strait, the most insecure strait in the world. Therefore, for the European Union or America the import risks from Middle East area will be smaller compared to that of China and Japan.

In addition, pipeline is one of the main forms of import transportation in EU and United States, while China and Japan depend mainly on shipping. For the United States, EU and Japan oil imports are undertaken basically by domestic fleet, while the ones that undertake China's oil import task are mainly foreign, so our oil import risk is higher than Japan, and EU and American transportation risk are relatively less.

Therefore, there is the certain limitation to the Neff (1997) method that weight coefficient of oil import risk only through correlated coefficient matrix of different areas outputs, so we adopt the AHP method to confirm the weight coefficient of the oil import risk in various areas. We sent 150 questionnaires altogether to the experts in the field of oil and gas, regaining 115 ones, among which 110 questionnaires are of effectiveness. Fig. 6-9 described the risk assessment system of oil import in detail, whose basic principle is: According to the objective judgment on reality, relying on the comparing scales and the judging principle, as well as the AHP method, we set the weights for strict consistency test, the higher the dimension of matrix n is, the bigger estimate bias is. Therefore, introduce a random index $R.I$ as the correcting value to test the consistency of the judging matrix with a more reasonable random consistency index, and the final result indicated each judging matrix satisfies the above consistency standards ($C.I < 0.10$, $C.R \leqslant 0.10$).

Calculate the risk weight coefficient of several import sources of several main oil importers and regions with the AHP method, shown as the Table 6-3.

Table 6-3 The risk weight coefficient w of oil import of main country

EU	SA	Russia	NA	WA	ME	U.S.	Mexico	Uncertain	Other
w	0.029	0.022	0.149	0.175	0.37	0.094	0.034	0.057	0.07
U.S.	ME	Mexico	Russia	NA	WA	Canada	SA	Europe	Other
w	0.322	0.039	0.143	0.195	0.122	0.019	0.028	0.069	0.063
Japan	China		SeA		WA		ME		Other
w	0.109		0.061		0.246		0.489		0.095
China	Russia		SeA		WA		ME		Other
w	0.051		0.117		0.217		0.537		0.078

Note: SA is South America, NA is North Africa, WA is West Africa, ME is Middle East, SeA is Southeast Asia.

As to our country, the oil import from the Middle East area, the risks are relatively high because of the area's political turbulence and its constant military conflict in some places. Over history, several heavy oil crises, the break-down to supply and the sharp rise in oil prices, are nearly all caused by the Middle East. In addition, our oil import from the Middle East will have to get across from the most insecure Malacca strait in the world. Almost all the oil deliveries are taken by foreign fleet, so generally the import risk from the Middle East area is the greatest. Then because of the unsteady political situation of West Africa political turmoil and strikes taking place occasionally, the foreign fleet undertaking the transport, and "the Malacca Strait" being the only route too, the import risk from this area is relatively high. Though the competition of oil import is fiercer in Southeast Asia, its political situation is relatively stable, and our country's fleet undertakes a larger percentage of transport task than when we import from Middle East and West Africa, so the risk is relatively slightly lower. It is generally thought that the risk from Russia is lower, mainly because it is not OPEC member, and OPEC's measures are difficult to impose on Russia that influences the world energy safety such as underproduction or petroleum embargo, etc. Besides, China and Russia keep friendly bi-relationships, who have common international political and military rivals, and China's oil deliveries have been mainly by railway from Russia, so the risk from Russia is the lowest. Other areas include: North and Southeast Africa, South America, Australia, and Kazakhstan, etc. Most of them are neither OPEC members nor the big exporters, so they have little effect on the world oil market, and relatively small competition for oil import, so the import risk is slightly lower.

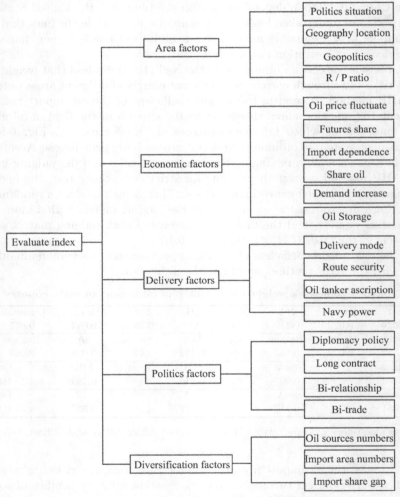

Fig. 6-9 The Risk Assessment Index System of oil import

6.3.4 Risk Assessment of Oil Imports in Some Major Oil Importing Countries

6.3.4.1 Source of Data and Pretreatment

All data are taken from *BP Statistical Review of World Energy* (2002, 2003, 2004). And then we calculated the share from several main areas, that is, the amount of oil imports

6.3 Risk Assessment of Oil Imports and Countermeasures

accounting for the total in that year in need of assessment.

6.3.4.2 Application of HHA Method to the Assessment of Oil Import Risk

Use Eq. (6-3) to calculate the oil import diversification index of main importers and regions during the period 2001–2003, and Si represents the share of a country's oil import from the i area. Fig. 6-10 shows results of calculation.

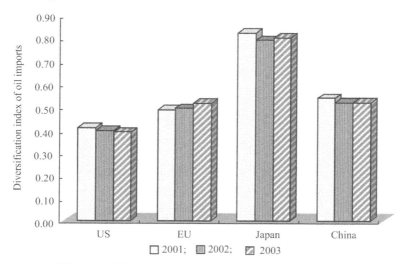

Fig. 6-10   Diversification index of the main oil importers

The results of calculation indicate that the diversification index of Japan's oil import is the highest, more than two times of United States, followed by China and European Union, and the lowest is United States (importing from more than 60 countries such as Saudi Arabia, Mexico, etc.). The oil import diversification index of United States presented the downward trend during the period 2001–2003, which is identical with the more sources of American oil import and the smaller gap of import quantity between oil import areas. The diversification index of European Union is rising, mainly because European Union is increasing the oil import from Russia constantly in recent years. Though European Union is seeking for import diversification constantly, oil import depending on Russia unduly made the index increase year by year. And the indexes of Japan and China dropped greater in 2002, then slightly rose in 2003, but still little below that of 2001. The result of the study fully proved the achievements of these countries' diversification policy in oil imports. But because the domestic oil demands suddenly increased, and the number of new oil sources and import quantity increased very limited, the import from some areas or countries suddenly soared, and then the diversification index rebounded. So there is a long road to China and Japan's oil import diversification.

According to the Eq. (6-4), we can calculate risk index of different oil importers and regions over 2001–2003. As Fig. 6-11 shows, the import risk of China is smaller than Japan, but much higher than European Union and United States, and it kept rising over 2001–2003. The risk of European Union presented a downward trend, and the risks of United States and Japan declined greatly in 2002, slightly went up in 2003, but still lower than that in 2001.

6.3.4.3   Discuss and Analysis of the Results

(1) European Union is constantly increasing supply from areas with less risk, reducing the import from ones with relatively high risk to realize import sources diversification.

As Figs. 6-12 and 6-13 show, EU increased supply from Russia with less risk by a large margin, from 181.0 million tons (account for 31.8% of total imports) in 2001 to 244.0 million tons (41.2%) in 2003, import share increasing by 9.4%, which lead to an increase in the diversification index of EU.

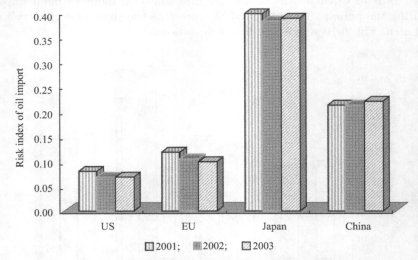

Fig. 6-11 Import risk index of the main oil importers

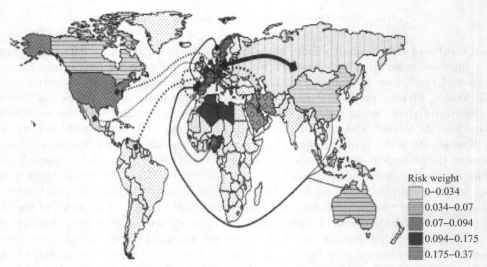

Fig. 6-12 Changes of the risk and import from oil sources of EU in 2001–2003
(Note: The full line is the increase of import and the dashed is the reduction of import)

① Because of the increasing import from the lower risk areas, the risk index of EU will be reduced. ② EU has reduced import from such areas as Middle East, North Africa by 21.9 million tons, 6.3 million tons separately, while increasing Russia's import. The import share of these areas has been reduced by 4.9% and 1.7% separately, and the risks are all relatively high to European Union, risk coefficient is 0.37 and 0.149. So the reduction has further lowered the oil import risk of EU. ③ EU has also increased the number of countries with low risk and the import from these areas further, the quantity rising from 3.1 million tons (account for 0.6% of total imports) in 2001 to 6.2 million tons (1.0%) in 2003. So the

6.3 Risk Assessment of Oil Imports and Countermeasures

oil import risk index of EU presented a downward trend over 2001–2003, with the risk being reduced gradually.

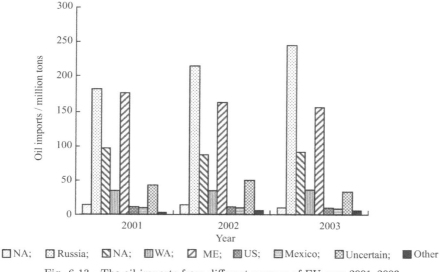

Fig. 6-13 The oil imports from different sources of EU over 2001–2003

(2) The United States reduces its import dependence on areas with large shares, and increases the import from those with low risk, to further perfect the diversification strategy of oil imports.

As Figs. 6-14 and 6-15 show, in order to reduce the dependence on some countries and regions, the United States reduced the import from areas with large shares and decreased the gap between the main oil import sources gradually during 2001–2003. The import from the Middle East which had the largest share and risk was reduced by 12.0 million tons, from 138.0 million tons (24.1% of total amount) in 2001 to 126.0 million tons (account for 20.8%) in 2003. The import from South America reduced by 6.0 million tons, from 126.0 million tons (22.0%) in 2001 to 120.0 million tons (19.9%) in 2003.

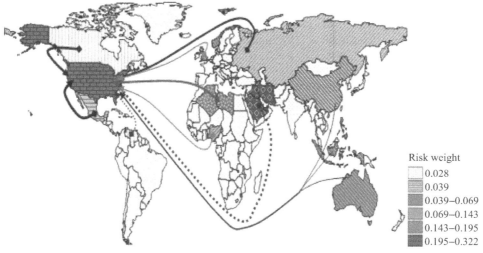

Fig. 6-14 Changes of the risk and import from oil sources of US in 2001–2003
(Note: The full line is the increase of import and the dashed is the reduction of import)

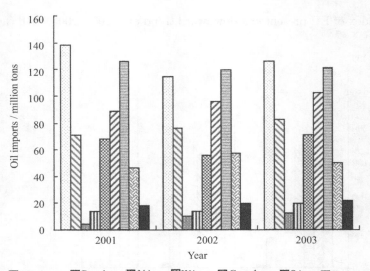

□ ME;   ◨ Mexico;   ▨ Russia;   ⊞ NA;   ▩ WA;   ▨ Canada;   ▨ SA;   ▨ Europe;   ■ Others

Fig. 6-15   The oil imports from different sources of United States over 2001–2003

① The gradually reduction of share gap between the main oil sources, especially reduction of import from areas with high risk, makes the United States decrease not only the diversification index of its oil import, but also the import risk. ② United States is planning to increase import from those areas with relative low risk (Canada, Mexico, etc.), whose risk coefficient is 0.019 and 0.039 respectively. The import from Canada and Mexico increased by 14.0 million tons and 10.7 million tons respectively over 2001–2003, the share increasing by 1.6% and 1.2% respectively. At the same time United States still expands the numbers of oil import sources actively, importing from more than 60 countries now, which further reduces its import diversification index and import risk. ③ United States increased the import from Mexico and Europe by 5.4 million tons and 10.8 million tons respectively in 2002, and the import share had increased by 1.3% and 2.1% respectively compared with the last year. The import from Europe in 2003 reduced by 6.9 million tons compared with 2002, and the share of Europe and Mexico had dropped by 0.1% and 1.9% respectively compared to that of the last year. Because the risks of Mexico, Europe are relative low, reducing the import from Mexico and Europe will strengthen the import risk of United States. ④ United States reduced the import from West Africa which had relative high risk by 12.6 million tons in 2002, while increased by 15.3 million tons in 2003. The import share of West Africa also declined from 11.9% in 2001 to 9.8% in 2002, then went back to 11.7% in 2003 again, causing the import risk index of United States to be slightly higher than that of the last year in 2003. Therefore, although the import diversification index of United States reduced year by year, its risk happened to rebound in 2003.

(3) Japan reduces share of traditional oil import sources, and increases the import of new sources, to try hard to realize the diversification of oil imports.

As Figs. 6-16 and 6-17 show, ① Japan is an energy-importing country. In order to reduce the oil import risk and ensure energy security, Japan's import from the traditional sources (the Middle East, Southeast Asia) reduced by 13.4 million tons and 5.9 million tons respectively in 2002, the import share dropping by 3.2% and 2.0% respectively compared with that in 2001. But in 2003, the import from Middle East and Southeast Asia increased by 13.0 million tons and 0.5 million tons respectively, the share from the Middle East going up by 1.3% compared with that in 2002. The risk from the Middle East is the highest, so reducing its import and share will decrease the risk, and vice versa. ② Reducing the oil import quantity and share from the traditional sources will disperse risk and decrease

6.3 Risk Assessment of Oil Imports and Countermeasures

the diversification index effectively, and vice versa. ③ Japan increased its import from West Africa which had higher risk by 4.0 million tons over 2001–2003, and the import share increased by 1.4%. Its import from China keeps the same basically, and the share had all been 1.6% for three years in succession, so the diversification index and import risk have all been keeping at a high level all the time. ④ Japan increased import from new sources by 9.7 million tons in 2002, import share increased by 4.0%, while reduced import from them by 2.6 million tons in 2003, import share reduced by 0.9%. Because of the high risk from the Middle East and West Africa area and the low risk from new import sources, both the diversification index and import risk index present the trend of upward after being downward in 2001–2003, but still lower than the level of 2001.

Fig. 6-16 Changes of the risk and import from oil sources of Japan in 2001–2003
(Note: The full line is the increase of import and the dashed is the reduction of import)

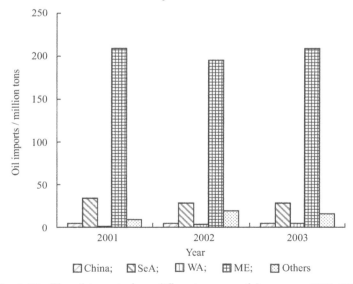

Fig. 6-17 The oil imports from different sources of Japan over 2001–2003

(4) China paid insufficient attention to the risk of new oil sources, with the oil import risk index being on the rise, so there is a long way to go for China's oil import diversification.

As Figs. 6-18 and 6-19 show, since the oil demand of China increased rapidly in recent years with the fast development of economy, while the domestic oil output increased slowly, oil import and dependence went up all the way, the oil import dependence increased from 16.7% in 1999 to 39.1% in 2003; however, with the increasing of imports, China paid insufficient attention to the risk from different sources.

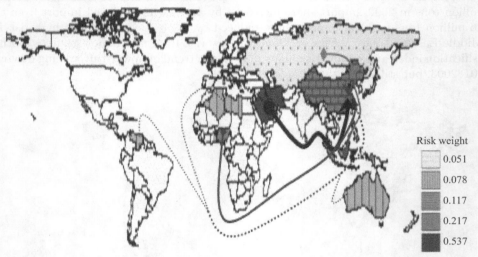

Fig. 6-18 Changes of the risk and import from oil sources of China in 2001–2003
(Note: The full line is the increase of import and the dashed is the reduction of import)

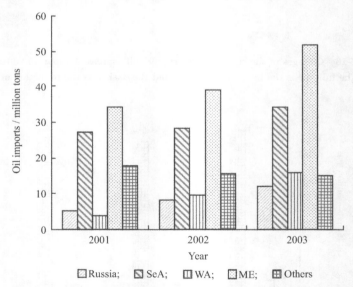

Fig. 6-19 The oil imports from different sources of China over 2001–2003

① Compared with West Africa, it brought much lower risk to China to import oil from Russia, the risk coefficient of the two sources being 0.217 and 0.051 respectively, so it is wise to increase import from Russia theoretically. While in fact the import from Russia only increased 6.6 million tons in 2001–2003, in contrast the import from West Africa increased 11.9 million tons, increasing by 3.3% and 7.9% respectively. Although half of import from West Africa is shares of oil, getting across the Malacca Strait raised the transporting risk.
② Though the import from Southeast Asia which had relative low risk increased year by

year during the period 2001–2003, import proportion is reduced year by year, dropping from 30.8% of the total in 2001 to 26.5% in 2003, reduced by 4.3%, so it is also difficult to reduce oil import risk of China in this kind of adjustment. ③ The import risk from other sources was lower than Southeast Asia, but both the import quantity and share reduced year by year in 2001-2003, and the import decreased 2.9 million tons and the share from 20.2% in 2001 to 11.6% in 2003, dropping 8.6%. So reducing the quantity and share from sources with less risk has increased the import risk. ④ The import from Middle East which had the highest risk over 2001–2003 increased by 17.6 million tons, though import share in 2002 was the same to that in 2001, and increased by 1.7% in 2003. Raising the import and share from areas with higher risk will undoubtedly strengthen the import risk of China. ⑤ Because the import shares of Russia and West Africa increased in 2002, that of Southeast Asia and other sources were reduced, Middle East having not changed, and the number of China's oil sources increased constantly, so the diversification index dropped. But the import share of the Middle East went up 1.7% in 2003, so the risk index rebounded in 2003, slightly higher than the level of the last year. ⑥ China is paying insufficient attention to the risk from every oil source, which leads to increase the import and share from areas with relatively high risk (the Middle East, West Africa) and reduce that from areas with low risk (other sources and Southeast Asia). Therefore, although the number of China's oil sources is increasing year by year, the import risk index is on the rise.

(5) Though European Union and United States both import oil mainly from nine countries and regions, the import diversification index of European Union is higher than that of United States, and is also on the rise. The main reason is that the import share of EU from Russia is up to 41.2%, while the share of United States's largest import source is only 20.8%. China and Japan both import oil from five main countries and regions. However, the diversification index of Japan is higher than China, mainly because its import share from the Middle East is up to 79.3%, and China's is 40.4%. So increasing new import sources, reducing the dependence on major import sources is the best way for reducing the diversification index and import risk index.

6.3.5 Results

Through calculation and comparative analysis as above, we can get the following main conclusions:

(1) To general merchandise, the smaller the diversification index is, the smaller the risk is, but because of the particularity of oil imports, its diversification index is not in direct relation with import risk, which is verified by the rising trend of European Union's oil import diversification index and the dropping import risk index by year over 2001–2003.

(2) The Middle East is the most insecure area for oil supply in the world, but no oil importer is able to get rid of relying on its imports shortly, for both its oil reserves and export quantum take the absolute predominance.

(3) Increasing the number of oil import sources blindly can not reduce the import diversification index and import risk. The facts of oil imports of United States and European Union has fully proven that the best way to realize the import diversification and to reduce import risk is decreasing share gap between the main sources, and increasing the import and share from new sources with low risk, and adopting diversified import modes.

(4) China's energy demand gap and limited import sources shortly make us have to increase the import and share of risky areas like the Middle East, so only China had upward oil import risk of several main oil importers over 2001–2003.

(5) The adjustment to national oil imports of United States, the European Union and Japan during 2001–2003 reduced the market share of OPEC members, thus Russia is in a very important position in the world oil market now. A new pattern for world oil market

is forming gradually, and every main oil importer and region is decreasing dependence on import from OPEC gradually to realize its oil import diversification.

6.4 Policy Suggestions on China's Energy Security

National macroeconomic policies are to guide economic development, technological progress, industrial structure and consumer behavior, just like a baton. A scientific and objective national policy can activate an industry. Similarly, a series of scientific and objective energy policies can accurately guide the whole society's energy consumption behavior, consumption structure, industrial structure and technological progress. Therefore, based on the previous analysis and quantitative results, combined with the current situation of China's energy security, we put forward several relevant policy recommendations, hoping to assist policy-makers to formulate national energy security policy with reference information.

6.4.1 Energy Diplomacy Policy

(1) Implement the "international cooperation" energy strategy, and actualize the diversification of energy diplomacy.

In the period of high oil prices, energy competition is not merely an economic issue but also a political issue. Because of some non-economic factors, Sino-Japanese and Sino-Indian energy competition appear to be even more subtle. In fact, faced with a severe energy shortage situation, China has begun to pay more attention on energy diplomacy. The senior leaders frequently visit countries and regions that have energy cooperation and have the forthcoming cooperation with China. At present, China has conducted energy cooperation with the Middle East, South America, Africa, Central Asia, Russia, and Kazakstan, ASEAN and other major oil-producing regions and countries. China's energy diplomacy and oil imports have been basically showing a "diversification" pattern.

(2) Implement political and energy integration diplomatic policy to occupy an advantageous position in a new world energy pattern.

The economy and national security are closely related to energy security. Energy security is a prerequisite and guarantee for national security. The government should enhance attention on energy security. In the complex international energy geopolitical environment, as a major oil importing and consumption country, China should follow the example of Japan to take energy diplomacy as a national focus on diplomatic activities, initiatively establish friendly unions and strategic partnerships with the major energy producers, and strengthen high-level exchanges and friendly contacts, learn actively from the experience of Sino-Russia "AN-DA" pipeline, and win chips from the politics for China's energy enterprises overseas investment and cooperation. China will combine oil imports strategy with the independent and peaceful diplomatic policy organically, pay more attention to the relationship with Russia and Caspian region countries, strengthen the oil pipeline negotiations and cooperation with those countries, increase the oil pipeline transport and land transport proportion in total oil imports, improve the transport capacity of domestic fleet companies, actualize the diversification oil import channels and reduce the potential oil imports risks, and lower the risk of oil imports.

6.4.2 Oil Import Policy

(1) Strengthen oil import risk analysis; implement the diversification of import sources.

When policy makers frame the strategic plan for oil imports, they must fully consider the import source risk index, attach importance to developing cooperative relations with new born oil-exporting countries, cooperate through the market and diplomatic means to evade import risks, increase the number of sources of imports as much as possible, increase the imports and share in the risk relatively lower countries, and decrease imports and share in higher risk areas, so as to gradually actualize the diversification of oil imports. At present, China's oil imports come from about 30 countries or regions such as: the Middle East, West Africa, Southeast Asia, and several other countries, far below that of United States which imports oil from more than 60 countries with a diverse channel. In order to reduce China's energy supply security risks, oil imports should achieve diversification, to gradually reduce the imports and share of the Middle East, and to increase Russia, Caspian region's oil imports, thus lowering marine oil transport proportion to reduce oil imports supply security risks.

(2) Adopt market and diplomatic means to evade the imports risk; strengthen regional energy cooperation.

With the new pattern of international energy gradually being formed, the Sino-United States, Sino-Japan, Sino-India and other countries' oversea oil resources competition is becoming more intense. At the same time, when summing up the experience and lessons, we should reinforce regional energy cooperation, and use the neighboring countries' crude oil processing capacity, so that the two countries share crude oil imports risks, achieving a mutually beneficial win-win.

(3) China's crude oil and refined oil exports should be unified to prevent excessive energy flowing out.

In the market economy environment, it is understandable for oil production enterprises to pursue high profits by increasing oil exports. However, China's three major oil companies are all state-owned enterprises; the government has the right to unify management of their oil exports. So China should establish a scientific and reasonable mechanism for the oil imports and exports, avoiding the abnormal phenomena that the oil exports sustain increase when domestic oil market supply has a shortage, and ensure the nation's oil supply security.

(4) Make subsidy policies and measures for petrochemical enterprises to protect the supply of refined oil.

As China's existing petroleum products price has not yet fully converged with the international refined oil price, consequently, there appears the phenomenon that purchase prices and sales prices are inverse in domestic oil market. To change petrochemical enterprises' production losses in the period of high oil prices, the government should make corresponding subsidy policies and measures, such as the profits of upstream oil production departments to bear the processing loss of downstream enterprises. Meanwhile, the government should also mandate the downstream petrochemical enterprises not to reduce production, or limit that production, when the oil price is at the high level, so as to protect the nation's refined oil supply security.

6.4.3 Strategic Petroleum Reserve Policy

(1) Accelerate the pace of national strategic petroleum reserves system, and steadily implement oil futures.

The SPR is one of the most effective measures that can ensure the oil-importing countries' energy supply security. In 2004, the Chinese government formally adopted the resolution of establishing a national strategic petroleum reserves project. The first period project includes four reserve bases, in which the largest base is Zhenhai base, which is expected to be completed by the end of 2006. The rest were completed in 2008. China not only has no available strategic petroleum reserves, but also buys oil in cash rather than in futures. So current priority should be accelerating the pace of oil futures, and using market means to evade the oil price risk. In addition, a perfect national strategic petroleum reserves system also includes relevant laws, regulations and policies. China should combine its own national conditions and characteristics, actively take reference from foreign experience of

SPR, such as Germany's funds collected manner of SPR and Japan's management style, gradually improve China's strategic petroleum reserves system and establish oil and gas resources strategic reserves on marine and land-based, implement the preferential policies for SPR, and increase commercial strategic reserves, consequently enhancing the size of national strategic petroleum reserves.

(2) Establish the warning mechanism for oil supply security, and prevent people's psychological panic about oil crisis.

China should start monitoring of petroleum products market, the warning of oil supply, emergency measures, and so on, so as to form a perfect oil supply security and warning system (such as if the oil supply fell below the 80% of plan amount, the warning system will be activated, if less than 70%, take urgent measures and contingency plans etc.). We should control the crisis of oil supply in the budding stage, minimize the impact on economy that is brought by oil prices fluctuation, and prevent people's psychological panic about oil crisis.

6.4.4 Conservation and Renewable Energy Policy

(1) Encourage the development and application of energy conservation technology, and make full use of renewable energy.

There is a vast potential for China's energy conservation, for China's energy intensity is one to two times higher than the developed countries' average level. The government should pay attention to frame specific plans and clarify the energy consumption standards, goals and policy measures for various sectors, especially focusing on the key sectors' energy conservation work. At the same time, the government should encourage the application and generalization of energy conservation technology, and develop energy conservation and environmental protection cars, and energy conservation houses and public buildings. Meanwhile China needs to optimize the terminal energy structure continually, and establish and perfect the incentive mechanisms from taxes, prices and sectors' policies. China also should make full use of hydropower, wind power, biomass energy, solar energy and other renewable energy resources, and promote the development of renewable energy resources.

(2) Increase R&D investment of key technology in upstream oil and gas industry, in order to accelerate the pace of technological evolution.

International experience shows that the commercial investment in the key technology of upstream exploration, development in oil and gas industries, is the key point that increases oil and gas reserves and outputs. Therefore, China should strengthen investment in research and development of key technology in upstream oil and gas industry, and improve the level of oil and gas exploration and exploitation technologies. In order to provide strong technical support and improve the technological level of oil and gas enterprises, we should reduce the costs of energy exploration, drilling, production and processing, and improve the comprehensive utilization of energy efficiency.

(3) Vigorously develop clean coal technology, improve the efficiency of coal use, and achieve the economic growth mode change.

Clean coal technology is the bridge to future energy. China's coal resources are very rich, and China's coal consumption accounted for about 65.0% of the total energy consumption. Therefore, clean coal technology in China has a broad prospect. China should take the energy transformed road where electrification is the core, turn coal into clean electricity, cleaner coal gas, etc. China should optimize terminal energy consumption structure, improve the efficiency of coal use, and ensure the national energy security.

Due to the pressure of environmental protection and greenhouse gas emissions, China should gradually optimize its coal-based energy consumption structure; develop natural gas and renewable energy, and achieve the economic growth mode change. At the same time, China should optimize the energy economic structure, reduce energy intensity, implement the preferential policies that are beneficial to the development of renewable energy, and promote renewable energy development and utilization.

6.4.5 Off-Shore Oil Policy

(1) Protect the maritime energy resources, and open new oil transportation routes.

When U.S. is trying to control the world's oil resources as much as possible, other countries are also doing the same for their own future energy security, economic development and national security. So the international energy new order is gradually formed. Therefore, when in disputes involving energy resources, China's government should not only guarantee the national offshore oil and gas resources from being violated, but also be actively involved in overseas oil resources development and cooperation. China should open new oil transport routes (such as the China-Burma oil routes, etc.) to reduce the oil imports amount through the Malacca Strait. At the same time, China should strengthen energy cooperation with the surrounding regions and the Caspian oil exporting countries (Russia, Kazakhstan, etc.), sign long-term energy trading contracts, and lay oil and gas transportation pipelines, to reduce the risk to oil supply.

(2) Enhance exploration technology, develop naval power, and protect the marine oil and gas resources.

Since the 1960s, China has exploited the maritime oil and gas resources. The success ratio of marine oil and gas exploration and recovery ratio of oil and gas have been improved continuously because of scientific research and technical innovation. So China's oil and gas outputs are rising year by year. According to the statistics, in 2005, the new discovery of oil geological reserves about 200 million cubic meters, and 1.5 billion cubic meters of natural gas in Bohai Bay. So Bohai Bay is becoming China's third largest oil-gas field. Besides the Bohai Bay, China's maritime oil production bases also include the west of China South Sea, east of China South Sea and China East Sea three main oil-gas fields. By the end of 2004, China's maritime proven recoverable oil reserves was more than two billion barrels of oil equivalent. Therefore, enhancing the maritime oil and gas resources exploration and exploitation technology is important to keep oil and gas production increasing stably. It has very important strategic and practical significance. Meanwhile China should strengthen the Navy, protect China's oil imports shipping routes, safety and the security of the marine oil resources.

6.5 Summary

We introduce the definition and connotation of energy security, and analyze carefully the contemporary status and hidden troubles of China's energy security based on China's energy resource reserve and demand. With the rapid growth of oil import dependence, China began to establish SPR, but it will take huge capital, so the studies of optimal SPR level become one of the most debatable questions. We establish an optimal decision model based on the decision tree method, and quantify the optimal SPR level to meet China's economy development in 2005, 2010 and 2020. Our results show that more China's SPR is not better if we consider that oil fluctuation will cause economic loss. When oil price is in lower level, the SPR of 60–90 days' oil import is optimal for China's economic growth and energy security, which accords with the SPR level required by IEA. When oil price is in higher level, the SPR of 30–60 days' oil import is optimal. Therefore, international oil price will directly have an impact on the SPR level.

As one of main oil importers, there are many impact factors affecting oil import. Based on the integrated method of quantitative and qualitative methods, we systematically analyze the diversification index and risk index of China's oil imports. The results show that it will not decrease effectively the diversification index and risk index of China's oil imports if we blindly increase the number of oil import sources. The facts of U.S. and European Union suggest that the optimal solution to realize diversification and reduce oil import risk is to decrease the oil import gap among different import sources, i.e., decrease the import from regions whose import risk is relatively high, and increase the import and share from regions whose import risk is rather low. China should also adopt these measures to decrease oil import risk.

Based on the above analysis and discussions, we present some policy implications:

(1) Energy diplomacy policies. We should implement the "international cooperation" energy strategies, actualize the diversification of energy diplomacy, push the integration diplomacy of politics-energy, and take an advantageous position in the new energy situation.

(2) Oil import policy. We should strengthen oil import risk analysis; implement the diversification of the imports sources; evade the imports risk through the market and diplomatic means; strengthen regional energy cooperation, manage uniformly China's crude oil and refined oil exports; prevent excessive energy export; make subsidy policies and measures of petrochemical enterprises, ensure the supply of refined oil.

(3) Strategic petroleum reserve policy. We should accelerate the pace of national strategic petroleum reserves system, steadily implement oil futures, establish the warning mechanism for oil supply security, and prevent people's psychological panic about oil crisis.

(4) Conservation and renewable energy policy. We should encourage the development and application of energy conservation technologies; make full use of renewable energy; increase R&D investment of key technology in upstream oil and gas industry; accelerate the pace of technological evolution; develop vigorously clean coal technology; improve the efficiency of coal use and achieve the economic growth mode change.

(5) Off-shore oil policy. We should protect the maritime energy resources; explore new oil transportation routes; enhance exploration technology and naval power to protect the marine oil and gas resources.

CHAPTER 7

Energy Technology and Its Policy

In the 21st century, energy systems, on the one hand, will satisfy the increasing energy demand to fuel economic growth, and on the other hand will be designed to protect the environment. In the development of the economy, energy and environmental system, energy technology plays a critical role. Policy makers should be aware of the driving forces, source, paces and direction of energy technological progress. Through studies of the world's energy technology system and China's energy policy, this chapter tries to answer the following questions:

- What phases are experienced by the world's energy technological advances? In each phase, what are the driving forces?
- What are the dynamics of international energy R&D expenditures? What trends can be traced for the technology portfolios and distribution among countries?
- What experiences could be referenced to develop China's energy technology policy?
- What's the roadmap for China's energy technology substitutions?
- What about China's renewable energy policies?

7.1 Paradigm Transitions of Energy Technological Change

Technological change is not only an engineering phenomenon, but also a complex social process involving technical, economic, social and institutional factors in a mesh of interactions (Perez, 2004). There are at least three major traditions of research on the impact of change in the economic environment and on the rate and direction of technological change (Ruttan, 1996): The demand-pull tradition emphasizes the relative importance of market demand on inducing technological advances; the theory of induced innovation focuses on impacts of the relative factor endowments and prices on technological innovation; and the third tradition is to understand the stability in factor shares in spite of the very large substitution of capital for labor.

In 1960s and 1970s, the debate between demand-pull and technology-push was hot, which ended by the criticism on demand pull theory from Mowery and Rosenberg (1979) and Dosi (1982). Although demand pull theory got criticism that the concept of demand used in many of empirical studies in 1960s and 1970s has been so broad or imprecise as to embrace virtually all possible determinants, Rosenberg (1974) insists that market demand plays a central role in technical change. The beginning of oil industry clearly confirms the importance of market demand (lamp fuel). Dosi (1982) argues that extreme forms of technology-push approaches, allowing for a one-way causal determination (from science to technology to the economy) fail to take into account the intuitive importance of economic factors in shaping the direction of technical change.

The concept of technological paradigms and trajectories, capturing some common features of the procedures and direction of technical change, were introduced by Giovanni Dosi in 1982. Under this framework, it is a comprehensive and systematic analysis to investigate the pace and direction of technological change at micro-technological level. Later on, diverse streams of research focused on technological change at the levels of sectors, regions, and nations. Freeman and Perez (1988) explored at an even higher level synthetically defined *techno-economic paradigms* to include broad elements into the analysis.

It is clear now that both the supply and demand factors play an important role in technological advances, and their relative impacts on technological advances vary in different times and in different industries. In the history of a specific industry, different factors that drive technological progress also vary in time. Based on the differences of factors that mainly affect technological advances in energy industry, three phases of the history of energy technological change are identified as shown in Table 7-1.

Table 7-1 Techno-economic paradigms in energy

Phase	Paradigm	Driving Factor
Phase I (Before 1859)	Natural transitions	Sci. & tech.
Phase II (1859–1992)	Hydrocarbon lock-in and induced innovation	Market forces
Phase III (1992–?)	Toward clean energy and sustainable 3E system	Public policy

Borrowing the concepts of paradigm-based theory, we identify these three paradigms in energy technological change, trying to make this kind of identification coincide with the development of energy economics and main determinants of energy technological change.

7.1.1 Phase I: Natural Transitions (before 1859)

In the ancient time before agricultural civilization, the most important material to the human is food, in some sense the digestible energy. Men hunted animals while women gathered wild fruits for food. The use of fire is a significant invention that greatly promoted human's life quality during this period. And this lifestyle proceeded for quite a long time. As the population increased rapidly, the hunter-gatherer lifestyle no longer provided adequate food and was not sustainable, because the Malthusian Limit was reached. At this time, technical progress was needed to transmit the development orbit to another. Only after humans began domesticating plants and animals, and harnessing animals (e.g. cattle and horses) for energy use, the long agricultural civilization period began. During this civilization period, humans exploited and utilized several energy sources, such as wind, coal, and hydro (for example, the earliest windmills were applied to work in Persia around 200 BC), with biomass as the primary energy source. Humans got through the long period of agricultural civilization.

Coal is the most abundant fossil energy source[①] and it has the longest history which has been used since the cave man. Coal has been an industrial commodity since the 5th century when Romans occupied Britain and began burning "the best stone in Britain". But it was the Industrial Revolution that played a major role in expanding the use of coal significantly after the invention of Watt's steam engine that made it possible for machines to do work. The radical technological end-use innovation, the steam engine powered by coal, initiated a grand transition of primary energy source from biomass to coal. More and more sectors, such as transportation, iron and steel, and electricity generation, were set up and were beginning to use coal. A transition from biomass to coal, which is also the transition from agricultural civilization to industrial civilization, occurred. Gradually, coal dominated the world's energy consumption and became the primary energy source. It was quite a long period before the great *Industrial Revolution* that took place in the 18th century, which brought human beings into the era of industrial civilization and changed the world and people's lifestyle dramatically.

Advances in science and technology played a critical role in the transitions during this phase, and fueled the transitions. Not only technologies like fire use, invention of wheel and steam engine, but also the development of natural science such as Newton's three laws, Ben Franklin's understanding of the nature of electricity and Maxwell's equations, paved the way

---

① World estimated recoverable coal reserve is 1,081,279 million short tons (2004), widely distributed around the globe. Source: EIA, *International Energy Annual*, 2002.

7.1 Paradigm Transitions of Energy Technological Change

for the future of energy technology. Fig. 7-1 depicts the significant scientific and technology events in energy history.

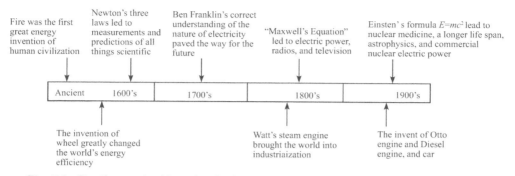

Fig. 7-1 Significant scientific and technology events in energy history. Based on EIA(2005)

Fig. 7-2 shows the shares of world primary energy supply from 1850 to 2000. The substitution of one primary energy source for another clearly tells the outline of energy technology dynamics. It is clear to see from Fig. 7-2 that before oil became prevailing, the global energy system was simple with competition and substitution of just two main energy technologies–biomass and coal. Although energy system as a whole is characterized by transitions from one dominating energy source to another: the development of each primary energy source follows a process of S-shaped curve and then was substituted by another (Grubler, et al., 1999), energy systems were getting more complex at the turn of 20th century and more factors affected the procedure of technological change. The paradigm of energy technology (more precisely, the main driving forces behind energy technological progress) changed greatly since the beginning of oil.

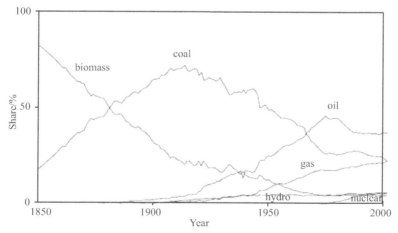

Fig. 7-2 Shares of world primary energy supply, 1850–2000
(Source: Sanden, Azar, 2005)

7.1.2 Phase II: Hydrocarbon Lock-in and Induced Innovation (1859–1992)

Modern oil industry began its history in the efforts to find the raw material for lamp fuel to substitute for increasingly expensive whale oil in the United States. In 1859, the first oil well drilled by Edwin L. Drake ended the whaling industry. Kerosene was accepted by the lamp fuel market and became the main product of crude oil. But the early period of oil industry was in a state of disorder and anarchy, the price of oil fluctuated wildly. It soared or fell quickly, depending on the market being flooded with an abundant supply or starved by a strong demand. In 1859 for example, the crude oil barrel was sold for 20US$ and two years later only for 52 cents. This was ended by 1900 under the domination of John D Rockefeller's Standard Oil Trust[②] and the oil industry had achieved a measure of stability (Piece, 1996).

The technological change of refining was quite crucial to the oil industry at the beginning, but in the early period, oil refining was a simple process by evaporation and had very low entry barrier. The beginning of 20th century saw a series of far-reaching changes on the verge which had greatly reshaped the oil industry. The dramatic changes in market pulled the technological changes in oil during this period. Although oil industry was founded to provide lamp oil, electricity sharply changed the lighting market structure: kerosene lamps lost the market to bulbs rapidly. As the first car was invented by Ford, the automobile was becoming a significant market for gasoline which shifted the oil industry from kerosene supply for lighting market to gasoline supply for the automobile industry, and the batch-type thermal cracking process in 1913 raised the distillation yield considerably. The refining technological changes were initiated by the development of automobile industry. Another major change was the development of catalytic cracking (first operated in 1936) which resulted in a substantial improvement in the quality of the gasoline. By 1942, the continuous catalytic cracking process had been commercialized and dominated the entire refining industry (Martin, 1996; Piece, 1996).

With the use of oil diffused increasingly in almost all the industries, by the 1960s the world consumption of oil exceeded coal, and oil became the new primary energy source. In the years prior to 1973, the demand for energy expanded exponentially. From 1950 to 1973, the U.S. aggregate energy use grew by 3.5% per year, roughly matching real GNP growth (3.7%), with demand shifting toward petroleum (4.3% annual growth) and electricity (7.7%) and away from coal (1.0%) (Sweeney, 1984). From nearly 80% of total energy consumption in 1929, the share of coal declined to slightly over 55% in 1955 and to 31% in 1973 (Lin, 1984). These trends, reflected throughout the world, were encouraged by flat or gradually declining energy prices.

Although being facilitated by a steep decline in the price of oil relative to coal, the replacement of coal by oil and gas occurred mainly in public mass transportation, power generation, and other industrial uses, and was based on technological and institutional factors. It took a century for oil to grow from its market share of 1% of world primary energy supply in the 1870s to the peak at 46%, during which period oil has formed a techno-institutional complex and created hydrocarbon lock-in (Unruh, 2000).

Because of the disparity of oil resource distribution, the world oil supply and then energy supply is influenced greatly by political reasons especially the unstable politics in Middle East region. Because this region's oil production in 1973 accounted for more than 50% of world total oil production, and its proven oil reserves is greater than that of total of the other regions (685.6 thousand million barrels, around two-thirds of world total at the end of 2002), the outside powers found their oil interests and their oil ambitions in the Middle East, thus helping a process which has made the region a center of international tension over long periods of time (Odell, 1968). The 1973 Yom Kippur War and 1979 Iran Revolution (followed by the Iraq and Iran war) caused two oil crises which greatly changed the energy

---

② John D Rockefeller's Standard Oil Trust is one of the most famous industrial organizations ever. The Trust, working to a vertical concentration to the whole industry, controlled much of the production, transport, refining, and marketing of petroleum products in U.S. In 1911, Standard Oil was declared a monopoly and broken up.

7.1 Paradigm Transitions of Energy Technological Change

sector and even the whole world economy (Goldstein et al., 1997). The crude oil prices[③] surged up dramatically: from 13.68 (2003 US$ per barrel) in 1973 they jumped to 43.38 in 1974 and then peaked at 80 in 1980. And later in 1990 the Iraqi invasion of Kuwait caused a short-term and slight oil price jump. Fig. 7-3 shows the oil prices fluctuation from 1860 to 2003, and the effects of "oil shocks" caused by oil supply interruptions in Middle East. The high prices fluctuation in the early period of oil was because the production varied significantly depending on the richness of oil wells.

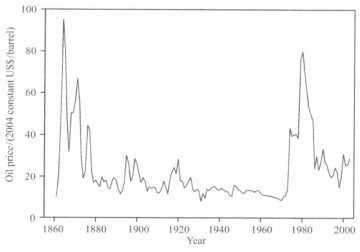

Fig. 7-3 Real oil prices, 1860–2003
(Data source: BP, 2005)

The oil crises of the 1970s had significant and pervasive impacts on energy market and government energy policy. High oil prices forced the firms to seek alternative energy resources and urged energy conservation. Aggregate statistics show there have been significant reductions in energy use, referred to as energy conservation, and significant shifts from one energy carrier to another, referred to as inter-fuel substitution (Sweeney, 1984). During this phase, the energy technological changes were mainly induced by market force—the high oil prices. The theory of induced innovation, introduced by Hicks (1932), states that changes in relative factor prices should lead to innovations that reduce the need for the relatively expensive factor. These changes directly led to two major waves in energy technological advances (Criqui et al., 2000):

(1) Improvement of energy conservation and energy use efficiency.

The world economy as a whole responded to the energy crisis by adjusting per capita energy consumption (Goldstein et al., 1997), at the same time, on the demand side, great efforts were carried out to find ways to save energy and promote energy use efficiency.

Fig. 7-4 depicts the trends of world energy supply and energy consumption from 1973 to 2001. On the demand side, the 1973 and 1979 energy crises both led to reductions in per capita energy consumption immediately. On the supply side, efforts were given to seek alternative technologies for energy supply and energy saving to reduce the need for the relatively expensive oil. Countries allocated much more resources to carry out research, for example, the IEA countries' public energy R&D expenditures increased dramatically when the oil prices surged up caused by the two oil crises, see Fig. 7-5. World energy

---

[③] Arabian Light posted at Ras Tanura. Source: *BP Statistical Review of World Energy*, 2004.

intensity (energy consumption per GDP) began to decline dramatically since late 1970s, which suggests improvements of energy conservation and energy use efficiency after the oil shocks.

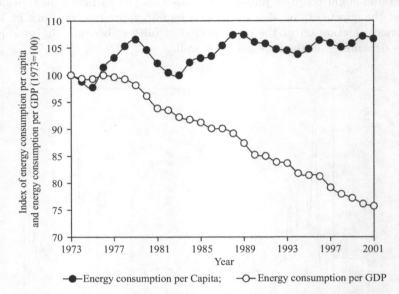

Fig. 7-4 Trends of world energy supply and energy consumption, 1973–2001

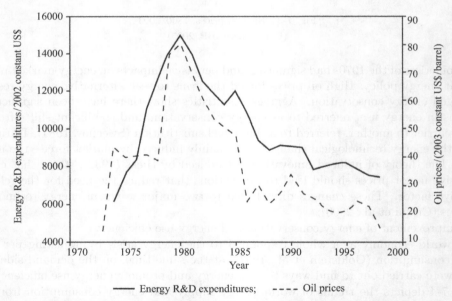

Fig. 7-5 Public energy R&D expenditures of IEA countries vs. oil prices

(2) Diversification of energy technologies.

Diversification of energy technologies is also an adjustment from the supply side. Energy firms already began to diversify their technologies in the 1960s. The big electro-mechanical engineering firms (such as General Electric, Westinghouse and Siemens) turned to develop technologies for thermal electricity generation toward light-water reactors, meanwhile the oil corporations began to invest in other technological directions like coal liquefaction, nuclear power, photovoltaic and other renewable technologies. This is because of their pessimistic opinions of the oil future (Martin, 1996). But it was the oil crises in the 1970s that forced the governments to re-orient their energy (technology) policies concerning security reasons and long-term goals. Countries, especially with high reliance on imported oil and high sensitivity to security matters like Japan, commenced a long-term program of substituting oil with nuclear, coal, and new energy technologies which aimed to reduce the dependence on imported oil (Lesbirel, 1988). As an important precursor of technological innovation, the R&D expenditures change in advance, governments allocated much more resources to technological innovation in energy as the oil prices surged up, and the timing of expenditures adjustments on energy R&D coincided with energy price changes in the highly volatile price of oil. Public energy R&D expenditures have shown a rising trend of technological variety from early 1980s (Zhang et al., 2006).

During this phase, hydrocarbon lock-in and induced innovation are the symbols of energy technological change. Market demand played a critical role in the beginning of oil industry (detailed above), and gradually oil formed hydrocarbon lock-in. Induced innovation of energy was caused by relative expensive factor (oil prices) in global energy market. So, market forces are the key driving factor of technological progress in this period.

7.1.3 Phase III: Transition toward Clean and Sustainable Energy System (1992–)

Energy is a precondition not only for economic development, but also essentially for social well being and decisive for environmental performance (EU, 2002). In view of the deep and broad relationships that energy has with society, economies and environment, sustainable development calls for an integration of these systems. But sometimes, the incompatible economic and environmental goals make it difficult to determine the best tradeoff between economic development and environment protection. For example, serious economic limitations and insurmountable technological challenges may definitely appear under strict legislation standards. On the contrary, it is obvious that a sharply growing economy consumes more energy and produces more pollutants.

Environmental concerns with the local and regional problems of urban smog and acid rain already started. These concerns led to some governmental regulations. Before the United States. Congress passed the Air Pollution Control Act of 1955, many states of United States. and local governments had passed legislation dealing with air pollution. In 1963, the *Clean Air Act* in the United States., which was gradually amended from then on, regulated SO_2 and NO_x emissions and others. During the decade of the 1980s, in part because Reagan's administration placed economic goals ahead of environmental goals, the legislation was not amended and its performance was not satisfying.

The human induced environmental degradation, such as global climate change, has gained great attention in the globe since the turn of 1990s. And environmental concerns have moved from mainly local or national issues to international or global issues. In 1988, recognizing the problem of potential global climate change, the World Meteorological Organization (WMO) and the United Nations Environment Program (UNEP) established the Intergovernmental Panel on Climate Change (IPCC), aimed at assessing the scientific, technical and socio-economic information relevant to understanding the risk of human-induced climate change. In 1992, the United Nations Conference on Environment and Development was held in Rio de Janeiro, Brazil—a conference popularly known as the "Rio Earth Summit". Most countries joined an international treaty—the United Nations Framework Convention on Climate Change (UNFCCC)—to begin to consider what can be done to reduce global warming and to cope with inevitable temperature increases. The UNFCCC calls for "stabilization of greenhouse-gas concentrations in the atmosphere at a level that would prevent dangerous anthropogenic interference with the climate system." In 1997, governments agreed to an addition to the treaty, called the *Kyoto Protocol*[④], which has more powerful (and legally binding) measures.

Although the concept of environmental protection and sustainable development was not new, it was after the 1992 UNFCCC convention that most countries began to (or intended to) transfer to the strategy of clean energy technologies and sustainable 3E system. The transition to clean energy technologies and sustainable 3E system has just begun and may have decades to go. Industrialized economies have become locked-in to fossil fuel based energy and transportation systems through path dependent processes driven by technological and institutional increasing returns to scale (Unruh, 2000).

In order to obtain such an ambitious goal of sustainable 3E system, governments have a critical role to play, often in co-operation with the private sector, in stimulating energy technological innovation (energy technology research, development and demonstration, RD&D) and removing barriers to rapid and wide deployment of clean and efficient energy technology (IEA, 2000). It is definitely clear that this transition will not occur with only market selection and physical technological progress of new energy technologies due to the effects of market failure, lock-in and path dependence. Government intervention is necessary to guarantee the direction of energy technological change and prevent the energy system from market failure. Governments adopted supply-push policy (e.g., technology research, development and demonstration) to create and stimulate niche markets for new technology and offer an opportunity for technology learning. On the other hand, demand-pull method (e.g., market creation) was applied to guide the direction of technological advance. Various supporting policies have been implemented to promote the use of renewable energy, such as Non-Fossil Obligation (NFFO) in the UK (Mitchell, 1995), feed-in tariffs scheme in some European countries (Meyer, 2003) and green power exchange and green pricing in U.S. and others (Ackermann, et al., 2001).

In this period, traditional energy technologies had no significant improvement although they still dominate the world energy system, while the major characteristic may be energy technology innovation, diffusion and commercialization in renewable energy technology and clean technology towards sustainable 3E system being driven mainly by governmental policies.

7.2 Oil Shocks and Energy R&D Expenditures Reaction Patterns

Technological advances in energy are the key to the capacity of the world to address energy-related challenges, such as a reliable and affordable energy supply, and the great risks of global warming. Expenditure on research and development (R&D) has long been positively associated with the pace and quality of technological innovation in many sectors of human endeavor. In the field of energy, it's a widely held view that spending on R&D is an important precursor to the technological advances required to secure sufficient, safe and environmentally acceptable energy supplies, to use them more efficiently (WEC, 2001).

Although R&D plays an important role in technological advances, unfortunately it has been widely observed and recognized that expenditures on energy R&D had been declining world-wide from the early 1980s (Criqui et al., 2000; Dooley, 1998; Dooley and Runci, 1999; Dooley et al., 1998; Hoffert and al, 1998; Margolis and Kammen, 1999a, b; Sagar and Holdren, 2002). The sharp cut in energy R&D expenditure has caused much concern. Many scholars have investigated the energy R&D efforts, especially the energy R&D expenditures.

---

④ The Protocol's major feature is that it has mandatory targets on greenhouse gas emissions for the world's leading economies which have accepted it. The agreement also offers *flexibility* in how countries may meet their targets.

7.2 Oil Shocks and Energy R&D Expenditures Reaction Patterns

Margolis and Kammen (1999a, 1999b), using the U.S. patents held by government agencies and governmental energy R&D expenditures, find that ① energy technology funding levels have declined significantly during the past two decades throughout the industrial world; ② The United States. R&D spending and patents, both overall and in the energy sector, have been highly correlated; and ③ the R&D intensity of the United States. energy sector is extremely low when compared to other sectors. Popp (2002) uses the United States. patent data held by private firms from 1970 to 1994 to estimate the effect of energy prices on energy-efficient innovations and finds strongly positive effects of energy prices on innovations.

Margolis and Kammen (1999a, 1999b), among others, argue that the cutbacks in energy R&D are likely to reduce the capacity of energy to innovate, and the trends of declining funding levels are particularly troubling given the need for increased international capacity to respond to emerging risks such as global climate change. The declining trend of investments in energy R&D is very clear, and it is partly driven by changes occurring as a result of the deregulation of many advanced industrialized nations' energy sectors (Dooley, 1998). Dooley further concludes that long-term energy R&D, and in particular cleaner, environmentally-preferred advanced energy supply R&D, is unlikely to be supported by individual utilities in a competitive, deregulated utility market. Some others (Blumstein and Wiel, 1998; PCAST, 1999) discuss the relationship of public and private investments in energy R&D and the changing private energy R&D paradigm under the deregulation and restructuring in energy sector and the global new business environment.

In the context of complex policy arrangement, deregulation and restructure in energy sector and the global new business environment, there raises great interest in estimating how much to spend on energy R&D is optimal and on which technologies. In order to quantify the benefits provided by continued federal renewable electric R&D in the United States., Davis and Owens (2003) use "real option" pricing techniques to estimate the value of renewable electric technologies in the face of uncertain fossil fuel prices and then conclude that the level of renewable electric R&D funding is sub-optimal. Miketa and Schrattenholzer (2004) analyze the optimal R&D support for an energy technology, using a stylized optimization model of the global electricity supply system, in which cost reductions of electricity supply technologies are assumed resulting from the accumulation of capacity and R&D.

In fact, due to different energy resource gifts and different levels of economic development, the dynamics and portfolios of public (governmental) energy R&D expenditures varied from country to country and varied also from technology to technology. This section aims to shed some light on the issue of public energy R&D dynamics and portfolios in the detailed country dimension and technology dimension. As to the dynamics, we are interested in the patterns of public energy R&D funding responding to oil shocks in the 1970s and afterwards, which partly reflect the changes of energy technology policy. On other hand, the allocation of energy R&D input reflects the priorities of energy technology fields. Two dimensions are taken into account, the technology dimension and the country dimension, in this chapter: ① in the technology dimension, the dynamic pattern of energy R&D that resources allocation among different technologies is focused on, and ② in the country dimension, we are interested in the question: How different are the countries in terms of priority of energy R&D funding?

The IEA survey of government energy technology R&D expenditures is the only consistent international data set which includes public investment in energy R&D of IEA countries from 1974–2001 (IEA, 2003). Approximately 96% of the industrialized world's public sector energy R&D (in 1995) is carried out in only nine countries: Canada, France, Germany, Italy, Japan, the Netherlands, Switzerland, the United Kingdom, and the United States (Dooley, 1998; IEA, 2003). These nine countries' aggregate energy R&D expenditures are great enough, so, they are chosen as representatives in this analysis.

7.2.1 Adjustment from Demand–Side and Supply–Side to Oil Shocks

To understand why oil crises happened and why they have that strong effect on world energy and economic systems, the first step is to review the world energy structure and oil prices before and after oil crises.

The world consumption of commercial energy from 1955 to 1973 increased by 143%, comparing with an 80% increase in the preceding 26 years, 1929–1955 (Lin, 1984). The accelerated growth of energy consumption occurred in conjunction with both an acceleration in the tempo of economic growth and a decline in the real price of oil and energy. During this period, the replacement of coal by oil and gas occurred in public mass transportation, power generation, and other industrial uses. The share of coal declined from nearly 80% of total energy consumption in 1929 to slightly over 55% in 1955 and to 24.8% in 1973, and the share of oil and gas reached 61.2% (IEA, 2004).

During the late 1960s and 1970s, the world oil supply and then energy supply was influenced greatly by political reasons especially the unstable politics in the Middle East region. Because this region's oil production in 1973 accounted for more than 50% of total world production, and its proven oil reserves are greater than that of the sum of other regions (685.6 thousand million barrels, around two-thirds of the world total at end of 2002). So the outside powers have found their oil interests and their oil ambitions in the Middle East, thus helping a process which has made the region a center of international tension over long periods of time (Odell, 1968). The 1973 Yom Kippur War and 1979 Iran Revolution (followed by the Iraq and Iran war) caused two oil crises which greatly changed the energy sector and even the whole world economy (Goldstein et al., 1997). Fig. 7-6 shows the real oil prices from 1945 to 2003 and the effects of the two oil shocks on the oil prices.

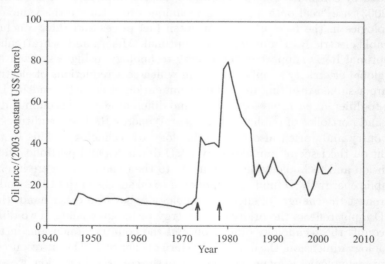

Fig. 7-6   Real crude oil price, 1945–2003

The first and second oil shocks drove oil prices to rise sharply and dramatically, while the Kuwait War in 1990 was just a mini-shock leading the oil price slightly higher for only six months. The two oil crises in 1970s were caused by the high oil share of energy consumption and high dependence on Middle East oil. From then on, governments began to adopt a strategy of diversification not only in energy technologies but also in importing strategy, in order to reduce the risk of dependence on imported oil from the Middle East.

The world economy as a whole responded to the energy crisis by adjusting per capita energy consumption (Goldstein et al., 1997), while the energy technological system responded

7.2 Oil Shocks and Energy R&D Expenditures Reaction Patterns

by adjusting energy technology policies. The theory of induced innovation, introduced by Hicks (1932), states that changes in relative factor prices should lead to innovations that can reduce the need for the relatively expensive factor. As an important precursor of technological innovation, the R&D expenditures change in advance. The timing of expenditures adjustments on energy R&D coincides with energy price changes in the highly volatile prices of oil, as shown in Fig. 7-7. Governments allocated much more resources to technological innovation in energy as the oil prices surged up. While the oil prices decrease of the 1980s again altered energy R&D budgets.

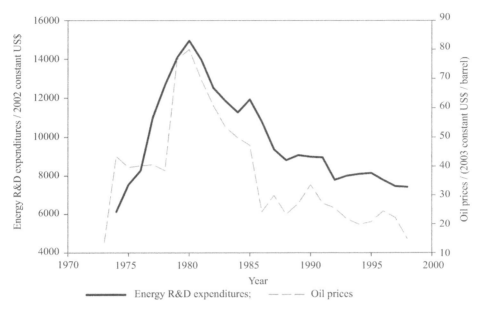

Fig. 7-7 Governmental energy R&D funding, IEA total

Analogous to the cobweb model (see Fig. 7-8) of the theoretical relationship between price and demand, the relationship between oil price and public energy R&D expenditures appears to be similar. When oil price increases, following the 1973 and 1979 oil interruptions, the governments sharply increased the energy R&D funding to seek cheaper and safer energy options, while following the price decreases of the 1980s, the energy R&D funding decreased. In fact, adjusting consumption (demand) is the reaction to price changes from the demand side, while adjusting R&D policy to allocate more resources is to seek affordable energy supply from the supply side.

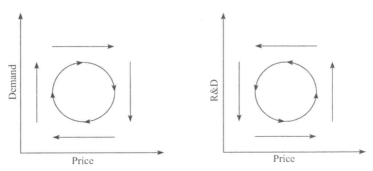

Fig. 7-8 Phase diagram of the cobweb model

Fig. 7-8 shows the cobweb model of the demand and R&D adjustments according to price changes. The system usually follows a circular path. In fact, with time lags in the responses of these variables, or with "shocks" that suddenly change the trajectory, the circular path may be either clockwise or counterclockwise. Beside the circular path converging to the equilibrium center, the system has a possibility of diverging away from it. In the following Figs. 7-9–7-12, the horizontal axis shows the nominal oil price of the Middle East, while the vertical axis shows the nominal public energy R&D funding in national currency respectively. These figures show three types of reaction patterns or trajectories, namely the clockwise trajectory, the counterclockwise trajectory and the diverging trajectory, of the world energy technology system faced with the oil shocks.

7.2.2 Energy R&D Reaction Patterns to Oil Shocks

7.2.2.1 Clockwise Trajectory

The clockwise trajectory implies that the governmental energy R&D expenditures increased sharply immediately after the first oil shock in 1973, while the second oil shock in 1979 just made the expenditure rise a little or even decrease. The United States., Germany and the United Kingdom are the typical countries that follow this clockwise trajectory. Because of the big proportion of the three countries' expenditures in IEA countries (around 50%), the IEA total expenditure also follows the same pattern. The trajectories are shown in Fig. 7-9.

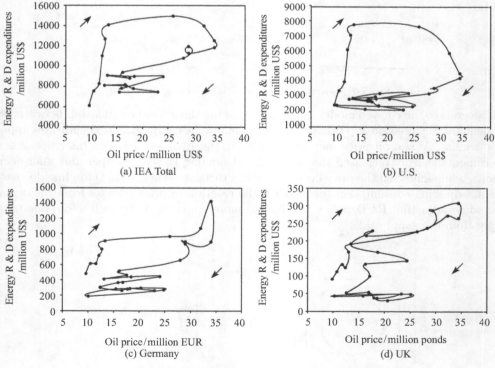

Fig. 7-9 Clockwise trajectories

This pattern says that the first oil shock has greater impact on public energy R&D expenditure than the second one. The reason of why this pattern appears in these three countries lies in their energy structure and economy sensitivity to oil supply. Fig. 7-10 depicts the percentage changes of oil consumption in some IEA countries during the two oil shocks. The percentages are changes of oil consumption in two years after the oil shock compared to oil consumption in the year of oil shock.

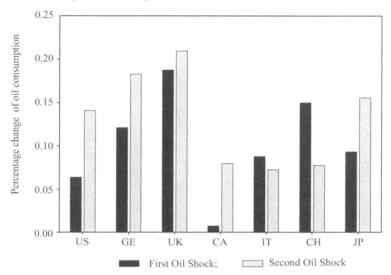

Fig. 7-10 Percentage change of oil consumption during the two oil shocks

The percentage changes of oil consumption in the United States., Germany and the UK are greater than other countries, in other words, their economies responded to the oil shocks by adjusting the oil consumption sharply because they are sensitive to oil supply. Although the percentage reduction of oil consumption in the United States. was lower than that in German and the UK, the United States. absolute oil consumption was so huge which was nearly twice of the sum of Germany and the UK's consumption. So, they responded to the first oil shock very quickly and increased energy R&D budgets significantly.

There is an interesting fact: these three countries' energy R&D budgets follow the clockwise trajectory, which suggests they reacted quickly and allocated more resources on energy R&D, so the second oil shock should shock their economies less. But, in fact when the second oil shock came, their percentage reductions of oil consumption were greater than that during the first shock. To explain this phenomenon, the outcome of their energy R&D activities after the first oil shock should be investigated. Unfortunately, because of the unavailability of related data, we do not discuss this topic here. Anyway, it is reasonable to guess that the outcome of energy R&D activities or their applications are not satisfactory.

7.2.2.2 Counterclockwise Trajectory

Another similar circle, but counterclockwise trajectory appeared in countries like Italy and Canada, shown in Fig. 7-11. This counterclockwise pattern is explained by the significance of the second oil shock in 1979 rather than the 1973 one. These governments reacted when first shock came, but not as strongly as the countries like the United States., Germany and the UK mentioned above. But when they experienced the second, more efforts and fundings were put on energy technology progress.

The percentage reductions of oil consumption in these two countries are not as significant as that in the United States., Germany and the UK (see Fig. 7-11). In Canada, the first oil shock only lowered down its oil consumption by less than 1% which seems like normal fluctuation of oil consumption. But its government energy R&D expenditures increased right after the first oil shock. When the second oil shock came, Canadian oil consumption reduced 8% in 1981 comparing to 1979. The government energy R&D budgets rise again more sharply.

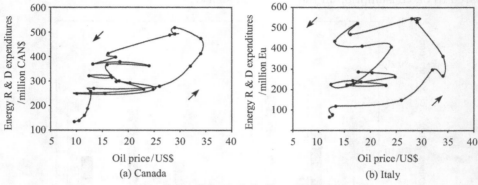

Fig. 7-11   Counterclockwise trajectories

In the Italian case, the government energy R&D expenditures data are not available until 1977, but it does not matter. The energy R&D budget in Italy in 1977 is comparatively small, so no matter how much were the budgets before 1977, the trajectory won't change. The Italian economy reduced oil consumption by around 8% during both the two oil shocks. The energy R&D expenditures was raised quite significantly during the second oil shock and even after the oil prices fell. It seems that the Italian government energy R&D expenditures have more of a time lag than other countries'. If we suppose the lag is three years, then the trajectory of Italy is like the UK's.

It is clear that it is dependence on oil and the reaction time of governments that make the circle trajectory clockwise or counterclockwise. Eventually, after the two oil shocks in the 1970s, public energy R&D funding decreased to a smaller range, and the smaller oil shock caused by the *Kuwait war* made no such great effect as the former two.

7.2.2.3   Diverging Trajectory

But in some countries especially those which are greatly dependent on imported energy and with restricted energy resources, the two oil shocks made the R&D funding diverge. As shown in Fig. 7-12, the cases of Japan and Switzerland illustrate a different pattern: The energy R&D expenditures grew even after the oil price declined in 1980s, which eventually diverged to a higher level.

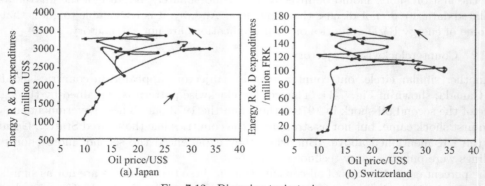

Fig. 7-12   Diverging trajectories

The reason of why this diverging pattern appears in Japan and Switzerland is obvious: they have no oil resources and are highly dependent on imported oil. So, even after the oil shocks, their government energy R&D expenditures rose too.

7.3 Energy R&D Expenditures in Technology Dimension

The above-mentioned patterns of the relations of R&D efforts to oil shocks reflect different energy structure and different levels of dependence on imported oil in the countries. Although the government energy R&D patterns vary from country to country, it is true that the world as a whole cut back the R&D investments in energy. It is very difficult to evaluate the proper investment level considering the requirement of safe energy supply and risks of global climate change. But the decline in energy R&D investment seems a big challenge to a sustainable energy system and a stable economy.

7.3.1 Entropy Statistics

Nevertheless, the sizes of energy R&D budgets do not tell the whole story. Sagar and Holdren (2002) point out that: an undue focus on the sizes of R&D budgets under-emphasizes the character of the energy R&D portfolios in terms of payoff horizons, risk levels, and fuel sources (i.e., allocation of the input).[5] In this sector and the next sector, we apply entropy statistics to measure the variety of public energy R&D investment among technologies and countries, aiming at shedding light on the allocation of governmental energy R&D input.

Entropy statistics could offer a methodology to analyze the varied patterns in frequency distributions, which have been applied in the analysis of distributions of R&D activity, in the measurement of market concentration using market shares and in the variety and competition of technologies (Frenken et al., 2004; Saviotti, 1988). With entropy as a measure of variety (or, on the contrary, concentration), one can characterize the flexibility of portfolios of countries and firms. The higher the technological variety (Jacquemin and Berry, 1979) being explored, the more flexibility a firm, an organization, or a country has to react to uncertain future developments, and the lower the risk level it may have. The applications of entropy as an indicator of variety have been made in the area of market concentration, R&D activities, etc.

Similar to (Frenken et al., 2004), we classify R&D expenditures in a two-dimensional intensity matrix. Thus, the X-dimension represents m technologies ($i = 1, ..., m$) and Y-dimension concerns n countries ($j = 1, ..., n$). Then, the two-dimensional intensity matrix e_{ij} describes the intensity of energy R&D expenditures belonging to energy technology i and country j. According to IEA R&D statistics, energy technologies are classified into 12 1-digit categories: Conservation, Oil and Gas, Coal, Solar, Wind, Ocean, Biomass, Geothermal, Hydro, Nuclear (Fission and Fusion), Power and Storage and Other technologies and research. Thus, we construct a 12×9 R&D intensity matrix, with X-dimension representing the 12 technologies and Y-dimension the 9 main IEA countries. Table 7-2 summarizes the meaning of the entropy values calculated by the formulation.

The marginal intensities of energy R&D funding could be obtained by the following:

$$e_{i\cdot} = \sum_{j=1}^{n} e_{ij} \qquad (7\text{-}1)$$

$$e_{\cdot j} = \sum_{i=1}^{m} e_{ij} \qquad (7\text{-}2)$$

The entropy of distribution of R&D funding among technologies is given by:

---

[5] In Sagar and Holdren (2002), they argued three aspects that are being under-emphasized: (1) the character of the energy R&D portfolio in terms of payoff horizons, risk levels, and fuel sources (i.e., allocation of the input), (2) the effectiveness of energy R&D efforts in terms of technological advances for a given expenditure (i.e., input-output relationships), and (3) the effectiveness of implementation and diffusion of new technologies (i.e., utilization of the output).

Table 7-2   Interpretation of entropy value

	Value (with m technologies and n countries)	
	Low	High
$H(X)$	Few technologies dominate R&D funding (less variety in technologies)	More technologies share a nearly equal funding intensity (more variety in technologies)
$H(Y)$	One single country dominates R&D funding (less variety among countries)	More countries share a nearly equal funding intensity (more variety among countries)
$H(X|Y_j)$	Country j funds only a few technologies (less variety of technologies in country j)	Country j funds technologies with nearly equal share (more variety of technologies in country j)
$H(X|Y)$	Narrow funding portfolio of countries (less average technological variety in all countries)	Wide funding portfolio of countries (more average technological variety in all countries)
$H(Y|X_i)$	R&D funding on technology i dominated by one or few countries (less national variety active in technology i)	More countries are more or less equally involved in funding technology i R&D (less national variety active in technology i)
$H(Y|X)$	Few countries fund technology R&D (less average national variety active in technologies)	More countries fund technology R&D (high average national variety active in technologies)

$$H(X) = -\sum_{i=1}^{m} e_{i.} \ln e_{i.} \qquad (7\text{-}3)$$

The value of $H(X)$ captures the energy R&D funding variety presented in the portfolios of R&D funding among energy technologies in the country as a whole. The portfolio will range between a minimum entropy, with one single technology completely dominating R&D spending, and a maximum one, with all technologies having an equal share in total R&D expenditures:

$$H_{\min}(X) = -\sum_{i=1}^{m} 1.0 \ln 1.0 = 0 \qquad (7\text{-}4)$$

$$H_{\max}(X) = -\sum_{i=1}^{m} \left(\frac{1}{m}\right) \ln \left(\frac{1}{m}\right) = -\left(\frac{m}{m}\right) \ln \left(\frac{1}{m}\right) = \ln m \qquad (7\text{-}5)$$

Similarly, the entropy of energy R&D funding among IEA countries is given by the following Eq. (7-6), which indicates the variety of countries that fund different energy technologies.

$$H(Y) = -\sum_{j=1}^{n} e_{.j} \ln e_{.j} \qquad (7\text{-}6)$$

Conditional entropies can enable one to measure the distribution of energy R&D funded on technologies by country j that captures the variety of technologies of country j [Eq. (7-7)], and the distribution of energy R&D among n countries in technology i which captures the variety of countries active in technology i [Eq. (7-8)].

$$H(X|Y_j) = -\sum_{i=1}^{m} \left(\frac{e_{ij}}{e_{.j}}\right) \ln \left(\frac{e_{ij}}{e_{.j}}\right) \qquad (7\text{-}7)$$

$$H(Y|X_i) = \sum_{j=1}^{n} \left(\frac{e_{ij}}{e_{i.}}\right) \ln \left(\frac{e_{ij}}{e_{i.}}\right) \qquad (7\text{-}8)$$

Average conditional entropy is defined as the weighted average of conditional entropies. The following Eqs. (7-9) and (7-10), respectively, represent the weighted average of conditional entropies of all individual countries, and of all individual technologies.

7.3 Energy R&D Expenditures in Technology Dimension

$$H(X|Y) = \sum_{j=1}^{n} e_{\cdot j} H(X|Y_j) = -\sum_{j=1}^{n} e_{\cdot j} \sum_{i=1}^{m} \left(\frac{e_{ij}}{e_{\cdot j}}\right) \ln \left(\frac{e_{ij}}{e_{\cdot j}}\right) = -\sum_{j=1}^{n} \sum_{i=1}^{m} e_{ij} \ln \left(\frac{e_{ij}}{e_{\cdot j}}\right) \quad (7\text{-}9)$$

$$H(Y|X) = \sum_{i=1}^{m} e_{i\cdot} H(Y|X_j) = -\sum_{i=1}^{m} e_{i\cdot} \sum_{j=1}^{n} \left(\frac{e_{ij}}{e_{i\cdot}}\right) \ln \left(\frac{e_{ij}}{e_{i\cdot}}\right) = -\sum_{i=1}^{m} \sum_{j=1}^{n} e_{ij} \ln \left(\frac{e_{ij}}{e_{i\cdot}}\right) \quad (7\text{-}10)$$

7.3.2 Energy R&D Expenditures in Technology Dimension

The decline of public energy R&D investment from the early 1980s has been widely recognized in the literature. As shown in Fig. 7-13, during the 1970s the total IEA public energy R&D expenditures rose sharply, affected by the sharply rising oil prices during the world oil crisis. When the oil crisis ended, the energy R&D expenditures fell too. Fig. 7-13 shows the technological entropy $H(X)$ at global level and the average conditional technological entropy at country level $H(X|Y)$. Both the technological entropy and the average conditional technological entropy rose till 1981 in the period of the oil crisis, and sharply fell in 1982 and 1983 which may be interpreted by the end of the oil crisis. Then gradually, the entropies rose again from 1983 on. The rising trend of technological entropy $H(X)$, from early 1980s, represents a rising technological variety, in other words, the energy technologies have received more and more equal R&D funding. The rising trend of average conditional entropy $H(X|Y)$ also demonstrates the technological variety at national level.

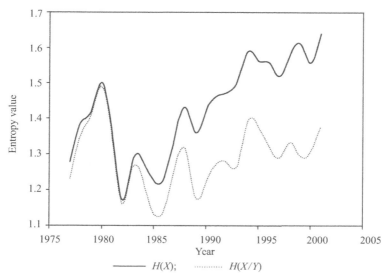

Fig. 7-13 Entropy of energy R&D distribution at global level and national level

But note that 12 energy technologies are considered in this analysis, thus the maximum theoretical technological entropy $H(X) = \ln(12) = 2.485$, the maximum technological entropy in the real series is less than 1.7. This is because different technologies are different in importance to the energy system and economy. For example, fossil fuels and nuclear energy are more important than the renewable technologies thus more budgets were allocated to these technologies, e.g., the expenditure on nuclear R&D is so huge that it possesses more than 50% of total energy R&D investment nearly in all the years concerned. So, the relative trends of entropy value say something while the absolute entropy value makes no sense.

As to different technologies, the conditional entropy ($H(Y|X_i)$) trends of the distribution in countries are shown in Fig. 7-14. More countries invested in the conservation, oil and gas

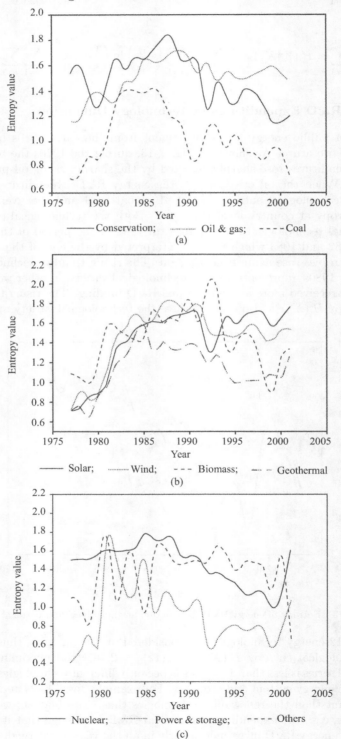

Fig. 7-14 Conditional entropy of country distribution

7.3 Energy R&D Expenditures in Technology Dimension

and coal technologies during the oil shocks in 1970s, but this trend flipped from the middle of the 1980s which means expenditures in these technologies were narrowed to fewer countries (as shown in Fig. 7-14(a)). The similar trends appear in the nuclear, power and storage and the other energy technologies as shown in Fig. 7-14(c). The situation of renewable energy technologies is different. From Fig. 7-14(b), there are obvious sharply increasing trends of interest in renewable energy technologies in more countries. Furthermore, the entropy values do not decline sharply when the oil shocks ended, which suggests a persistent interest of these countries in the renewable energy technologies.

7.3.3 Energy R&D Expenditures in Country Dimension

We now turn to the energy R&D expenditures in the country dimension. Fig. 7-15 depicts the trends of entropy of country distribution in energy R&D expenditures at the global level and the aggregate technology level. The entropy of country distribution in energy R&D $H(Y)$ increased during 1977–1985, and then gradually declined from 1985 on. The declining trend of entropy of country distribution means the distribution of energy R&D investment was gathered into fewer countries. And the average conditional entropy of country distribution at technology level $H(Y|X)$ shared the same trend. These trends tell that fewer countries were more and more dominating in energy R&D investments from 1985 on. Within the countries concerned in this book and even in a broader coverage, the United States and Japan hold a quite large proportion of world energy R&D expenditures.

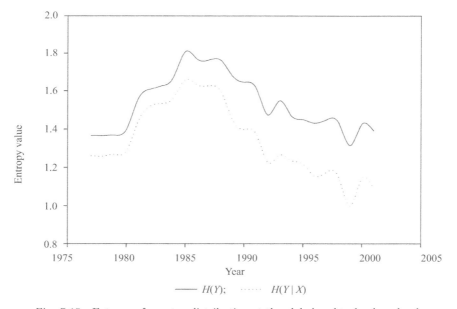

Fig. 7-15 Entropy of country distribution at the global and technology level

Let's turn to more detailed entropy. The following Fig. 7-16 shows the conditional entropy of energy R&D expenditures at some specific country levels. The conditional entropy of technology distribution in Canada, the United States., Japan and France shown in Fig. 7-16(a) does not vary too much during all the years concerned, which suggests that the diversification level of energy R&D funding seems to remain stable. While most of the European countries diversified their energy R&D gradually since the oil shocks, as shown in Fig. 7-16(b).

Comparing Figs. 7-13 and 7-15, on one hand, shows the rising entropy of trend of energy R&D in technology dimension, and the declining entropy of that in the country dimension.

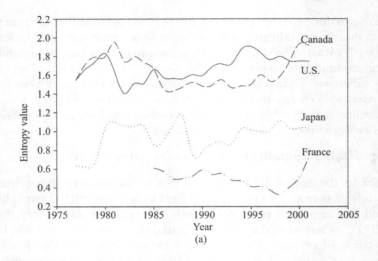

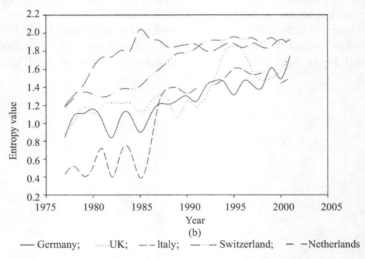

Fig. 7-16 $H(X|Y_j)$ of different countries

The following Fig. 7-17 depicts the structure change of public energy R&D expenditures for nine selected IEA countries in 1980, 1990 and 2000.

The structural change in energy R&D varies among these countries. Some sharp changes occurred in the 1980s, while others in the 1990s, both in technology and country. For example, the nuclear R&D percentage in Italy, Netherlands and Switzerland fell dramatically in the 1980s, while Canada, the United Kingdom and the United States. in the 1990s, and France and Japan even hold a constant percentage of total R&D investment during the whole period. Overall, the structure of public energy R&D changes dramatically and some countries have eliminated their broad categories of energy technology R&D and shifted towards fewer favored technologies. The following lists the priorities in these countries, which is obtained from Fig.7-17 by listing major invested technologies. The following Table 7-3 lists the priorities in these countries.

As shown in Table 7-3, nuclear dominated the energy technology R&D in all these countries in 1980, and fossil fuel possessed the second position in a few countries. In 1990, nuclear still held the largest proportion in most countries, while conservation and renewables exceeded

7.3 Energy R&D Expenditures in Technology Dimension

nuclear in Netherlands and fossil fuel exceeded in United States. Most countries, except France, favor other technologies beside nuclear. For example, Germany favors renewables, Italy favors power and storage, Japan and the United States. favor conservation, and France favors nuclear persistently.

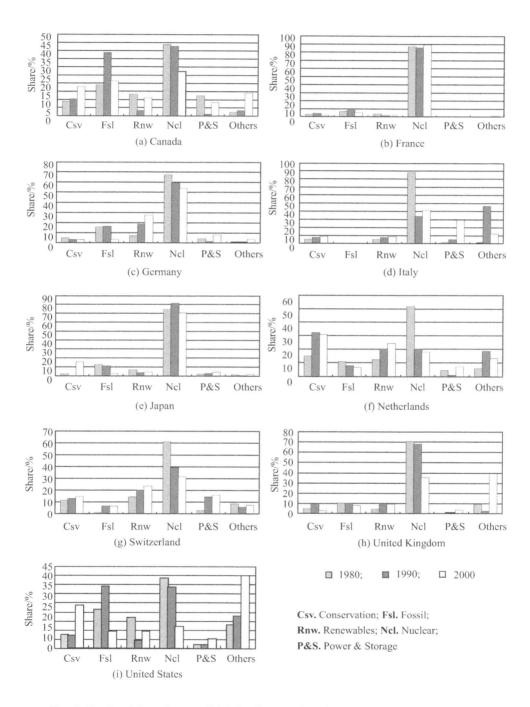

Fig. 7-17 Portfolios of energy R&D funding in selected countries, 1980, 1990 and 2000

Table 7-3 Energy technology priorities

	1980	1990	2000
Canada	Nuclear; Fossil Fuel	Nuclear; Fossil Fuel	Nuclear; Conservation; Fossil Fuel
France	Nuclear	Nuclear	Nuclear
Germany	Nuclear; Fossil Fuel	Nuclear; Renewables	Nuclear; Renewables
Italy	Nuclear	Nuclear; Others	Nuclear; Power & Storage
Japan	Nuclear	Nuclear; Fossil Fuel	Nuclear; Conservation
Netherlands	Nuclear	Conservation; Renewables; Others	Conservation; Renewables;
Switzerland	Nuclear	Nuclear; Renewables	Nuclear; Renewables
United Kingdom	Nuclear	Nuclear	Others; Nuclear
U.S.	Nuclear; Fossil Fuel	Fossil Fuel; Nuclear	Others; Conservation

7.4 The Substitution Routes of Energy in China and the Policy Analysis on Renewable Energy

Under the current situation of energy production and technology level, the dominant energy in China is still conventional energy led by fossil fuels. The economic structure dictates oil and coal together as the primary energies. However, along with the decreasing reserves of fossil energy, as well as the threat of green-house-gases and environmental pollution to sustainable development from combusting of fossil fuel, fossil energy will certainly be substituted for renewable energy.

Under the current situation, because of the dual restrictions from technology and costs of renewable energy, fossil energy will remain dominant in China for a long time. Therefore, facing the scarcity of oil resources, a more feasible solution is turning coal, which is relatively abundant, into something more efficient and clean, to replace oil. Considering the trend of energy technology, it is reasonable to foresee that in the future century, the evolvement of energy in China will be: in the early half century, seeking alternative solutions to oil resource under the fossil energy framework; in the later half, substituting fossil energy for renewable energy with relative technologies (Zhu, 2001).

7.4.1 Alternatives to Fossil Energy

The current foremost body of Chinese energy is fossil energy which is dominated by coal and oil. This situation is decided by the proportion of exploitable-reservation of fossil energy in the entire energy body. The exploitable-reservation of fossil energy accounts for 63.5% of the entire energy. Additionally, although there are large amounts of hydro that technically can be used, the water power stations require a vast amount of investment and have a long investment cycle. The water resource which can be explored accounts for much less than 36.5% economically. This kind of energy structure determines that fossil energy is and will be dominant for a long time.

The scarcity of oil in China is an austere reality. The increasing dependence on imported oil for the economic development is becoming a more and more important factor to national security. *On the other hand*, consider the great consumption of oil all around the world. The increase of oil price becomes an inevitable trend, as well as the intensity of oil supply. Therefore, the substitution for oil becomes necessary. Under the situation that renewable energy cannot offer an exact solution, seeking a more abundant reserve of fossil energy to take the place of oil becomes an imminent need.

Therefore, *the current main problem of fluctuation* is how to realize the substitution of coal, which is an abundant reserve, for oil. The fossil energy mainly includes coal, oil and gas. Along with technology development, another new methane resource–natural gas hydrate–has gradually captured people's attention and become a potential clean fossil energy.

From the viewpoint of different kinds of fossil energy, the substitution of oil can be realized through coal, gas, and natural gas hydrate. Which way to take, however, is decided by the development and utility of technologies.

(1) Natural gas alternatives.

It is a strategy of many countries to substitute oil with relatively abundant reserves of natural gas. Gas to liquid technology (GTL) is the determinant key factor of a successful alternative. Currently, GTL is a hot-topic all around the world and many countries are endeavoring to develop the technology to get qualified liquid fuel. Some countries have already industrialized GTL technology. It is estimated that in the future 5-10 years, industry products such as gasoline, diesel and chemical products made from natural gas or other fuel gas would emerge in the international markets (Xu, 2004). In China, because the scale of natural gas exploration and supply is still small, GTL technology has gained little attention. From the view of an alternative fossil energy strategy, GTL should become one of the developmental focuses of China in the near future.

(2) Natural gas hydrate alternatives.

Natural gas hydrate is a crystal compound made of water and natural gas under certain temperature, pressure, gas saturation, salinity and pH value. It is commonly called "combustible ice" or "flammable ice". The main component of natural gas hydrate is methane, thus it is sometimes called methane hydrate (CH_4-nH_2O). in Russia and Canada, the permanent frozen earth zone in the Arctic area and ocean around the world, large amounts of natural gas hydrate have been found. According to the calculation of Kvenvolden in 1988, the organic carbon in natural gas hydrate counts for 53.3% of all the organic carbon in the world, while that of the sum of coal, oil and natural gas is 26.6%. According to the United States. data, the entire energy in natural gas hydrate in the world is two to three times of that of coal, oil and natural gas (Chen, Fan, 2004). The study and exploration of natural gas hydrate has already been carried out in China, and enormous mineral resources have been detected in the East and South Sea of China. It is estimated that the energy of natural gas hydrate in the South Sea of China alone accounts for half of the oil resource reservation in China. Therefore, the exploration and utility of natural gas hydrate will become an important energy alternative. Now United States., Japan and other developed countries are actively working on the development of natural gas hydrate exploration and utilizing technologies; and all these countries plan to apply the technologies to the market. In China, now there has not enough technology to exploit the natural gas hydrate, but from a total energy development perspective, it should be adopted into the integrated energy strategies and policies.

(3) Coal alternatives.

Now the most well-developed and feasible alternative energy technology in China is coal alternatives. The key determinant of coal alternatives is the advancing of coal processing and utilizing technologies. These technologies include two aspects: liquefaction of coal and clean coal technology. The aim of clean coal technology is to improve the efficiency of coal utility and control the emission of pollutants in a relatively short time. Therefore, the application of clean coal technology depends on two elements: efficient clean power generation technology and colligated processing technology of commercial coal. In developed countries, the main consumption use of coal is generating power, because generation can realize efficient utility of coal, and be helpful to control environmental pollution. So, improving the proportion of coal to generate power is an inevitable choice to optimize the end energy structure and improve the efficiency of coal consumption in China. There are still gaps between current coal generation technology in China and advanced technology worldwide, so developing efficient and clean coal generation technology is an important part of clean coal technology. Beside power generation, there are many other industries and lifestyles that also consume a lot of coal. Therefore, developing integrated processing technology of commercial coal, improving its heat efficiency through the ways of washing, filtration, shaping, smashing and additives, becomes another key point in the clean coal technology. The liquidation of coal will be discussed as a case study in the next section.

7.4.2 Case Study: Liquidation Technology of Coal in China

Based on the analysis above, in the progress of energy alternatives in China, substituting fossil energy with renewable energy is an inevitable trend. Also the development of renewable energy should be a long term strategy emphasis of energy technology in China. However, in the following several decades, the situation in China, even in the world, will be still trapped in fossil energy dominance. Therefore, the alternative fossil energy framework should be the key point of short-to-mid term energy technology strategies and policies. Among the fossil energies, because the relative weak development of natural gas technology in China, the development of natural gas hydrate is consequently difficult to foresee. Therefore the coal alternative technology which mainly stems from coal liquidation, coal gasification and coals clarification, becomes the most economical and practical choice in China. In this section, strategy analysis would be carried out on coal liquidation technology, as well as corresponding policy suggestions.

7.4.2.1 SWOT Analysis on Coal Liquidation Technology in China

(1) Strength of coal liquidation technology in China.

The strength of coal liquidation technology in China mainly lies in the abundant reserves and the vast energy market.

The coal resource in China is rich, and the producing capacities and potentials are both great. These situations make coal the leading energy in China and more important when compared with oil and natural gas; the same situation makes coal a relative stable, reliable and cheap energy resource. According to the calculation of National Mineral and Resource Department, the proven exploitable reserves of coal around China is 2,040 trillion tons. Therefore, the abundant coal resource can provide enough support to the industrialization of coal liquidation technology.

Meanwhile, the stable growth of the Chinese economy creates a great demand for energy in the future. Under the current energy demand structure which stems from the economic situation, the demand for liquid fuels cannot be taken by the solid and gas fuels. The coal liquidation technology can provide qualified gasoline, diesel and aviation kerosene, which means the substitution for oil can be realized through coal liquidation technology without changing the current end demand structure.

Additionally, the mature situation of international related technology along with accumulated national research is another advance for the development of coal liquidation technology in China. The coal liquidation technologies have been studied and applied for several decades in other countries, and more than 20 years in China. Now the demonstration projects which are based on national technology and with the enhancement of imported advanced international technology have been launched. These projects set good bases for the development and application of the coal liquidation technology.

(2) Weakness of the coal liquidation technology in China.

Conversely to the advantages of abundant reserves, growing market and mature technologies, the development of coal liquidation technology in China also faces some restrictive conditions. For the development and application of any new technology, extensive funds are necessary. One of the main characteristics of coal liquidation technology industrialization is the great primary investment. According to the results of pre-feasible research carried out by related departments together with United States, Japan and Germany, setting up a direct-liquidation plant with the producing capability of 1 million tons oil per year requires more than RMB 80 trillion yuan to 100 trillion yuan. Such a huge investment is very difficult for current Chinese coal producers, most of who can not even secure the safe production process now. Therefore, the financing of coal liquidation projects would become the most important key point.

On the other hand, although the international coal liquidation technology is well developed, there still is a distance to industrialized application in China, and part of the key technology is still to be imported. This situation brings risks to its development in China. Lacking independent intellectual properties on key technologies will not only increase high costs of importing technology, but also increase the question about whether China can get an advanced technology. Consequently the market competency of coal liquidation products will be affected.

(3) Opportunity of coal liquidation technology development in China.

The current opportunity of coal liquidation technology development in China can be presented in three aspects: increasing international oil price, increasing demand for energy, and increasing awareness of environmental protection.

Compared with the stagnancy of coal liquidation technology due to low oil price, the high oil price brings opportunities to the relatively high cost alternative technology. The coal liquidation products face the same situation. According to the results of some feasibility assessments, the cost of oil which is made from coal liquidation is around RMB 1.4 thousand yuan per ton to 1.6 thousand yuan per ton, and the cost of oil made from indirect liquidation of coal is RMB 2 thousand yuan per ton. It can be deduced that, if the international oil price keeps stable at 20–25US\$ per barrel, the Chinese coal liquidation industry will get a reasonable profit. However, the oil price has already reached 28.83US\$ in 2003, and grew sharply in 2004. In early February of 2004, the price was over 30US\$ and in late March it reached 40US\$ per barrel. In August 2005, the oil price reached 70US\$ per barrel. The growth of oil price reflects the economic principle that along with the exploitation of resources the scarcity will lead to growth of prices. In the future decades, although it will fluctuate in some scale, the general trend of increasing of oil price is inevitable. This high oil price market environment brings vast opportunity to the development and application of coal liquidation technology. The economic profits will absorb more funds coming to the R&D of related technologies.

As mentioned above, the great energy demand in China is the basis and advantage of coal liquidation technology development. Meanwhile, this demand market keeps growing at a stable fast speed. All these situations provide good opportunities for technology development. The growth of energy demand will enlarge the gap between supply and demand, and further increase the demand for coal liquidation products. Besides, the economic growth will provide substantial assurance to the scienctific and technology progress. Along with the enhancement of national investment into R&D, the efficiency of coal liquidation will be improved, and the cost will be reduced.

Along with the increase of environmental awareness of the Chinese public and government, at the energy-end consumption structure, the proportion of coal burned directly will be reduced, which requires a more safe and clean consumption method. Also the transfer from coal to electricity does not meet the market requirement for liquid fuel. When coal is exhausted, the demand for energy will also form the driver for coal liquidation technology.

(4) Threat to coal liquidation technology development in China.

Along with the opportunities of increasing oil price, increasing market and environmental awareness, the development of coal liquidation technology also inevitably faces some negative factors and risks. The range of oil price undulation becomes larger and larger. The increasing price brings opportunity, while the inevitable downside and adjustment will bring potential risks and shocks to coal liquidation technology. The resilience of the technology is an important factor. In the meantime, the price of coal in China is also increasing. Along with the increasing cost of safe production and the shares on environment management, the trend of growth of coal price will become significant, which will also decrease the competition of coal liquidation products and brings risks to production.

Beside the economic risks, there are also technical risks. Although the coal liquidation technology is mature in the world, it has existed in China for only 20 years. Without any experience in industrialization and commercialization, the risks in the process of enlarging scales are inevitable. Additionally, whether the capacity of R&D can assure the advances of technology is also a problem. Once the technology has lagged, the competition will be lost also. Along with the R&D of new technologies, whether there will be other more low-cost and clean new energies is another threat to the development of coal liquidation technology.

7.4.2.2 Policy Suggestions on Promoting the Development of Coal Liquidation Technology

Based on the SWOT analysis for coal liquidation technology, the fast increasing energy demands, relatively abundant coal resources and increasing oil importing risks all point to the fact that coal liquidation technology occupies an important strategy status in Chinese energy alternative routes. The Chinese government should be aware of the strategy significance of this technology in the short-to-mid term dominant energy alternatives and economic development, and put the development of this technology to a high strategy level, following with policy guiding and supporting.

(1) Fortify government investment, and encourage participating of enterprises.

The Chinese government should clarify the strategy status of the coal liquidation technology in the entire energy technology system, fortifying the support with both funds and policies, and taking this technology as key points to be invested. Meanwhile, some policies, such as favourable tax rate, tax drawback on investment to R&D and low interests for loans, should be adopted to encourage energy enterprises and other investors to participate in the R&D of coal liquidation technologies, so as to promote the progress of technology development, and form the technology basis for scaled production.

(2) Promote international cooperation, introduce advanced technology.

The government should promote international cooperation and introduce advanced technology to realize the fast development of technologies. An effective way of promoting international cooperation is to make use of the price advantages of coal resources and labor to attract foreign enterprises setting joint ventures with Chinese energy enterprises. This way can, on the one hand, resolve the big amount of demanded funds, and on the other hand set demonstrations with the international advanced technologies and then drive the development of related national technologies and industries.

(3) Encourage industrialization, and share risks with corporations.

Those existing demonstration plants and projects in national enterprises should be protected and actively fostered. Facing the fluctuating international oil price and the increase of domestic coal price, government could set special risk funds to protect these projects, as well as share the risks due to the fluctuation. Meanwhile, the government could make governmental purchases of those coal liquidation products or make some allowance policies to promote the competitive and survival capacity of these enterprises. These policies could provide a favorable environment for the development and application of the technologies.

7.4.3 Policy Analysis on Renewable Energy Technology

As a primary and non-renewable energy, fossil energy resources will be reduced along with the increase of exploitation. Consequently the prices will go up and the competition for the resource will be exaggerated. This situation forms a potential threat to economic and political security around the world. On the other hand, the combustion of fossil energy is a main leading reason for green house gas emissions. The green house gas emission affects the global climate and environment severely. Although there has not been a global agreement yet, along with the increasing environment degradation and impacts on the social economic activities, the human society will inevitably take more and more strict measures on controlling the green house gas emission. Because nowadays the new energy technologies which can take the place of fossil energy still cannot support the economic development and the needs of living yet, fossil energy will still take the position of dominant energy for a relatively long time. However, from a long term view, the energy alternatives are a strategy direction of future energy industries development. Therefore, in the meantime of promoting the alternative of coal and natural gas to oil, China has to speed the R&D on new energy technologies, to form technology reservation and set technical basis for the new era of energy alternatives. The routine of dominant energy alternatives should be a cross lapped process, in which the general trend should be alternating from fossil energy to non-fossil energy, from non-renewable energy to renewable energy (as shown in Fig. 7-18).

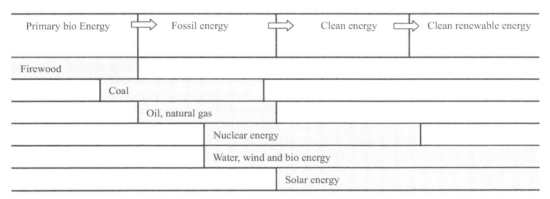

Fig. 7-18 Alternative routines for dominant energy

7.4.3.1 Chinese Policies on Renewable Energy

Currently, renewable energy has insufficient competition when compared to fossil energy. Because the chief energy system has stemmed from fossil energy, the development of renewable energy has to be supported by government. In many countries there are regulations and laws aiming at encouraging and supporting the development of renewable energy. These supporting policies can be categorized as: supporting R&D, governmental demonstration of projects and economic promotion measures.

(1) Supporting R&D.

The investment in renewable energy technology in China, compared with developed countries, is small. Now there still are no complete statistics on R&D investment. According to some estimations, in 2000 the energy R&D investment accounts for only 0.045% of GDP, while this number for 15 EU countries is 2% (Ragwitz, Miola, 2005). The investment in renewable energy R&D is even lower. In the period of the sixth and seventh five-year plans (1980–1999), the accumulated investment in national key projects related to renewable energy R&D is only RMB 20 billion yuan.

(2) Government demonstration projects.

The development of renewable energy utilities in China, especially large-scale application projects, is closely related to the promoting of government demonstration projects. From the 1980s, the development of renewable energy in the rural areas is promoted through four main government projects (Zhou, 1996): integrated project on rural economic development and rural energy development, small hydro power in rural electricity promotion, renewable energy development in reducing poverty projects and the efficient combustion stoves promotion in the rural women releasing project. In the Table 7-4, the main government demonstration projects in the 1990s are shown.

Table 7-4　Projects on renewable energy from Chinese government

	Project name	Supporter	Description
1996	Bright Project	The National Development and Reform Commission	Providing renewable power for 2 million people by 2010
1996	Riding-wind Plan	The National Development and Reform Commission	Actualizing the ratio of fans made in China 60%–80%
1997	Double-Adding Project	The National Commission of Economy and Trade	Accelerating the self-producing rate of wind power equipment, and providing RMB 9000 billion yuan favourable loans
2000	The Tenth Five-year Plan	The National Development and Reform Commission	Assuring the installed capacity of wind power reaches 1500MW by 2005
2000	Plan for development of renewable energy industry	The National Development and Reform Commission	Assuring the installed capacity of wind power reaches 7000MW by 2015
2002	Intensive Bright Project	The National Development and Reform Commission	Providing RMB 18 thousand billion yuan for solar power and wind power, to accelerate the progress of power supply in remote areas

(3) Economic promotion measures.

• Allowance policy for investment.

From 1987, the Chinese government set a special interest-subsidy loan for the rural energy projects. This loan is mainly used in mid-to-big scale marsh gas projects, solar-heat utility, reconstruction and promotion of wind power. The government allowance is a conventional measure of economic promotion in China, and now the allowance for renewable energy is mainly used in the R&D, demonstration projects building, and the development of industrialization of technologies.

• Favorable tax policy.

China sets a low tax rate for the renewable energy equipments importing. Because many domestic renewable energy projects utilize favorable loans or donations from foreign governments and institutions, this policy assures that these investments can avoid the tariffs and value-added tax. To resolve the problem that the actual tax of wind power is higher than fire power, the taxation rule on wind power should be adjusted to half of the value-added tax.[6] Because renewable energy is a high-tech industry, some local governments, such as Inner Mongolia and Xinjiang, take the favorable policy of income tax exemption. Additionally, the renewable energy power projects in western areas can get preferential tax according to the "Western development favorable tax policy".

• Favorable price policy.

The favorable price policy is mainly focused on the electricity generated from renewable energy, which is now mainly wind power. The policy assures the market for renewable energy power and provides favorable prices for this power. In 1994, the government issued "Regulation on the management of wind power marketing", to clarify the favorable price policy of "return the capital, pay the interest, get reasonable profits" to wind power generation. Also this regulation requires the electricity network to purchase the wind power totally. This policy is a feed-in tariffs policy, which is the main policy in China currently to attract the investment in renewable energy generation.

---

[6] On December 1, 2002, the National Financial Department and the National Revenue, "Announcement on the value-added tax of resource integrated projects and other products".

7.4.3.2 Problems in the Renewable Energy Policy

In China the development and application of renewable energy have received a lot of attention, especially the development in rural and remote areas. In recent years, along with the promotion of governmental demonstration projects, there has already formed a basis for the scaled industrialization production. However, because of the limitation of economic development and some insufficiency of policy making, there are still gaps between the current status of renewable energy development and its potential. These gaps are as below:

(1) Insufficient support for R&D investment and marketing.

Renewable energy is an intensive industry on both technology and funding, which decides that the supports for renewable energy development should be strengthened from both technology R&D and marketing. On the one hand, the intensified input to renewable energy R&D from the government can resolve the problem of insufficient motives for its development, and improve the entire social welfare level; on the other hand, strengthen cultivating the market of renewable energy is a pass to fill the technical void of fossil energy.

(2) Insufficient accord between policies.

Because in the government sections there is no special department for the development of renewable energy, many policies and demonstration projects are from different departments. The diverse purposes of different departments make their policies out of accord. For example, to reduce the cost of wind power investment, in 1998 the tariff for importing wind power set was canceled. But this policy set difficulty to the national wind power set manufacturers because of the introduced unfair competition (Liu, et al., 2002). On the other hand in recent years, to promote the scaled industrialization of wind power, some related departments are working actively on supporting the domestic production of wind power equipment, and some demonstration projects have been set up. The diverse aims of these two policies impact the development of domestic enterprises both actively and negatively, thus weakening the efficiency of these policies. On the other hand, the development and market cultivation of renewable energy needs long term support from the government, which makes the accord in the long term also a notable problem.

(3) Insufficient system for renewable energy project investment.

For a long time, the main funds source of renewable energy projects, especially those large scale demonstration projects, is from three channels (Liu, et al., 2002; Yang, et al., 2003): ① appropriate funds from government; ② donations and loans from foreign governments and international institutions; ③ FDI. The singularity of investors for renewable energy projects, as well as the lacking of financing channels, limits the development of renewable energy.

The issuing of the "Renewable Energy Law", which legally puts emphasis on protecting power pricing, eliminates the potential uncertainties which the investors would face. Therefore, more and bigger investors will be absorbed into the exploitation and application of renewable energy, and the development of commercial application and industrialization of renewable energy will be greatly promoted.

7.5 Policy Suggestions for China's Energy Technology

The main drivers of technological advances in energy changed from the development of science and technology to market force, and then shifted to clean and sustainable energy. Today, the factors that affect change in energy technology are getting more complex.

(1) Public policy shall lead and support the technological advances in energy.

In the field of energy, governmental intervention is necessary and important. Public policy is responsible for not only the long-term social welfare, but also for the improvement of the market efficiency (IEA, 2003). Shifting towards clean and sustainable energy system shall be pulled by public policy because of market inefficiency, which suggests the importance of public policy in the future energy technological advances. First, in the global trends of deregulation and restructure of energy sector and the background of global environmental issues, firms' R&D activities are determined by the return ratio of R&D. This can not assure the realization of long-term social welfare, while government shall be responsible for the social welfare. Second, governmental energy R&D expenditures should direct the R&D activities and technological progress, as Mansfield and Switzer (1984) reveal that governmental energy R&D spending could bring incentives and direct firms to complementary R&D studies. And finally, because of the phenomenon of lock-in and externality that commonly appeared in the energy sector, clean and sustainable energy technologies could not yet compete with mature traditional energy technologies. Thus, public energy policies are needed to support the new energy technologies to create a fair environment that may solve the problem of market failure.

(2) Governmental energy technology policy should not be influenced greatly by the market and should avoid myopic behaviors.

Over more than 30 years, oil shocks have been the biggest events that have affected the energy economy in the world. They not only changed the world's energy economic situation, but also changed the direction and pace of energy technological progress. To respond to the shocks, different reactive patterns have appeared, because different countries have different economic levels, different dependency on oil, different energy resources and different expectations on progress and results of oil shocks. The upsurge of oil prices has pushed the nations to adopt diversifying portfolios of oil imports, and also induced (energy) technological innovation.

Although the needs for fighting global climate change are pressing, it is impossible to transfer to clean and sustainable energy system by the market mechanism itself. Governmental intervention is critical in the transition. Governmental R&D expenditure, as an policy instrument, will play an important role. However, from the mid 1980s, governmental R&D expenditure was getting smaller. Furthermore, the behavior of governmental energy R&D is influenced by the market, which is a dangerous myopic behavior especially with the background of rising risks of global climate change.

(3) Technological advance is an historical process, so energy technology policy shall be in line with a country's resource advantages.

The trend of globalization is irreversible. Most of the energy firms are multi-national corporations, and help the diffusion of technology. While we compare the status and development of energy technology of each country, it is easier to find the diversification of energy technology in different countries which is determined by the resource and energy policy.

① Diversification of energy technology R&D policies in different countries. First, because of differences of resource, economic development level and energy consumption structure, countries' energy R&D expenditures reacting to oil shocks take different trajectories. Second, different countries' energy technology priorities became diversified. After the oil crisis, Germany and other many European countries prefer new energy R&D, Italy prefers electricity and its storage R&D, France prefers nuclear energy R&D, and the United States. and Japan prefer energy saving technologies.

② Institutions are the only explanatory factor that leads countries to different energy technology paths (Martin, 1996). In France, after small electricity firms merged into Electricite de France (EDF), EDF led France on the road of *all-electric, all-nuclear* and France is locked into nuclear technology. In Denmark, the government established a niche market for wind power which makes Denmark one of the leaders of wind power technology.

Technological progress is a historical and systemic process, and is relevant to the resource gifts and technological ability of a country. Our study also reveals the differences of the paths of energy technology could be explained only by the national innovation systems.

(4) Comprehensive and harmonious energy policy is needed to support the transition to clean and sustainable energy system.

The support of energy technology from public policy is not possible from only one or two governmental administrations, while it needs support from multigovernmental agencies. Thus, governmental public policy needs to form a comprehensive framework. The harmonious support framework is another important issue. First, from the angle of time scale, policy supporting the transition to clean and sustainable energy system should be long-term oriented. And secondly, from the point of sectors, because there is no single administration responsible for energy administration in China, some of the energy policies are not in line with each other. In one word, we need comprehensive and harmonious energy policy from different administrations to support the energy technological advances.

7.6 Summary

In this chapter, the transition of energy techno-economic paradigms is first investigated. Based on different driving forces, three energy techno-economic paradigms are identified, saying science and technology driven, market driven and public policy driven. Now global technological change is shifting to the era of public policy driven, which suggests the factors that influence the technological change are increasing more and more, and the governmental public energy policy will play a critical central role in the future energy technological progress.

Although public policy will be a key for the energy technological advances, especially in the process that we transition to a clean and sustainable energy system, unfortunately the chapter does not find an optimistic status of current energy R&D expenditures. Through the research on international energy R&D spending responding to the oil shocks, it is found that most of the IEA countries are influenced by the market when they invest in energy R&D. This is a myopic investment behavior, especially with the background of global climate change.

The distribution of governmental energy R&D expenditures in technological and national dimensions shows the trends of technological diversification and concentration in national dimension. These trends are of help for optimizing energy structure and assuring energy supply, and provide more technology options to fighting global climate change. After the oil crisis, together with the trend of technological diversification, concentration of energy R&D spending to fewer countries is another obvious trend. As the overall energy R&D budget is declining, only several countries dominate the energy R&D activities globally.

With the trends of the development of world's energy technology, and considering China's fundamental realities, this chapter tries to analyze the substitution roadmap and policy. We argue that, at present the key issue of energy substitution in China is to realize the substitution of coal for oil. Renewable energy is the direction of future energy development. We also investigate the development of renewable energy and the implementation of renewable energy technology policy in China, and argue that China shall further strengthen the policy support for renewable energy and make scientific decisions on policy design. Finally, based on the above research, implications for China's energy technology development are drawn.

CHAPTER 8
China Energy Outlook

On the basis of international and domestic energy development background-analysis, we conducted research focusing on a number of important and hot issues of China's current energy strategy and policies and reached a series of conclusions and policy recommendations drawn from the research.

Based on previous research, at this stage, in order to achieve a sustainable energy development, guarantee the energy demand for building a harmonious society and ensure China's energy safety, we suggest focusing on the following urgent issues: stabilizing the energy production, promoting energy conservation in key industries, expanding energy varieties and diversifying energy supply channels, optimizing the energy consumption structure, strengthening the study and macro-control of the energy market, eluding the risk of energy supply and so on. Therefore, China's energy development strategy should follow the basic principles of "technology priority, multiple developments, environment friendly, economic security". The guidelines for China's future energy strategy and policies are:

- Stabilizing domestic production, exploiting oversea resources, and ensuring economic growth.
- Optimizing economic structure, improving energy efficiency, and controlling aggregate demand.
- Strengthening technology innovation, developing clean energy, and protecting the ecological environment.
- Facing global markets, consummating the institutional mechanism, and achieving harmonious development.

8.1 Total Energy Consumption: Large Amount, Rapid Growth, Large Regional Differences

Since 2003, international oil prices have remained high. However, high oil prices have not slowed China's energy consumption. In 2005, China's total energy consumption was 2.22 billion tce, growing by 9.5% compared with that of 2004.

In 2005, China's primary energy production was 2.06 billion tce, growing by 9.5% compared with the last year. The production of raw coal was 2.19 billion tons, growing by 9.9%. Crude oil's production was 0.18 billion tons, growing by 2.8%. Total generating capacity in 2005 was 2.47 trillion kWh, growing by 12.3%. At the end of 2005, the capacity reached the highest history level, 0.5 billion kW.

As the world's second largest consumer of oil, according to the data supplied by China's National Development and Reform Commission, China's net oil imports were 136.17 million tons in 2005, of which net crude oil imports were 118.75 million tons, with 1.2% growth. Because of the rapid economic and energy consumption growth rate, for a long time China will continue maintaining a strong momentum of growth in energy consumption.

In the third chapter, combined with the actual situation in China, we build six scenarios with different technical and economic development-bases, with the assumptions of foremost factors that affect energy demand, and we conduct scenario analysis of the eastern, central and western regions' energy needs in 2010 and 2020 under various scenarios, as shown in Figs. 3-3 and 3-4. Regional energy forecast results show that the regional differences of China's future energy demand are large; moreover, this will be an increasing trend.

Meanwhile, in the third chapter, the forecast results show that: in 2020, the raw coal demand will be 2.196–2.357 billion tons; the crude oil demand will be 266–286 million tons. However, the domestic raw coal production will be 1.871 billion tons and crude oil production will be 180 million tons. Therefore, in 2010, China's coal production will be unable to meet the domestic demand of economic development, as well as a higher crude oil shortfall. In 2020, the raw coal demand will be 2.482–3.334 billion tons; the crude oil demand will be 424–569 million tons. However, the domestic raw coal production will be 2.161 billion tons and crude oil production will be 180 million tons in 2020. The gap between raw coal and crude oil supply and demand will increase further. Therefore, in the next 5–15 years, China's energy supply and demand contradictions will be even serious. Energy conservation, multiple developments, increasing oil supply capacity, and other issues will be our focus from the perspective of energy security.

8.2 Oil Imports Become Diversified, Imports Risk Reduces Gradually, and SPR Scale Increases Incrementally

Following implementation of China's oil imports diversification strategy, imports of oil from the Middle East will gradually decrease. The smooth completion of Sino-Kazakhstan oil pipeline and the formal signing of the contract of Russian Far East oil pipeline-Daqing branch indicate the initial implementation of China's oil import channels diversification strategy. Kazakhstan and the Caspian Sea's oil will continue to flow to China Xinjiang through the Sino-Kazakhstan oil pipeline. At the same time the Russian Far East region's oil and gas resources will first flow to northeastern China through Daqing branch pipeline. By then, there will be at least 50 million tons of crude oil imported through east and west pipelines, which will greatly reduce the risk to crude oil imports in China.

The Chinese government's wise decision of implementing "international cooperation" strategy, effectively lowers China's oil imports' risk. In 2005, China's overseas oil and gas development and utilization of resources made a breakthrough progress. CNPC bidding for Kazakhstan oil PK oil companies succeeded. The three major oil companies' CNPC, SINOPEC, CNOOC's share of oil in the Middle East, North Africa and other regions, have increased. China gradually implements its oil imports channels diversification strategy. All these measures reduce the risk of imported crude oil successfully. Therefore, China's future oil imports would show the diversified pattern and the imports risk will be reduced gradually.

National strategic petroleum reserve (SPR) can effectively reduce the short-term oil supply disruptions and oil price volatility on the economy, and protect national oil supply security. In March 2004, China officially launched its national strategic petroleum reserve project. The goal of the first project is to establish the equivalent of 20 days of oil reserves size, the largest reserve base in Zhenhai in Zhejiang Province, and the sixteen tanks of the first project were completed in August 2005. The other three reserve bases will be completed in 2008. In 2010, China is expected to achieve a 30 days' combined size of the reserves. Although China's strategic petroleum reserves grow rapidly, compared with Japan's 171 days, the Republic of Korea's 67 days, IEA member countries' 90 days' size of the reserves, there is still a great gap. Therefore, the scale of China's national strategic petroleum reserve will continue expanding; the capacity for resisting the risk of oil supply interruption and emergency response will increase steadily. However, achieving a guarantee of oil supply security still has a long way to go.

8.3 Coal Supply and Demand Are Basically Balanced, Clean Energy Is Developed, and Consumption Patterns Are Diversified

China has abundant coal resources, and is one of the few coal-based energy consumption countries in the world. China's coal not only meets the self-sufficiency needs, but is also exported on a large scale. With the rapid growth of China's energy consumption, pressure on supply is greater than ever before. Because many coal safety accidents happened, China has further intensified its closures of small coal mines. By the end of 2005 China closed more than 4000 small coal mines that did not meet the nation's safety standards. Therefore, China's speed of coal output growth will be slowed down. In the third chapter, we use the multi-regional input-output model to predict the future of China's coal demand under different contexts. The forecast results indicate that in 2010, demand will rise to 2.196–2.387 billion tons, and in 2020 to 2.482–3.334 billion tons. Meanwhile, we also use the system dynamics model to predict China's coal production under different contexts. The forecast results indicate that in 2010 China's coal output will be about 2.28–2.54 billion tons, and in 2020 about 2.53–2.69 billion tons. Obviously, China's coal demand growth rate is significantly higher than the growth rate of output; therefore, in 2020 China may turn from a coal exporting country to a net importer.

The steel industry is an energy-intensive industry; China's steel industry is a typical coal-intensive industry. Coal occupies as much as 70% of the steel industry's total energy consumption. Therefore, to improve the efficiency of China's steel industry's coal use is important and practically significant to improve the efficiency of China coal's comprehensive use. On July 20, 2005, the Nation's Development and Reform Commission has formally promulgated the "iron and steel industry development policy" and emphasized that the steel industry should be developed as following: "with sustainable development and the recycling economic concept, strengthen the environmental protection, improve the overall efficiency level of using resources, pay more attention to energy conservation and lessening consumption". The document clearly put forth that, in 2010, the comprehensive energy consumption per ton of steel and the comparable energy consumption per ton of steel will drop below 0.73 tce and 0.69 tce. respectively; in 2020 will drop below 0.7 tce and 0.64 tce, respectively. As showed in Fig. 8-1, from 1994 to 2003, comprehensive energy consumption per ton of steel in China's steel industry has dropped from 1.52 tce to 0.79 tce, with a drop of 48%; particularly in 2001–2003, there was a markedly increased drop. Therefore, the future coal comprehensive utilization efficiency of China will be further enhanced.

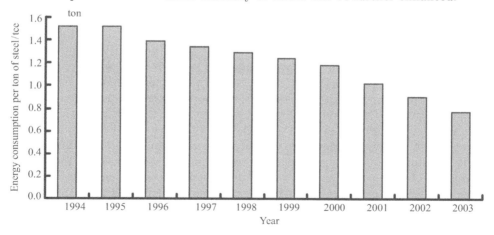

Fig. 8-1   1994–2003 the comprehensive energy consumption per ton of steel in China

In order to promote the renewable energy's development and utilization, increase energy supply, optimize the energy consumption structure, coordinate the energy use and environmental protection, and achieve the economic and social sustainable development, on February 28, 2005, China has formally adopted the "People's Republic of China Renewable Energy Law", which includes wind, solar, hydro, biomass, geothermal, ocean energy, and other non-fossil energy, and came into effect on January 1, 2006. The formulation and implementation of "Renewable Energy Law" will greatly promote China's renewable energy development. China's future energy consumption will gradually turn to the clean, efficient non-fossil energy development.

In the past two years, China's renewable energy use was developing at an average speed over 25% per year. Statistics show that, by the end of 2004, China's total installed hydropower capacity had reached 1.1 million kilowatts, wind power in the grid and the total installed wind power capacity had reached 76 million kilowatts, solar PV had reached 6 million kilowatts, the use of solar water heaters accounted for more than 40% of the world, and the rural households had built more than 110 million methane-generating pits.

In the third chapter, we use the multi-region input-output model to predict the future natural gas demand in China under different scenarios. The results show that, in 2010, China's demand for natural gas will be about 33.785–45.61 billion cubic meters, and in 2020 the demand will be about 79.757 billion–123.75 billion cubic meters. According to objectives displayed in "China's new energy and renewable energy development outline (1996–2010)", in 2010 the China's wind power capacity will reach 1–1.1 GW, the total use of solar energy will reach 4.67 million tce, biomass power generation capacity in 2010 will be more than 300 thousand kilowatts, the total use of geothermal energy will reach 1.51 million tce, and the installed capacity of small hydropower and the generating capacity will reach 2,788 MW and 1,170 MW. According to the Nation's Development and Reform Commission's plan, by the end of 2020, the renewable energy's percentage in the primary energy will increase from 7% to 15%. By then, the total capacity of hydropower will reach 300 million kilowatts, the wind power capacity will reach 30 million kilowatts, solar PV 1 GW, and biomass fuels 50 million tons. Therefore, coal, oil and other polluting fossil energy will gradually be replaced by renewable energy and new energy alternatives. However, due to the constraints such as technology and production costs, if there is no governmental great support in policy, capital and technology, China's renewable energy industries are very difficult to form the industrialization and scale effect in a short term.

8.4 Energy Efficiency Is Improved Steadily, Energy Conservation Potential Is Tremendous, and Technological Progress Is the Key

China's latest formulation of the "11th Five-Year Plan" energy development planning, clearly stipulates: "In 2010, the energy consumption per unit GDP will be 20% lower than that in the end of the 10th Five-Plan". The target of reducing energy intensity by 20% in the next five years is mainly on the basis of the "Special Plan For Medium-and Long-term Energy Conservation" formulated in 2004. China is to strive for quadruple GDP, while only doubling energy consumption in 2020. Only in the "11th Five-Year Plan" and the next two five-year plans China will achieve the goal of the energy consumption reduction by 20%, make the per unit GDP energy consumption reduction by 50% in 2020, and thereby realize the double use of energy consumption while quadrupling the economic growth target. Facts show the that, in 1980–2000, China's GDP growth and energy consumption growth ratio is about 2:1, which is to say the economic growth quadrupled with energy consumption doubled.

Statistics show that in the past 20 years, China's energy efficiency increased rapidly. In 2003, the energy consumption per unit GDP was about one third of that in 1980s. In 1980, China's energy intensity was about 7.90 tce per RMB 10 thousand yuan of GDP (in 1990 prices); in 2005, the energy intensity was about 2.83 tce per RMB 10 thousand yuan of GDP, with an average annual energy intensity decrease of 4.0%. Therefore, reducing energy intensity by 20% in the "11th Five-Year Plan" period and reaching an average annual reduction by 4.4% can be realized basically. In fact, China's energy conservation still has a tremendous potential. Nowadays, the energy consumption per dollar GDP is about nine times that in Japan, four times that in the United Kingdom and three times that in America (Kanekiyo, 2005). Therefore, there is still much space and potential for China's energy utilization efficiency's further development. Building a harmonious society which was proposed in the 16th Chinese Communist Party Congress has provided a strong policy support for energy conservation and energy efficiency improvement. For a long time, China's energy works will mainly focus on energy conservation and consumption reduction.

We have used the Malmquist index decomposition method to conduct some researches. The results show that, from 1994 to 2003, the energy efficiency of China's iron and steel industry had been improving, with an average annual growth of 5.4%, the cumulative growth of 60.3%, and the largest increase was in 2001 and 2003. The rise of China's iron and steel industry's energy efficiency mainly relies on technological progress. During 1994 and 2003, the average annual rate of technical progress of the energy use was 9.1% and the cumulative growth had reached 118.2%. During this period, the infrastructure investment and renovation in China's iron and steel industry were enlarged, a lot of advanced technology and equipment were added, bringing about an increase in the utilization of energy and technological progress.

We believe that, in the future, China should take effective measures to continue enhancing contributions to energy-saving by technological progress. At this stage, the government departments need to enlarge the support to steel production research and development of energy conservation technologies, and continue to encourage enterprises to adopt advanced technology and equipment; need to speed up the elimination of energy intensive equipment and techniques, formulate and implement strict pollutant emission standards, and reasonably increase the steel industries' cost of the environment, promote enterprises to improve secondary energy use efficiency; make appropriate access conditions to the steel industry, to avoid duplication of low-level projects and waste of resources, make energy-efficient steel industry play a more prominent role in the whole of society's energy conservation.

Over the long term, technological innovation must be enterprises' spontaneous behavior, rather than the executive order forced upon by the administrator. Government departments should gradually focus on energy-saving contributions brought by technical efficiency improvements, promote the reform and reorganization of the steel industry, play a guiding role in the property market and in energy conservation, promote scaled economies that contribute to energy-saving, use actively the fiscal policy and tax policy to trigger steel enterprises to reduce energy consumption (while spontaneously relying on technological advancement and management improvement), which is more urgent in the Baoshan and other iron and steel enterprises at present.

8.5 Total Emissions Continue to Expand, and the Industrial Structure Goes toward a Carbon-Intensive Trend. The Impact of Consumer Behavior Should Not Be Underestimated

So far, the global climate change has been the most significant environmental problem, and it's one of the most complex challenges faced by human beings in the 21st century. The international negotiation on the mitigation of climate change not only has something to do with the survival of mankind, but also has direct impact on the sustainable modernization process of developing countries. China's coal-based energy consumption structure has led to a series of environmental problems, such as emissions of greenhouse gases, acid rain in the region, suspended particle pollution, etc. As the second largest emitter of carbon dioxide only next to the United States, China had a carbon dioxide emission of about 1076.04 million tons of carbon due to the use of fossil fuels and cement production process in 2003. However, per capital emissions was below the world's average level. Data from World Bank's database show that, during the past 25 years, China has reduced its growth rate of carbon dioxide emissions effectively from 431.7 tons per million dollars in 1978 (1990 international dollars) to 175.9 tons per million dollars in 2000 through economic restructuring, energy efficiency improving, alternative energy discovery, afforestation and family planning. However, with the constraint of the structure of energy consumptions, energy use technologies, capital and other factors, China's growth rate of carbon dioxide emissions is still much higher than the average level of the world's major developed countries of 135.1 tons per million dollars. Therefore, the control of carbon dioxide emissions is not optimistic.

The greenhouse gas emissions reduction in future poses a severe challenge to China's development and we are facing increasingly heavy international pressure. At present, China has high pollution emissions, beside carbon dioxide ranking No. 2 in the world. And other types of greenhouse gas emissions are topping the list, which accounts for a larger share of global emissions. On the one hand, it has a serious impact on the ecology and environment. On the other hand, it faces immense pressure of implementing international environmental conventions. In Chapter 3, our research with the multiregional input-output model shows that China's carbon dioxide emissions will continue to rise. The conclusion represents that emissions of 2010 will be up to 13.77–14.79 million tons of carbon and the data of 2020 will be 16.87–22.66 billion tons of carbon, and the total carbon dioxide emissions will surpass the United States in about 2025. In 2030, the average emissions will surpass the global average level. If the economy keeps growing rapidly as in the current situation, China may rank No. 1 in carbon dioxide emissions all over the world. With an increasingly strong call on developing countries to reduce emissions, China will face a lot of energy and environmental problems, which may eventually affect the political, economic and energy security.

In Chapter 5, we have used AWD methodology to show that from 1980 to 2003, the decline of China's energy intensity played a dominant role in the decreasing of carbon emission intensity. And the changes in the structure of energy end-use consumption inhibit the decreasing of carbon emission intensity, which represents the terminal structure of energy consumption in three major industries going in the carbon-intensive direction. Meanwhile, from 1987 to 2002, the industrial structure inhibited the decreasing of carbon emission intensity, which also shows that China's industrial structure was going in the direction of carbon-intensity from 1987. The decline of China's energy intensity and carbon emission intensity resulted from the reform and opening of China's economic system, improved management and technical progress. And improving energy efficiency will still be an important way for reducing carbon emissions. However, the industrial structure and the end-use energy consumption structure which play a dominant role in primary energy consumption and carbon emissions is going in the direction of carbon-intensive type, and we should pay adequate attention to this trend.

We applied STIRPAT model to show that the country's population proportion, especially the 15–64 year old population proportion affects the carbon dioxide emissions dominantly, followed by energy intensity, thus we can make sure that reducing energy intensity and guiding the population aged 15–64 in production and life can slow the growth rate of carbon dioxide emissions. On different income levels, the impact of population, economy and technology on carbon dioxide emissions differs. Therefore, when we formulate medium and long-term strategies to reduce emissions of carbon dioxide, we should fully consider the impact of population, economy and technology at the different development stages. In fact, during the year 1999 to 2002, about 26% of China's total annual energy consumption and 30% of carbon dioxide emissions were due to the living behaviors and other economic activities which were used to satisfy the demand of these behaviors of rural and urban residents.

Through the study of Chapter 5, we can find that there is tremendous energy saving space in residents' living sectors. Under the premise that the basic living standards are not lowered, we can save energy equaling to 21.763 million tons of standard coal in the housing, automobile, motorcycles, and household appliances sectors, which makes up 11.0% of total energy-use of the resident-sector in 2002, and is equivalent to an annual reduction in carbon dioxide emissions of about 1628.8 tons of carbon, especially in the following areas:

(1) Heating, cooling, lighting and other energy-saving constructions have a large energy-saving space.

On the 2002 level, if the per capita living expenditure of Chinese urban residents increases 100 RMB per year, it will lead to an indirect energy consumption of 20 million tons of standard coal. And carbon dioxide emissions will increase by 14.18 million tons of carbon, which accounts for 10.15% of the total energy consumption of indirect energy consumption of urban residents' living behavior, and makes up 9.5% of the total indirect carbon dioxide emissions caused by the acts of urban life. If we turn down the indoor air conditioner by 1°C in summer, then we can save electricity by 5% to 8%. Considering the number of air-conditioners in 2002, 0.56 million kWh of electricity can be saved. If the amount of airconditioner use decreases by an hour per day, then 8.7 million kWh of electricity can be saved, which is about 0.07% of the total household-using electricity, and can decrease carbon dioxide emissions by 16,700 tons of carbon.

(2) In the transportation sector, there is enormous potential for energy saving in household-own cars and motorcycles.

In the year 2002 and 2003, the annual growth rate of household passenger cars is 50.6%. With this growth rate, China's vehicle number per 100 families of urban residents will be 23.24 by 2010. According to fuel consumption per 100 km of cars and the average annual vehicle mileage at present, considering the number of urban families, their energy consumption will be 22.997 million tons of standard coal (15.6 million tons of gasoline). The CO_2 emissions will be up to 12.487 million tons of carbon, which is 26.4 times the gasoline consumption and carbon dioxide emissions for passenger cars in 2002. Acounting to the standard of an average vehicle fuel consumption of 3 liters per 100 km which was drawn by the Department of Energy of U.S. in 1993 and the environmental protection department of Germany in 1994, the gasoline-saving potential of China's passenger cars in 2010 will be 8.259 million tons and 52.9% of the energy can be saved. Then we can reduce 6.611 million tons of carbon emissions. Research shows that, the number of motorcycles in China will reach 447 million in 2010. And the annual energy consumption will be 98.29 million tons of standard coal (66.8 million tons of gasoline). If the average fuel consumption per 100 km of motorcycle can decrease by 1%, then the total gasoline consumption will be reduced by 982,900 tons of standard coal in 2010, and also carbon dioxide emissions can be reduced by 533,700 tons of carbon, which is equivalent to 4.39% of total energy used in private transport sectors in 2002.

(3) The energy efficiency of basic rural life is to be improved.

During the year 1999 to 2002, China's rural residents living direct energy use (commodity energy) was up 3.4% of the total energy consumption. Among it, the proportion of the coal was about 68%, mainly for cooking and heating. The general thermal efficiency of briquettes used for cooking was about 30%, equivalent to the standard of European and American countries in the early 1970s. Therefore, the efficiency of China's rural coal use has large space for improvement.

8.6 Energy Strategies and Policies Should Highlight International Cooperation, Diversification, and Sustainability

Through studies above, we believe that the guiding line of China's future energy development strategies and policies should be to stablize domestic production, develop overseas resources, ensure economic growth, optimize the economic structure, improve energy efficiency, control total demand, strengthen technological innovation, develop clean energy, and protect the ecological environment. Facing the world market, we should implement the system to achieve harmonious development. We should focus on the following three aspects.

(1) Further strengthen the "international cooperation" in energy strategy.

Since constrained by the reserves and production capacity of domestic oil and gas, in order to protect national security of energy supply, China has formulated and implemented an "international cooperation" energy strategy. In diplomacy, China's independent foreign policy of energy will be further improved, and strategic cooperation and trade with energy exporting countries will become more and more frequent, diversified energy foreign policy will remove the political obstacles for domestic enterprises to explore overseas oil and gas business; China will continue to formulate some policies in taxes and industry developing fields to support domestic enterprises developing the overseas oil and gas resources. China is always encouraging oil and gas enterprises to take active use of domestic and foreign resources and markets, which is called two resources, two markets. Further, we can improve the share of overseas oil to protect national security of energy supply. Strategically, the communication and cooperation between China and neighboring countries in energy information, energy imports, forecast of future energy market and energy use will be further developed. And the risk to China's energy use will gradually decrease. Meanwhile, we will further strengthen the cooperation with energy exporting countries in the fields of energy production and energy technology. Then the risk to the energy supply will decrease too. Accordingly, the overseas investment yield of domestic enterprises will increase. Therefore, in a long period of time, the "international cooperation" strategy will be a basic principle of China's energy development. And it is effective to protect China's security in energy supply and economy.

(2) Make efforts to promote "diversity" energy strategy.

China's "Medium and Long-term Energy Development Planning" explicitly points out that the energy strategy is "coal-based, diversified development", which is totally different from the energy consumption structure of "coal mainly" in the past. It indicates a great change in China's energy strategy. Due to the lack of oil and gas resources, abundant coal resources make "coal-based" structure a unchangeable reality in the short term. The key issue is that we can not maintain "coal own dominant" for a long term. Firstly, coal mining is getting more and more difficult. At present about 90% of coal resources are in the middle and western regions where exploitation and transportation is difficult. Secondly, due to technical constraints, the process of coal mining development and utilization will bring environmental pollution of land, water and atmosphere. Thirdly, with the change in the economic structure and the improvement of people's living conditions, more efficient, clean and convenient energy is needed, and coal doesn't have these three main advantages.

Moreover, China's "Medium and Long-term Energy Development Planning" refers to diversification strategy of energy consumption, which says: "accelerating the development of nuclear power, renewable energy and vigorously developing hydropower". Among them, the change of strategy in nuclear power and renewable energy from "moderate" to "accelerate" is the most noticeable, and vigorously developing hydropower is the most effective measure in short term to achieve the diversification energy consumption strategy. Therefore, considering energy security, environmental protection, optimization of the structure of consumption and so on, promoting diversification strategy is an inevitable result of energy development.

(3) Continue implementing the "sustainable development" energy strategy.

"Sustainable development" strategy is one of the most important factors to protect stability and development of nation and region. At this stage, we should strengthen our energy supply infrastructure and make energy use reasonable, formulate a long-term energy development plan, and make the development of China's energy resources scientific and orderly.

On the one hand, China's energy supply capacity will be further strengthened, particularly the overseas energy resources utilization. Since using Russia's abundant oil is one of the most positive choices of energy development, China will increase oil and gas imports from Russia. And investment in Russian and the Caspian region's stream petroleum industry in the field of oil exploration and development will be further expanded. In the coming few years, there will be greater development in energy trade and regional energy cooperation with the neighboring countries. Along with the start of the "Angarsk-Nakhodka line" and "China Daqing Extension" oil pipeline engineering, a modern energy transport network and mature regional energy market is just around the corner. Therefore, the Northeast Asian region, particularly China, will enhance its speaking rights on the world oil price.

On the other hand, China will effectively improve the conversion and reasonable use of energy. Technology transfer from developed countries, especially clean coal technology and CDM/ETS will play an important role. At the same time, a scientific and rational energy market planning, especially the formulation of tax policy on inter-regional fossil energy and non-fossil energy is also very important. The future of China's energy use technology will develop rapidly. Thus, technological progress will be the key point of sustainable energy development.

References

Agiobenebo T J. 2000. Market structure, concentration indices and welfare cost. Department of Economics, University of Port Harcourt, Nigeria & Department of Economics, University of Botswana, Botswana.

Akaike H. 1974. A new look at the statistical model identification. IEEE Trans. on Automatic Control, 1: 716–723.

AMR. 2005-12. Energy and economic database. Academy of Macroeconomic Research. http://www.amr.gov.cn:8000/showcedb/index.asp. (In Chinese)

An F Q, Liu J A, et al. 2005. Oil security holds together national vitals. Http://www.Sinopecnews.com.cn/shzz/2003-07/16/content_55887.html. (In Chinese)

Ang B W, Zhang F Q.2000. A survey of index decomposition analysis in energy and environmental studies. Energy, 25: 1149–1176.

Armington, Paul S A. 1969. Theory of demand for products distinguished by place of production. IMF Staff Papers. 16: 159–176.

Asafu-Adjaye J. 2000. The relationship between electricity consumption, electricity prices and economic growth: Time series evidence from Asian developing countries. Energy Economics, 22: 615–625.

Bach S, Kohlhaas M, Meyer B, et al. 2002. The effects of environmental fiscal reform in Germany: A simulation study. Energy Policy, 30(9): 803–811.

Bachman, Ingram. 1999. A Macroeconomic Response to Oil Price Shocks in Pacific Rim Economies. Philadelphia: Wharton Economic Forecasting Associates.

Bacon, Rockets R W, et al. 1991. The asymmetric speed of adjustment of UK retail gasoline prices to cost changes. Energy Economics, 13: 211–218.

Bettendorf Leon, Stephanie A van der Geest, Marco Varkevisser.2003. Price asymmetry in the Dutch retail gasoline market. Energy Economics, 25: 669–689.

Bidsall N. 1992. Another look at population and global warming. Population, Health and Nutrition Policy Research Working Paper, WPS1020, World Bank, Washington D.C.

Bin S, Dowlatabadi H. 2005. Consumer lifestyle approach to U.S. energy use and the related CO_2 emissions. Energy Policy, 33: 197–208.

Blumstein C, Wiel S. 1998. Public-interest research and development in electric and gas utility industries. Utilities Policy, 7: 191–199.

Borenstein S, Cameron C A, Gilbert R. 1997. Do gasoline prices respond asymmetrically to crude oil price changes? Q. J. Econ., 112: 305–339.

Boserup E. 1965. The conditions of agricultural growth. Aldine, Chicago.

Boserup E. 1981. Population and Technological Change: A Study of Long-Term Trends. Chicago: University of Chicago Press.

BP. 2002. BP Statistical Review of World Energy 2002, British Petroleum (BP), London.

BP. 2003. BP Statistical Review of World Energy 2003, British Petroleum (BP), London.

BP. 2004. BP Statistical Review of World Energy 2004, British Petroleum (BP), London.

BP. 2005. BP Statistical Review of World Energy 2005, British Petroleum (BP), London.

CDIAC. 2005. Global, regional, and national fossil fuel CO_2 emissions. http://cdiac.ornl.gov/trends/emis/meth_reg.htm.

Chen B S, Lai T W. 1997. An investigation of co-integration and causality between electricity consumption and economic activity in Taiwan. Energy Economics, 19: 435–444.

Chen H. 2004. The importance of studying international market for petroleum strategy. Price: Theory and Practice, 1: 16–17. (In Chinese)

Chen H, Deng Y S. 2001. Suggestion for Chinese oil price formation mechanism. International Petroleum Economics, 2: 19–20. (In Chinese)

Chen J S. 2005-12. Oil market review and 2005 prospect. Panorama Network. http://www.p5w.net/p5w/home/futures/200501050336.html. (In Chinese)

Chen S T, et al. 1996. Analysis on the relationship between Chinese energy and economic growth after 1990s. Chinese Energy, 12: 24–30. (In Chinese)

Chen Y, Fan S S. 2004. Natural Gas Hydrate—Research on Energy Development Strategy. Beijing: Chemical Industry Press. (In Chinese)

Chen W. 2005. The costs of mitigating carbon emissions in China: Finding from China MARKAL-MACRO modeling. Energy Policy, 33(7): 885–896.

Chen W Y. 2003. Carbon quota price and CDM potentials after Marrakesh. Energy Policy, 31(8): 709–719.

China Coal Branch Coal Trade Promotion Council. 2005-11-5. The forecast of coal. http//www.coalword.net.cn/mtzy.htm. (In Chinese)

China Energy Development Report Editor Committee. 2001. China Energy Development Report 2001. Beijing: China Measure Press. (In Chinese)

China Energy Research Association. 2001. Energy Policy Research, 1: 50. (In Chinese)

China International Futures Brokers Limited Network Division. 2005-12. 2005 Global economy will go from slowdown to the recession. http://www.szcifco.com/readnews.asp?siteid=154729. (In Chinese)

China Power Daily. 2005-9-24. China's power industry enters a new era in the tenth-five plan. (In Chinese)

Choi KiHong, Ang B W. 2003. Decomposition of aggregate energy intensity changes in two measures: Ratio and difference. Energy Economics, 25: 615–624.

Christodoulakis N M, Kalyvitis S C, Lalas D P, et al. 2000. Forecasting energy consumption and energy related CO_2 emissions in Greece: An evalution of the consequences of the Community Support Framework II and natural gas penetration. Energy Economics, 22(4): 395–422.

Clinch J P, Healy J D, King C. 2001. Modelling improvements in domestic energy efficiency. Environmental Modelling & Software, 16: 87–106.

Cole M A, Rayner A J, Bates J M. 1997. The environmental Kuznetz curve: An empirical analysis. Environment and Development Economics, 2: 401–416.

Commerce Department. Yearbook of China's Foreign Economics Relations and Trade 1997/98, 1998/99, 2000, 2001, 2002, 2003. Beijing: China Foreign Economic Relations and Trade Press. (In Chinese)

Constructing forehanded society international workshop background report group. 2005-12-12. China's energy conservation situation, problem and challenges in industry. http://www.cdrf.org.cn/2005web/hotnews/050614_gongyejieneng.htm. (In Chinese)

Crompton P, Wu Y. 2005. Energy consumption in China: past trends and future directions. Energy Economics, 27: 195–208.

Daly H. 1996. Consumption: value added, physical transformation and welfare//Costanza R, Segura Q, Martinez-Alier J. Geting Down to Earth: Pratical Applications of Ecological Economics: 49-59. Washington D.C.: Island Press.

David L G, Donald W J, Paul N L. 1998. The outlook for US oil dependence. Energy Policy, 26(1): 55–69.

Davis G A, Owens B. 2003. Optimizing the level of renewable electric R&D expenditures using real options analysis. Energy Policy, 31: 1589–1608.

Davis W B, et al. 2002. Contributions of weather and fuel mix to recent declines in US energy and carbon intensity. Energy Economics, 25: 375–396.

Dension E F. 1962. The Sources of Economic Growth in the United States and the Alternatives Before Us. New York: Committee for Economic Development.

Dension E F. 1985. Trends in American Economic Growth 1929–1982. Washington D.C.: The Brookings Institution.

Development Research Center of State Council, Tsinghua University, China Automobile Technology Research Center, Chinese Research Academy of Environment Sciences. 2003. Motor vehicle fuel economy background report. http://www.efchina.org/documents/China_Fuel_Efficiency_Background_CN.pdf. (In Chinese)

Dickey D A, Fuller W A. 1981. Likelihood ratio statistics for autoregressive time series with a unit root. Econometrica, 4: 1057–1072.

Dickey D A, Fuller W A. 1979. Distribution of the estimators for autoregressive time series with a unit root. Journal of the American Statistical Association, 74: 427–431.

Dieta T, Rosa E A. 1994. Rethinking the environmental impacts of population, affluence and technology. Human Ecology Review, 1: 277–300.

Dieta T, Rosa E A. 1997. Effects of population and affluence on CO_2 emissions. The National Academy of Sciences of the U.S., 94: 175–179.

Dooley J J. 1998. Unintended consequences: energy R&D in a deregulated energy market. Energy Policy, 26: 547–555.

Dooley J J, Runci P J. 1999. Adopting a long view to energy R&D and global climate change. Battelle Memorial Institute, Columbus, Ohio, U.S.

Dooley J J, Runci P J, Luiten E. 1998. Energy R&D in the industrialized world: retrenchment and refocusing. Battelle Memorial Institute, Columbus, Ohio, U.S.

Doroodian K, Roy Boyd. 2003. The linkage between oil price shocks and economic growth with inflation in the presence of technological advances: A CGE model. Energy Policy, 31: 989–1006.

Dosi G. 1982. Technological paradigms and technological trajectories: A suggested interpretation of the determinants and directions of technical change. Research Policy, 11: 147–162.

Duchin F. 1998. Structural Economics: Measuring Changes in Technology, Lifestyles and the Environment. Washington D.C.: Island Press.

Economia & Energia. 2001. Supply of an instrument for estimating the emissions of greenhouse effect gases coupled with the energy matrix. http://ecen.com/matriz/eee24/coefycin.html.

Edenhofer O, Jaeger C C. 1998. Power shifts: The dynamics of energy efficiency: Energy Economics, 20: 513–537.

Ehrlich P R, Holden J P. 1971. Impact of population growth. Science, 171: 1212–1217.

Ehrlich P R, Holden J P. 1972. One dimensional economy. Bulletin of Atomic Scientists, 16: 18–27.

Engle R F, Granger C W J. 1987. Cointegration and error correction: representation, estimation, and testing. Econometrica, 55: 251–276.

Engleman R. 1994. Stabilizing the atmosphere: Population, consumption and greenhouse gases. http://www.cnie.org/pop/CO_2/intro.htm.

Engleman R. 1998. Profiles in carbon: An update on population, consumption and carbon dioxide emissions. Population Action International, Washington D.C.

Eric D L, Wu Z, Pat D, Chen W, Gao P. 2003. Future implications of China's energy-technology choices. Energy Policy, 31(12): 1189–1204.

Erol U, Yu E S H. 1987. On the relationship between electricity and income for industrialized countries. Journal of Electricity and Employment, 13: 113–122.

EU. 2002. Science and Technology for Sustainable Energy. Office for official publications of the European Communities, European Union, Luxemburg.

Fan Y, et al. 2006. Analyzing Impact Factors of CO_2 Emissions Using the STIRPAT Model. Environmental Impact Assessment Review.

Farla J, Cuelenaerel R, Blok K. 1998. Energy efficiency and structural change in the Netherlands, 1980–1990. Energy Economics, 20: 1–28.

Feng W, Wang S, Ni W, Chen C. 2004. The future of hydrogen infrastructure for fuel cell vehicles in China and a case of application in Beijing. International Journal of Hydrogen Energy, 29(4): 355–367.

Former State Department of Geology and Mineral Resources. 1992. China Geology and Mineral Resources Yearbook 1992. Beijing: Geological Press. (In Chinese)

Former State Department of Geology and Mineral Resources. 1993. China Geology and Mineral Resources Yearbook 1993. Beijing: Geological Press. (In Chinese)

Former State Department of Geology and Mineral Resources.1994. China Geology and Mineral Resources Yearbook 1994. Beijing: Geological Press. (In Chinese)

Former State Department of Geology and Mineral Resources. 1995. China Geology and Mineral Resources Yearbook 1995. Beijing: Geological Press. (In Chinese)

Former State Department of Geology and Mineral Resources. 1996. China Geology and Mineral Resources Yearbook 1996. Beijing: China Geology and Mineral Resources Yearbook Editorial board. (In Chinese)

Former State Department of Geology and Mineral Resources. 1997. China Geology and Mineral Resources Yearbook 1997. Beijing: China Geology and Mineral Resources Yearbook Editorial board. (In Chinese)

Former State Department of Geology and Mineral Resources. 1998. China Geology and Mineral Resources Yearbook 1998. Beijing: China Geology and Mineral Resources Yearbook Editorial board. (In Chinese)

Freeman C. 1989. The Third Kondratieff Wave: Age of Steel, Electrification and Imperialism. Research Memorandum 89-032, MERIT, Maastricht, Netherlands.

Freeman C, Perez C. 1988. Structural crises of adjustment, business cycles and investment behavior//Dosi G. Technical Change and Economic Theory. London: 38–66.

Ghali H Khalifa, EI-Sakka M I T. 2004. Energy use and output growth in Canada: A multivariate cointegration analysis. Energy Economics, 26: 225–238.

Gielen D, Chen C. 2001. The CO_2 emission reduction benefits of Chinese energy policies and environmental policies: A case study for Shanghai, period 1995–2020. Ecological Economics, 39(2): 257–270.

Gielen D, Moriguchi Y. 2002. Modelling CO_2 policies for the Japanese iron and steel industry. Environmental Modelling & Software, 17: 481–495.

Glasure Y U, Lee A R. 1997. Cointegration, error-correction, and the relationship between GDP and electricity: The case of South Korea and Singapore. Resource and Electricity Economics, 20: 17–25.

Goldstein S, Joshua, Huang X, Akan B. 1997. Energy in the world economy, 1950–1992. International Studies Quarterly, 41: 241–266.

Granger C W J. 1969. Investigating causal relations by econometric models and cross-spectral methods. Econometrica, 37: 424–438.

Greene D L, Jones D W, Leiby P N. 1998. The outlook for US. oil dependence. Energy Policy, 26(1): 55–69.

Greening L A. 2004. Effects of human behavior on aggregate carbon intensity of personal transportation: Comparison of 10 OECD countries for the period 1970–1993. Energy Economics, 26: 1–30.

Greening L A, Davis W B, Schipper L. 1998. Decomposition of aggregate carbon intensity for the manufacturing sector: comparision of declining trends from 10 OECD countries for the period 1971–1991. Energy Economics, 20: 43–65.

Greening L A, Ting M, Davis W B. 1999. Decomposition of aggregate carbon intensity for freight: trends from 10 OECD countries for the period 1971–1993. Energy Economics, 21: 331–361.

Greening L A, Ting M, Krackler T J. 2001. Effects of changes in residential end-uses and behavior on aggregate carbon intensity: comparison of 10 OECD countries for the period 1970 through 1993. Energy Economics, 23: 153–178.

Griliches Z. 1996. The discovery of the residual: A historical note. Journal of Economic Literature, 34: 1324–1330.

Grubler A. 1998. Technology and Global Change. Cambridge: Cambridge University Press.

Grubler A, Nakicenovic N, Victor G D. 1999. Dynamics of energy technologies and global change. Energy Policy, 27: 247–280.

Gui W L. 2004. How to cope with entering WTO for China oil price supervision system. China Industrial Economics Perspective, 3: 23–28. (In Chinese)

Han J C. 2005. How much oil is required in year 2015. Energy of China, 27: 41–42. (in Chinese)

Han Z, et al. 2004. Research on change features of Chinese energy intensity and economic structure. Application of Statistics and Management, 1: 1–6. (In Chinese)

Han Z, et al. 2004. The cointegration and causality of Chinese energy economy. System Engineering, 22(12): 17–21. (In Chinese)

Han Z Y, Fan Y, Wei Y M. 2004. Research on the cointegration and causality between GDP and energy consumption in China. Int. J. Global Energy Issues, 22(2/3/4): 225–232.

Han Z Y, Fan Y, Wei Y M. 2005. Energy structure, marginal efficiency and substitution rate: An empirical study in China. The 3rd Dubrovnik Conference on Sustainable Development of Energy, Water and Environment Systems.

Harrison Lovegrove & Co., John S Herold, Inc. 2005. Global Upstream M&A Review 2005-7-25. http://www.hargrove.co.uk/uploadedfiles/HLC_GUMAR_Release.pdf

He G. 2005. China's future way to hydropower: Exploit acceleratory and sustainable development. Hydrowater Technology, 36(2): 5–8. (In Chinese)

He H, Wang K, Ming J. 2005-12. High oil prices may become the norm, the fourth approaching shadow of the oil crisis. http://finance.sina.com.cn/stock/t/20050418/12231526793.shtml. (In Chinese)

He J K, Liu B, Zhang A L. 2002. An analysis on China's CO_2 emission reduction potential. Transaction of Tsinghua University (Social Sciences), 17(6): 75–80. (In Chinese)

He J. 2002. China hydropower resource development situation and development strategy discussion on 21st century. Hydrowater Technology, 10(1): 1–10. (In Chinese)

He Q. 2005. Advantages and disadvantages of refined oil pricing mechanism. http://www.southcn.com/car/lmrd/200505111739.htm. (In Chinese)

He Y K. 2004. Six years of oil price reform. China Petroleum and Petrochemical, 5: 31–32. (In Chinese)

Helkie W L. 1991. The impact of an oil market disruption on the price of oil: A sensitivity analysis. The Energy Journal, 12(4): 105–116.

Hickman B G. 1987. Macroeconomic impacts of energy price shocks and policy responses: A structural comparison of lburteen models. Macroeconomic Impacts of Energy Shocks Elsevier Science B.V., North Holland.

Hicks J R. 1932. The Theory of Wages. London: Macmillan.

Hsu C C, Chen C Y. 2004. Investigating strategies to reduce CO_2 emissions from the power sector of Taiwan. International Journal of Electrical Power & Energy Systems, 26: 487–492.

Huang S C. 2005. World Coal Development Report 2004. Beijing: Coal Industry Press. (In Chinese)

Huatongren Data Center, 2005-12-12. China statistical data application supporting system (Government Edition), http://218.249.174.12. (In Chinese)

Hwang D B K, Gum B. 1992. The causal relationship between energy and GNP: The case of Taiwan. The Journal of Energy and Development, 12: 219–226.

IEA, 2003a. Creating Markets for Energy Technologies. Paris: OECD/IEA.

IEA, 2003b. IEA Energy Technology R&D Database. Paris: OECD/IEA.

IEA, 2004. Key World Energy Statistics. Paris: IEA/OECD.

Inja P, Paul L, Donald J, et al. 1999. Strategic oil stocks in the APEC region. Proceedings of the 22nd IAEE Annual International Conference, International Association for Energy Economists.

Institute of Industrial Economic Research, Chinese Academy of Social Sciences. 2005. China Industry Development Report 2005: China's Industry Development under Resource and Environment Constraints. Beijing: Economic and Management Press. (In Chinese)

IPCC. 1995. IPCC Working Group I Summary for Policymakers. Cambridge: Cambridge University Press.

Jacquemin A P, Berry C H. 1979. Entropy measure of diversification and corporate growth. The Journal of Industrial Economics, 27: 359–369.

Jenne C A, Cattell R K. 1983. Structural change and energy efficiency in industry. Energy Economics, 5: 114–123.

Jiang W, Xiao T. 2001. Energy efficiency: Market disability and government function. Energy Policy Study, 4: 6–11. (In Chinese)

Jiang Z H, 1998. National and Regional Population Forecast. Beijing: China Population Press. (in Chinese)

Jiao J L. 2005. Econometrical Models and Empirical Study on the Problems of Oil Prices (Doctoral Dissertation). Hefei: University of Science and Technology of China. (In Chinese)

Jin B S. 2005-12. Economic globalization on China's energy security pick. http://www.ce.cn/macro/home/jjsp/200509/08/t20050908_4647869.shtml. (In Chinese)

Johansen S. 1988. Statistical analysis of cointegrated vectors. Journal of Economic Dynamics and Control, 213: 231–254.

Jonthan E Sinton. 2001. Accuracy and reliability of China's energy statistics. China Economic Review, 12: 373–383.

Kanekiyo K. 2005. Energy outlook of China and Northeast Asia and Japanese perception toward regional energy partnership. The Institute of Energy Economics, Japan.

Kirchgasser G, Kubler K. 1992. Symmetric or asymmetric price adjustment in the oil market: An empirical analysis of the relations between international and domestic prices in the Federal Republic of Germany 1972–1989. Energy Economics, 14: 171–185.

Kraft J, Kraft A. 1978. On the relationship between energy and GNP. Energy Development, 3: 401–403.

LBNL. 2004. China Energy Databook v6.0. Lawrence Berkeley National Laboratory, CD.

Lenzen M. 1998. Primary energy and greenhouse gases embodied in Australian final consumption: An input-output analysis. Enegry Policy, 26: 495–506.

Lesbirel S H. 1988. The political economy of substitution policy: Japan's response to lower oil prices. Pacific Affairs, 61: 285–302.

Li G S, Jiao D Q, Su Y S. 2004. China's oil and gas resource proving prospective. Contemporary Oil and Chemistry, 12(8): 4–8. (In Chinese)

Li H Y. Impact of international crude oil price on China price level. http://www.china.org.cn/chinese/EC-c/654187.htm. (In Chinese)

Li J F. 2002. Analysis and study on international competition of Chinese three big oil enterprises. Chemical Techno-Economics, 5: 9–23. (In Chinese)

Li S T, Hou Y Z, He J W. 2005. Economic growth prospect from the 11th five year to year 2020. http://finance.sina.com.cn/g/20050510/01431572557.shtml. (In Chinese)

Li W K. How far does China break loose the tie of high oil price and strive for international oil pricing. http://news.xinhuanet.com/ fortune/2005-08/27/content_3409971htm. (In Chinese)

Li Z N, Ye A Z. 2000. Advanced Econometrics. Beijing: Tsinghua University Press. (In Chinese)

Lin C Y. 1984. Global pattern of energy consumption before and after the 1974 oil crisis. Economic Development & Cultural Change, 32: 781–802.

Liu S, Li X, Zhang M.2003. Scenario analysis on urbanization and rural-urban migration in China. Institute of Geographic Sciences and Natural Resources Research, Chinese Academy of Sciences, Beijing. Working Paper.

Liu Q, Nobuhiro O. 2002. Methods of producing multi-regional input–output table for China and related problems. Statistical Research, 9: 58–64. (In Chinese)

Liu W Q, Gan L, Zhang X L. 2002. Cost-competitive incentives for wind energy development in China: Institutional dynamics and policy changes. Energy Policy, 30: 753–765.

Liu Z. 2005-9-16. Zhang Guobao said China would develop renewable energy acceleratory. Xinhua News Agency. http://news.xinhuanet.com/fortune/2005-09/26/content_3546715.html. (In Chinese)

Lu W, Ma Y. 2004. Image of energy consumption of well off society in China. Energy Conservation and Managerment, 45(9-10): 1357–1367.

Lutkepohl H, Reimers H E. 1992. Impulse response analysis of cointegrated system. Journal of Economic Dynamics and Control, 1: 53–78.

MacKinnon J G. 1991. Critical values for cointegration tests//Engle R F, Granger C W J. Long-run Economic Relationships: Readings in Cointegration. Oxford: Oxford University Press.

Maddison A. 2005. http://www.eco.rug.nl/~Maddison/.

Malthus T R. 1967. Essay on the principle of population. Dent, London, U.K.

Mansfield E, Switzer L. 1984. Effects of federal support on company-financed R and D: The case of energy. Management Science, 30: 562–571.

Margolis R M, Kammen D M. 1999a. Evidence of underinvestment in energy R&D in the United States and the impact of federal policy. Energy Policy, 27: 575–584.

Margolis R M, Kammen D M. 1999b. Underinvestment: The energy technology and R&D policy challenge. Science, 285: 690–692.

Martin J M. 1996. Energy technologies: Systemic aspects, technological trajectories, and institutional frameworks. Technological Forecasting and Social Change, 53: 81–95.

Meyerson F. 1998. Population, carbon emissions and global warming: The forgotten relation at Kyoto. Population and development reviews, 24: 115–130.

Miketa A, Schrattenholzer L. 2004. Experiments with a methodology to model the role of R&D expenditures in energy technology learning process; first results. Energy Policy, 32: 1679–1692.

Miller R E, Blair P D. 1985. Input-output Analysis: Foundations and Extensions. New Jersey: Prentice Hall.

Ministry of Commerce of the P. R. C. China Commerce Yearbook 2004, 2005. Beijing: China Commerce Press.

Ministry of Land and Resources. 1999. China Land and Resources Yearbook 1999. Beijing: China Land and Resources Yearbook Editorial Board. (In Chinese)

Ministry of Land and Resources. 2000. China Land and Resources Yearbook 1999. Beijing: China Land and Resources Yearbook Editorial Board. (In Chinese)

Ministry of Land and Resources. 2001. China Land and Resources Yearbook 1999. Beijing: China Land and Resources Yearbook Editorial Board. (In Chinese)

Ministry of Land and Resources. 2002. China Land and Resources Yearbook 1999. Beijing: China Land and Resources Yearbook Editorial Board. (In Chinese)

Ministry of Land and Resources. 2003. China Land and Resources Yearbook 1999. Beijing: China Land and Resources Yearbook Editorial Board. (In Chinese)

Ministry of Water Resources. 2005-12-12. China's hydro energy reserves and exploitable hydro energy. http://www.mwr.gov.cn/weiye/fubiao/biao6.html. (In Chinese)

Mokyr J. 1990. The Lever of Riches: Technological Creativity and Economic Progress. Oxford: Oxford University Press.

Mork K A, Olsen O, Mysen H T. 1994. Macroeconomic responses to oil price increases and decreases in seven OECD countries. The Energy Journal, 15(4): 19–35.

Mu H, Kondou Y, Tonooka Y, et al. 2004. Grey relative analysis and future prediction on rural household biofuels consumption in China. Fuel Processing Technology, 85(8-10): 1231–1248.

Nakicenovic N, Victor N, Morita T. 1998. Emissions scenario data base and review of scenarios. Mitigation and Adaptation Strategies for Global Change, 3: 95–120.

National Development and Reform Commission. 2004. China medium and long-term energy conservation plan. http://www.china.com.cn/chinese/PI-c/713341.htm. (In Chinese).

National Planning Committee. 1999. White Paper on China's New and Renewable Energy. Beijing: China Planning Press. (In Chinese)

National Planning Committee. 2000. China's New and Renewable Energy. Beijing: China Planning Press. (In Chinese)

National Planning Committee, National Science and Technology Committee, National Economic and Trade Committee. 1995. New and Renewable Energy Development Plan (1996–2010). (In Chinese)

NDRC. 2004-11-10. Energy conservation plan on mid-long period. National Development and Reform Committee. (In Chinese)

NDRC. 2005-12. China's electricity capacity installation is over 500MW. http://zys.ndrc.gov.cn/xwfb/t20051229_55285.htm. (In Chinese)

Neff T L. 1997. Improving energy security in Pacific Asia: Diversification and risk reduction for fossil and nuclear fuels. The Pacific Asia Regional Energy Security, Massachusetts.

Nelson C R, Plosser C I. 1982. Trends and random walks in macroeconomic time series. Monetary Economics, 10: 139–162.

News Center of United Nations. 2005-11-28. The first session conference of Kyoto Protocol founder. http://www.un.org/chinese/News/fullstorynews.asp?newsID=4756. (In Chinese)

NGO. 2001. Supply of an instrument for estimating the emissions of greenhouse effect gases coupled with the energy matrix. Economia & Energia, http://ecen.com/matriz/eee24/coefycin.htm.

Ni W, Thomas B J. 2004. Energy for sustainable development in China. Energy Policy, 32(10): 1225–1229.

References

Odell R, Peter. 1968. The significance of oil. Journal of Contemporary History, 3: 93–110.

Offshore Oil engineering Co Ltd. 2005-12. 2004 Year Report. http://www.stock2000.com.cn/refresh/arch/2005/03/16/814671.htm. (In Chinese)

Pachauri S, Spreng D. 2002. Direct and indirect energy requirements of households in India. Enegry Policy, 30: 511–523.

Paik I, Leiby P, Jones D, et al. 1999. Strategic oil stocks in The APEC region. Proceedings of the 22nd IAEE Annual International Conference, International Association for Energy Economics.

Paul S, Bhattacharya R N. 2004. CO_2 emission from energy use in India: A decomposition analysis. Energy Policy, 32(5): 585–593.

PCAST, 1999. Federal energy research and development for challenges of the twenty-first century. President's Committe of Advisors on Science and Technology, Washington D.C.

Perez C. 2004. Technological revolutions, paradigm shifts and socio-institutional change//Eric, R. Globalization, Economic Development and Inequality: An Alternative Perspective. Edward Elgar, Northampton, MA, U.S.: 217–242.

Phillips P C B, Perron P. 1988. Testing for a unit root in time series regression. Biometrika, 2: 335–346.

Popp D. 2002. Induced innovation and energy prices. The American Economic Review, 92: 160–180.

Qu G. 1992. China's dual-thrust energy strategy: Economic development and environmental protection. Energy Policy, 20(6): 500–506.

Rees W E. 1995. Reducing the ecological footprint of consumption. The workshop on policy measures for changing consumption patterns. Seoul, South Korea.

Reinders A H M E, Vringer K, Blok K. 2003. The direct and indirect energy requirement of households in the European Union. Enegry Policy, 31: 139–153.

Rgwitz M, Miola A. 2005. Evidence from RD&D spending for renewable energy sources in the EU. Renewable Energy, 30: 1635–1647.

Robert S P, et al. 2003. Econometric model and economic forecasting (4th ed.). Translated by Qian X J. Beijing: Machine Industry Press. (In Chinese)

Robison, Yunez-Naude, Hinojosa-Ojeda, et al. 1999. From stylized to applied models: Building multisector CGE models for policy analysis. North American Journal of Economics and Finance, 10: 5–38.

Roca J, Alcantara V. 2001. Energy intensity, CO_2 emissions and the environmental Kuznets curve: The Spanish case. Energy Policy, 29: 553–556.

Rosenberg N. 1976. Technology and environment//Rosenberg N. Perspectives on Technology. Cambridge: Cambridge University Press.

Ruttan V. 1971. Technology and the environment. American Journal of Agricultural Economics, 53: 707–717.

Sagar A D, Holdren J P. 2002. Assessing the global energy innovation system: Some key issues. Energy Policy, 30: 465–469.

Sahoo K C. 2005. Carbon finance opportunities at the World Bank. http://www.iges.or.jp/en/cdm/pdf/india/activity01/india10.pdf

Said S, Dickey D. 1984. Testing for unit roots in autoregressive-moving average models of unknown order. Biometrika, 71: 599–607.

Savabi M R, Stockle C O. 2001. Modeling the possible impact of increased CO_2 and temperature on soil water balance, crop yield and soil erosion. Environmental Modelling & Software, 16: 631–640.

Saviotti P P. 1988. Information, variety and entropy in techno-economic development. Research Policy, 17: 89–103.

SBMR, 2005-11-22. China's energy resource is still increasing. State Bureau of Material Reserve, National Development and Reform Committee. (In Chinese)

Schipper L, Bartlett S, Hawk D, et al. 1989. Linking life-styles and energy use: a matter of time? Annual Review of Energy, 14: 271–320.

Schmalensee R, Stoker T M, Judson R A. 1998. World carbon dioxide emissions: 1950–2050. The Review of Economics and Statistics, 80: 15–27.

Shang X Y, Lu M, Deng Y S, et al. 2005-12. High oil prices challenges China's economic structural, alternative energy facing the development of opportunities. http://finance.sina.com.cn/g/20041018/15151086875.shtml. (In Chinese)

Shi A. 2003. The impact of population pressure on global carbon dioxide emissions, 1975–1996: Evidence from pooled cross-country data. Ecological Economics, 44: 29–42.

Shi D. 1999. Structure change is a main factor influencing Chinese energy consumption. Chinese Industrial Economy, 11: 38–43. (In Chinese)

Shi D. 2001. Review and evaluation on Chinese energy price policy after reformation. Energy Policy Study, 3: 30–37. (In Chinese)

Shi D. 2002. The improvement of energy efficiency in Chinese economic growth. Economic Research Journal, 9: 49–56. (In Chinese)

Shi D. 2003-3-31. Oil prices influence economy vitals. China Securities. (In Chinese)

Shi F Q, 2005. A preliminary analysis on China's energy consumption elasticity. Statistical Study, 5: 8–11. (In Chinese)

Silberglitt R, Hove A, Shulman P. 2003. Analysis of U.S. energy scenarios: Meta-scenarios, pathways, and policy implications. Technological Forecasting & Social Change, 70: 297–315.

SIMA. 2005. http://sima-ext.worldbank.org/query/.

Simon H. 1973. Technology and environment. Management Science, 14: 1110–1121.

Simon J L. 1980. The Ultimate Resource. Princeton: Princeton University Press.

Slow R M. 1957. Technical change and the aggregate production function. Review of Economics and Statistics, 39: 312–320.

Solveig G, Rudman R. 1973. Paramenters of technological growth. Science, 182: 358–364.

Song C X, Xu Q. 2004. Modern west economics–micro economics. Shanghai: Fudan University Press. (In Chinese)

Starr C, Rudman R. 1973. Parameters of technological growth. Science, 182: 358–364.

State Custom Agency. 2006. Custom statistical database. http://www.customs.gov.cn/Default.aspx?tabid=400. (In Chinese)

State Power Information Network. 2005. Electricity Resource Distribution. http://www.bshp.com.cn/newsp/zgdl/zyfb/zyfb.htm, 2005-12-12. (In Chinese)

State Statistical Bureau. 1987. China Energy Statistical Yearbook 1986. Beijing: Energy Press. (In Chinese)

State Statistical Bureau. 1990. China Energy Statistical Yearbook 1989. Beijing: China Statistical Press. (In Chinese)

State Statistical Bureau. 1991. China Energy Statistical Yearbook 1990. Beijing: China Statistical Press. (In Chinese)

State Statistical Bureau. 1992. China Energy Statistical Yearbook 1991. Beijing: China Statistical Press. (In Chinese)

State Statistical Bureau. 1998. China Energy Statistical Yearbook 1991–1996. Beijing: China Statistical Press. (In Chinese)

State Statistical Bureau. 1999a. Comprehensive Statistical Data and Materials on 50 Years of New China. Beijing: China Statistical Press. (In Chinese)

State Statistical Bureau. 1999b. The Input-Output Table 1997. Beijing: China Statistical Press. (In Chinese)

State Statistical Bureau. 2000. China Statistical Yearbook 1999. Beijing: China Statistical Press. (In Chinese)

State Statistical Bureau. 2001a. China Industrial Economy Statistical Yearbook 2000. Beijing: China Statistical Press. (In Chinese)

State Statistical Bureau. 2001b. China Energy Statistical Yearbook 1997–1999. Beijing: China Statistical Press. (In Chinese)

State Statistical Bureau. 2001c. China Statistical Yearbook 2000. Beijing: China Statistical Press. (In Chinese)

State Statistical Bureau. 2002a. China Industrial Economy Statistical Yearbook 2001. Beijing: China Statistical Press. (In Chinese)

State Statistical Bureau. 2002b. China Statistical Yearbook 2001. Beijing: China Statistical Press. (In Chinese)

State Statistical Bureau. 2003a. China Industrial Economy Statistical Yearbook 2002. Beijing: China Statistical Press. (In Chinese)

State Statistical Bureau. 2003b. China Statistical Yearbook 2002. Beijing: China Statistical Press. (In Chinese)

State Statistical Bureau. 2003c. China Statistical Yearbook 2003. Beijing: China Statistical Press. (In Chinese)

State Statistical Bureau. 2004a. China Population Statistical Yearbook 2003. Beijing: China Statistical Press. (In Chinese)

State Statistical Bureau. 2004b. China Energy Statistical Yearbook 2000–2002. Beijing: China Statistical Press. (In Chinese)

State Statistical Bureau. 2004c. China Rural Statistical Yearbook 2003. Beijing: China Statistical Press. (In Chinese)

State Statistical Bureau. 2004d. China Statistical Yearbook 2004. Beijing: China Statistical Press. (In Chinese)

State Statistical Bureau. 2005a. China Statistical Yearbook 2005. Beijing: China Statistical Press. (In Chinese)

State Statistical Bureau. 2005b. Main data communique on the first economic census. 2005-12-14. (In Chinese)

State Statistical Bureau. 2006a. Statistical communique on the 2005 national economic and social development. 2005-2-28. (In Chinese)

State Statistical Bureau. 2006b. Data revision communique on the historical gross domestic product. 2006-1-9. (In Chinese)

Stock J H, Watson M W. 1989. Interpreting the evidence on money-income causality. Econometrics, 40: 161–182.

Sun J W. 1998. Changes in energy consumption and energy intensity: a complete decomposition model. Energy Economics, 20: 85–100.

Sun J W. 1999. The nature of CO_2 emission Kuznets curve. Energy Policy, 27: 691–694.

Sweeney L, James. 1984. The response to energy demand to higher prices: What have we learned? The American Economic Review, 74: 31–37.

Tang Y. 2005-7-2. Sino-US energy policy dialogue. People's Daily. (In Chinese)

Tatom. 1993. Are there useful lessons from the 1990–1991 oil price shock? The Energy Journal, 4: 129–150.

The Administrative Center for China's Agenda 21, Tsinghua University, 2005. Clean development mechanism. Beijing: Social Sciences Academic Press. (In Chinese)

The People's Bank of China. 1999. China Finance Yearbook. Beijing: China Statistics Press. (In Chinese)

Thomas G Rawshi. 2001. What is happening to China's energy consumption. Energy Policy, 28: 671–687.

UNFCCC. 2005. http://cdm.unfccc.int/Projects/registered.html.

UN Statistics Database. 2005. http://unstats.un.org/ unsd/snaama/downloads/GDPconstantNC-countries.xls. (In Chinese)

Varian H R. 1992. Microeconomic Analysis (3rd ed.). New York: W. W. Norton & Company, Inc.

Wang C. 2005. China's industrial distribution of CDM projects. Sino-Canada CDM PDD development training conference. Beijing. (In Chinese)

Wang D. Rising of oil price: The shadow of world economy development. http: //finance.news.tom.com/1638/20050404-199626.html. (In Chinese)

Wang H J. 2001. The input-output analysis on economic structure change and energy demands. Application of Statistics and Management, 5: 27–30.(In Chinese)

Wang J C. 2003. Coal utilized cleanly and structure change. Coal Economic Research, 4: 6–12. (In Chinese)

Wang N J, Ji R S. 2003. Coal liquefaction technology economic analysis. Coal Economic Research, 29(10): 41–43. (In Chinese)

Wang Q Y. 2001. Energy data version 2001. Research of Energy Policy, 1: 1–104. (In Chinese)

Wang Q Y. 2003. Energy efficiency in China and comparing abroad. Energy Conservation and Environmental Protection, 8: 5–7. (In Chinese)

Wang T S, Shen L S. 2001. Economic models collection of quantity economy & technique economy. Institute of Chinese Social Science Academy. Beijing: Social Science Literature Publishing House. (In Chinese)

Wang X L, Meng L. 2001. A reevaluation of China's economic growth. China's Economic Review, 12: 338–346.

Wang X P. 2003. An ecomonic analysis of using crop residues for energy in China. Environment and Development Economics, 8: 467–480.

Wang Z, Zhao W Z, Gao S X, et al. 2005-12.21 century Beijing round table 23: nation's energy strategy. http://business.sohu.com/20041030/n222762368.shtml. (In Chinese)

Weber C, Perrels A. 2000. Modeling lifestyles effects on energy demand and related emissions. Enegry Policy, 28: 549–566.

WEC. 2001. Energy Technologies for the 21st Century. World Energy Council.

Wei P, Xie D. 1991. Econometric analysis and development policy of Chinese energy and economy growth rate. Jiangsu Coal, 1: 15–20. (In Chinese)

Wei Y M, et al. 2006. Progress in energy complex system modeling and analysis. International Journal of Global Energy Issues, 25: 109–128.

Wei Y M, et al. 2007. The impact of lifestyle on energy use and CO_2 emission: An empirical analysis of China's residents. Energy Policy, 35: 247–257.

Wei Y M, Jiao J L, Liang Q. 2004-11. Discussion on the rise of international crude oil price. Science News. (In Chinese)

Wei Y M, Liang Q M, Fan Y, et al. 2006. A scenario analysis of energy requirements and energy intensity for China's rapidly developing society in the year 2020. Technological Forecasting and Social Change, 73: 405–421. (In Chinese)

Wei Y M, Wu G. 2003. The problems and challenges caused by energy sustainable strategy for China. Science News, 16: 8–9. (In Chinese)

Wen L K. 2005-12. US embark on the journey of emerging markets (3), Chinese economy under the rising crude oil. http://www.gtja.com/zcb/22/yw2.htm. (In Chinese)

Wold S. 1995. PLS for multivariate linear modeling//van de Waterbeemd H. Chemometric Methods in Molecular Design. Methods and Principles in Medicinal Chemistry. Weinheim, Germany VCH.

Wu G, Liu L C, Fan Y, et al. 2005. Oil import risk analysis of main oil import countries in the world based on HHA. Cross-Straits Energy Economics Conference, China Taipei. (In Chinese)

Wu G, Liu L C, Wei Y M. 2004. The comparing abroad of energy security policies. China Energy, 26 (12): 36–41. (In Chinese)

Wu K, Li B. 1995. Energy development in China. Energy Policy, 23(2): 167–178.

Wu L, Kaneko S, Matsuoka S. 2005. Driving forces behind the stagnancy of China's energy-related CO_2 emissions from 1996 to 1999: The relative importance of structural change, intensity change and scale change. Energy Policy, 33(3): 319–335.

Wu Y J, Xuan X W. 2002. Economic Theory of Environmental Tax and Analysis of Their Application in China. Beijing: Economic Science Publishing House. (In Chinese)

Wu Z, He J, Zhang A, et al. 1994. A macro-assessment of technology options for CO_2 mitigation in China's energy system. Energy Policy, 22(11): 907–913.

Xiao T. 2001. Which influences the domestic product oil price. China Petroleum, 3: 42–43.

Xinhua News Agency, 2005. Yearbook 2005 of The People's Republic of China. Beijing: Xinhua Press. (In Chinese)

Xinhua Press. 2005. China's crude oil output in 2020 may remain at about 180 million tons. http://www.china5e.com/news/oil/200505/200505260043.html. (In Chinese)

Xue X M. 1998. Calculation and comparison study of CO_2 emissions from China's energy consumption. Environmental Protection, 4: 27–28. (in Chinese)

Xu Z, Cheng G, Chen D, et al. 2002. Economic diversity, development capacity and sustainable development of China. Ecological Economics, 40(3): 369–378.

Xu Z H. 2004. Fossil Energy and the Development of the Situation—Energy Development Strategy Research. Beijing: Chemical Industry Press. (In Chinese)

Xu Z R. 2005-12. The limited range of international oil prices fall. http://www.sinopecnews.com.cn/shzz/2005-02/28/content_225259.htm. (In Chinese)

Yabe N. 2004. An analysis of CO_2 emissions of Japanese industries during the period between 1985 and 1995. Energy Policy, 32(5): 595–610.

Yang H, Wang H, Yu H, et al. 2003. Status of photovoltaic industry in China. Energy Policy, 31: 703–707.

Yang H Y. 2000. A note on the causal relationship between electricity and GDP in Taiwan. Energy Economics, 22: 309–317.

Yan J N. 2005-12. Oil price soaring leads to a tension of China's securities market. http://www.nanfangdaily.com.cn/jj/20050905/dd/200509050123.asp. (In Chinese)

Yan L, Kong L. 1997. The present status and the future development of renewable energy in China. Renewable Energy, 10(2-3): 319–322.

Yang H, Wang H, Yu H, et al. 2003. Status of photovoltaic industry in China. Energy Policy, 31: 703–707.

Yang W M. 2002. Discussion on reform of China oil price system after entering WTO. Price Theory and Practice, 7: 21–22. (In Chinese)

Yang R G, Fan Y, Wei Y M. 2005. China's investment in coal supply analysis—a system dynamics model. Application of Statistics and Management, 24(5): 6–12. (In Chinese)

Yao R, Li B, Steemers K. 2005. Energy policy and standard for built environment in China. Renewable Energy, 30(13): 1973–1988.

Yao Y F, Jiang J H, 2003. The research of China energy requirement to satisfy development potential. http://www.iwep.org.cn/kechixu/meeting/zhongguopaifangxuqiu_yaoyufang.ppt. (In Chinese)

Ye Y, 1996. The Macro-economic Assessment of China's Greenhouse Gases Abatement (Doctoral dissertation). Beijing: Tsinghua University. (In Chinese)

Yi D H. 2002. Data Analysis and Eviews Application. Beijing: China Statistics Press. (In Chinese).

Yi N. 2001. The connection of China oil price and international market. China Investment. 11: 12–15. (In Chinese)

York R, Rosa E A, Dieta T. 2003. STIRPAT, IPAT and ImPACT: Analytic tools for unpacking the driving forces of environmental impacts. Ecological Economics, 46: 351–365.

York R, Rosa E A, Dietz T. 2002. Bridging environmental science with environmental policy: Plasticity of population, affluence and technology. Social Science Quarterly, 83: 18–34.

Yu D F. 2005-4-5. Geothermal exploitation: A seductive cake. http://www.clr.cn/front/chinaResource/read/news-info.asp?ID=52346. (In Chinese)

Yu E S H, Choi J Y. 1985. The causal relationship between electricity and GNP: An international comparison. Journal of Energy and Development, 10: 249–272.

Zhang A L, Li J F. 2004. Analysis on interregional input-output models. Journal of Systems Engineering, 19: 615–619. (In Chinese)

Zhang F. 2005-12. The impact on China's economy of the international oil price rise. http://www.p5w.net/p5w/home/report/macroscopical/200502250924.html. (In Chinese)

Zhang H. 2000. Oil price system must be reformed after entering WTO. China Petroleum and Chemical Industry, 8: 26–29. (In Chinese)

Zhang H, Zhou S H. 2002. Chinese oil consumption and economy development in new era. Oil Warehouse and Service Station, 2: 10–13. (In Chinese)

Zhang J T, et al. 2006. An empirical analysis for national energy R&D expenditures. International Journal of Global Energy Issues, 25: 141–159.

Zhang J T. 2006. Paradigm transitions of technological change in the energy sector and policy implications. International Journal of Foresight and Innovation Policy.

Zhang M Y. 1999. Small samples of model and its causal relationship in the analysis of macro-economic. Systems Engineering—Theory & Practice, 11: 110–114. (In Chinese)

Zhang Z X. 2000. Can China afford to commit itself to an emissions cap? An economic and political analysis. Energy Economics, 22: 587–614.

Zhang Z X. 2000. Decoupling China's carbon emissions increase from economic growth: An economic analysis and policy implications. World Development, 28(4): 739–752.

Zhang Z. 2003. Why did the energy intensity fall in China's industrial sector in the 1990s? The relative importance of structural change and intensity change. Energy Economics, 25: 625–638.

Zhai G M. 2002. Oil and gas resources in China in the 21st century vision. Xinjiang Petroleum Geology, 23(4): 271–278. (In Chinese)

Zhao D Y. 2005-12. Oil prices hike affects China's stock market valuation. http://www.21our.com/readnews.asp?id=568304&str_id_star=1&str_id_end=927. (In Chinese)

Zhao L X, Wei W X. 1998. Study on energy and economic growth models. Forecasting, 6: 32–34.

References

Zhou F Q. 2001. Expectation and countermeasure suggestion for oil supply and demand in China. International Petroleum Economics, 5: 5–8. (In Chinese)

Zhou F. 1996. Development of China renewable energy. Renewable Energy, 9: 1132–1137.

Zhou Z Y, Tang Y G. 2004. China's natural gas situation and its status in the energy structure. China Mining, 13(6): 17–28. (In Chinese)

Zhou Z Y, Zhang K, Tang Y G. 2003. Characteristics and trends of China's proven reserves of oil Xinjiang Petroleum Geology, 24(4): 356–359. (In Chinese)

Zhu C Z. 2001. Energy substitution, electricity substitution and efficiency. Study on Coal Economy, 8: 12–14. (In Chinese)

Zhu C Z. 2004. The role of electricity industry in energy efficiency and conservation. Journal of Energy Saving and Environment Protect, 9: 11–14. (In Chinese)

Zhu X. 1999. China Mining Situation (the first Volume). Beijing: Science Press. (In Chinese)

Zi Y. 2005-12. There can't be a fourth oil crisis. http://www.p5w.net/p5w/home/furtures/200509270845.html.

Index

A

Academy of Macroeconomic Research Workgroup of State Development Planning Commission, 77
ADF unit root test, 35
Africa, 148, 167
Aggregate energy intensity, 44
AHP method, 234
Air conditioning, 200
Air Pollution Control Act of 1955, 255
Akaike information criteria, 34
Anthropogenic CO_2 emissions, 208
ARMA. *See* Multivariate models, 150
Armington assumption, 126
ASEAN, 244
Asia, 164
Asian Financial Crisis, 45, 124
Asymmetric Vector Error Correction Model (AVECM), 119
Auto-covariance, 32
AVECM. *See* Asymmetric Vector Error Correction Model, 119
Average temperature change, 172
Aviation kerosene, 272
AWD decomposition, 179
Azerbaijan, 167

B

Bacon's model, 119
Baoshan, 285
Bi-directional causality, 32
Biomass combined heat and power (CHP), 177
Bohai Bay, 247
BP Statistical Review of World Energy, 228
Brent crude oil, 101
Brightness Project, 13
British gasoline prices, 119
Burma, 167

C

Canada, 32, 235
Carbon dioxide, 7, 180
Carbon emission intensity, 286
Carbon tax revenue, 214
Caspian region, 244
Causality
    Causality Test. *See* Granger
CDM. *See* Clean Development Mechanism
CDM projects, 175
CEDAS. *See* China energy demand system
Central Asia, 167, 244
CERs. *See* Certified emission reduction
Certified emission reduction, 7
CGE. *See* Computable General Equilibrium
Chemical Import and Export Corporation of China, 161
ChevronTexaco Corporation, 6
China-Burma oil routes, 247
China energy demand system, 73
China Fuel Efficiency Background Report, 199
China International Petrochemical Industry United Co., Ltd., 161
China International Petroleum Chemicals Co. Ltd, 113
China National Petroleum Corporation, 6, 159
China Petrochemical Corporation, 159
China Petroleum, 123
China Petroleum Chemical, 123
China South Sea, 247
China Statistics Yearbook, 10, 198, 228
China United Oil Co., Ltd, 161
China's Oil and Gas Industry Annual Report, 224
China's rural regions, 198
China's National Development and Reform Commission, 13, 281
Carbon Dioxide (CO_2)
    Abatement, 171
    Concentration and global climate change, 172
    Control, China, 158
    Emission intensity, 222
    Fossil fuel combustion, 208
    Lifestyles, 207
    Tax, 125
    Technology efficiency, 196
    Mitigation, 171
    Global GDP, 147
    Income levels, 188
Chinese Academy of Sciences, 76
Chinese carbon emissions intensity, 178
Chinese oil pricing mechanism
    Reform, 158, 169
CHP. *See* Biomass combined heat and power
Civil biogas, 13
CLA, 194, *See* Consumer Lifestyle Approach
Clean Air Act, 262
Clean Development Mechanism, 7
Clockwise trajectories, 260, 262

CNOOC, 161
CNPC. *See* China National Petroleum Corporation, 6, 282
CO_2. *See* Carbon dioxide, 12
Coal briquettes, 200
Coal liquefaction, 254, 302
Coal production, 5
Coal reserves, 3, 221
    Mine construction, 87
    Proven reserves, 2, 8
    Regional, 2
Coal supply, 71
Coal-Bed Methane, 176
Coalmine security, 12
Cobweb Model, 259, 260
Cointegration, 28, 31, 118
Cointegration and causality, 31
Cointegration test, 32, 168
Commodities' flow path, 127
Computable General Equilibrium (CGE), 123
Construction in Energy Statistics Yearbook, 195
Consumer Lifestyle Approach (CLA), 196
Conversion efficiency, 2
Counterclockwise trajectories, 262
Crude oil imports, 15, 245
Crude oil price fluctuation, 104

D

Daily Brent, 151
Daqing crude oil, 101
    Crude oil price, 108
Decision tree, 225
    Optimal Petroleum Reserves, 225
Decomposition of energy intensity, 43
Development Research Center of the State Council (DRC), 125, 199
Dickey and Fuller developed DF method, 32
Diesel oil, 114
Diesel oil accumulative adjustment function, 122
Direct energy use of urban residents, 199
Disaster and climate, 142
Diverging trajectory, 260, 262
Diversification index, 241
Domestic oil fleet, 234

E

EB projects, 174
EC. *See* Energy consumption
ECE. *See* United Nations Economic Commission for Europe
Econometrics, 32
Economic growth, 3
Economic Market Stabilization, 144
Economic outputs, 47
    Subindustries, 49
Economic structure, 17, 20
    Trends in China, 28
Economy, 1
    Development, 1, 7
    Irregular economic development, 18
    Regions, 2, 85
    Scenarios, 71
    Society, 1
EIA. *See* Energy Information Agency
Elasticity of demand, 131
Electricity, 11
    Marginal efficiency, 58
Electricity production, 11
Eleventh (11th) Five-Year Plan, 284
Elman network, 151
Emitter-pay, 213
Energy conservation, 6
Energy consumption, 1, 3, 4, 7, 13, 14, 15, 16, 29, 34, 284
    Decline, 3, 30
    Different industries, 16
    Environment, 1
    GDP, 3
    Growth rate, 1, 45
    Industrial sectors, 16
    Per capita, 3
    Per unit GDP, 194
    Priority to coal, 14
    Production, 1, 3
    Provinces and cities, 20
    Regions, 2, 6
    Structure, 1, 4
    Sub-industries, 49
    Various countries, 4
    World and major countries, 3
Energy efficiency, 1, 17, 42
    Trend, 10, 34
Energy Efficiency Standards, 72, 176
Energy import dependence, 15
Energy Information Administration, 97
Energy Information Agency, 228
Energy intensity, 3, 4, 27, 83
    Subindustries, 49
Energy production, 4
    Coal, 3
    Crude oil, 4

Index

Hydropower, 4
Natural gas, 2
Policy implications, 18, 41
Energy Research Institute, 182, 198
Energy security
Definition, 220
Energy security, China, 12
Energy structure, change, 42
Energy technology priorities, 270, 278
Entropy of energy, 264, 265, 267
Entropy statistics, 263
Error Correction Model, 32
EU, 95
Expected value (EV), 226
Extended Linear Expenditure System (ELES), 126
Exxon Mobil Corporation, 148

F

Fast-moving-consumer goods (FMCG), 197
FDI, 277
Feed-in Tariffs scheme, 256
Ferrous metal ores, 51
Fertilizers and pesticides, 129
Fluctuation
Petroleum price, 95
Fluctuation of energy supply, 42
Forecasting, 71
Foreign stock markets, 114
Foreign-sourced advanced technology, 42
Fossil energy, 3
Alternatives, 270, 293
Fossil fuel-related CO_2 emissions, 183
FPMGA method, 151
France, 4, 268
Futures, 101, 141
Futures-weighted, 153

G

GARCH. See Multivariate models
Gasoline, 114
Gasoline, domestic, 101
Gasoline/diesel, 116
GDP. See Gross Domestic Product
GDP, income levels, 80, 186
General Electric, 254
Genetic algorithm, 150
Geologic prospecting investment factor (ICGP), 90
Geological prospecting, 87
Geothermal resources, 3, 10
Germany, 3

Global CO_2 emissions, 172
Global economy, 95
International petroleum, 95
Global optimizing method, 151
Global warming, 7, 185, 208
GPMGA. See Pattern-matching technique
Granger causality test, 32, 109
Green Power Exchange, 256
Green pricing, 256
Greenhouse gases, 171
Gross Domestic Product(GDP), 3, 164
Growth of regional population, 79
Energy requirements, 71
Growth rate of China, 59
Guangzhou, 223
Gulf of Mexico, 148

H

Heavy-industrialization, 57
HHA. See Hirschman-Herfindahl Agiobenebo
High-tech industry and Information & Communication, 166
High-tech industrialization, 57
Hirschman-Herfindahl Agiobenebo (HHA) method, 234
Home energy use, 197
Household appliances, 287
Hurricane Ivan, 153
Hurricane Katrina, 223
Hydro fluorocarbons (HFCs), 171
Hydroelectricity, 14
Hydropower, 4, 5, 8, 9, 12, 23, 24, 176, 177, 181, 245, 284

I

IEA. See International Energy Agency
IEA forecast, 149
IET. See International Emissions Trading
IMF report, 103
Import risk index, 219
Oil importers, 5, 219
Impulse response analysis, 110
Index Decomposition Analysis (IDA), 177
Indirect energy use, 199
Residents' lifestyle, 171
Indonesia, 5, 167
Industrial Revolution, 3, 250
Industrialization, 1, 57
Input–output model, 72
Multi-regional, 72
Integrated energy intensity, 48
Intergovernmental Panel on Climate Change

(IPCC), 172
International crude oil price, 95
International diesel oil, 115
International Emissions Trading, 7
International Energy Agency, 6, 219, 224
International energy competition, 6
International market risk, 95
International oil price, 11, 95
Inter-regional IO, 73
I/O model. See Input-output model
IPAT. See STIRPAT model
Iran, 98
Iraq war, 105
IRIO. See Inter-regional input-output model
Italy, 4, 257

J

Japan, 5, 147, 224
JI. See Joint Implementation
JI projects, 176
Joint Implementation, 7
Journal International Petroleum Economics, 115

K

Kazakstan, 244
Kvenvolden, 271
Kyoto Protocol, 7, 29, 171, 173, 175, 176, 208, 209, 211, 218, 219, 256
 Post-Kyoto, 173

L

Laspeyres method, 43
LDQ. See Logarithm of DaQing
Libya, 167
Light-water reactors, 254
Lipschitz index, 96, 100
Liquidation technology, coal, 272
Logarithm of DaQing, 107
LPG, refinery gas, 181

M

Malacca Strait, 223, 233
Malmquist index, 285
Malthusian Limit, 250
Malthusian perspective, 195
MAPE. See Mean Absolute Percentage Error
Marginal efficiency, 58
Maxwell's equations, 250
ME. See Marginal efficiency
ME of coal, 58
 of electricity, 65
 of petroleum, 64
Mean Absolute Percentage Error (MAPE), 152
Medium and Long-Term Science and Technology Plan, 166
Methane (CH_4), 171
Methane capture projects, 177
Mexico, 103
Middle East, 29, 100
 Crude oil, 112
 War, 99
Mining and processing of ferrous metal ores, 50
Morlet wavelet, 96
MRIO. See Multiregional IO model, 73
MRS, 58
Mtces, 29, See Metric tons of carbon equivalents
Multiregional IO. 73
Multistep prediction, 150
Multivariate models, 145, 150

N

Nation's Development and Reform Commission, 283, 284
National Demonstration Project of Straw Gasification, 13
National Development and Reform Commission (NDRC), 116
National energy security, 12, 219
National Information Center, 149
National macroeconomic policies, 244
National meteorological stations, 10
Natural gas, 2, 181
Natural gas
 Hydrate, 271
 Production, 5
 Reserves, 2, 8
 Proven, 2
 Trade, 6
NDRC. See National Development and
Reform Commission, 119
Netherlands, 73
New policies and technologies, 42
New York, 6, 159
New York Mercantile Exchange, 6, 29
Nigeria, 5, 148
Nitrous oxide (N_2O), 171
Non-Fossil Obligation (NFFO), 256
North Sea Oil Field, 101
Nuclear electricity, 14, 67
Nuclear energy, 166, 278

Index

Nuclear power, 4, 288
NYMEX, 113, *See* New York Mercantile Exchange

O

Ocean Trade Corporation of China, 161
OECD, 107
Oil
    Crises, 95
    Exploitation, 106
    Fluctuation, 95
    Fund, 165
    Imports, 124, 221
    Price risk, 106
    Reserves, 1
    Proven reserves, 24
    Terrestrial reservoir, 7
Oil shock, 224, 278
Oman, 167
OPEC, 5, *See* Organization of Petroleum Exporting Countries
OPEC countries, 102
OPEC Policies, 142
OPEC Price Zone, 103
OPEC region, 104
Organization of Petroleum Exporting Countries (OPEC), 4

P

Paasche method, 43
Partial Least Squares, 186
Pattern-matching technique, 150
Pattern modeling and recognition system, 150
People's Republic of China Renewable Energy Law, 283
Perfluorocarbons (PFCs), 169
Petrochemical sector, 140
Petrochina, 159, 161, 167
PetroKazakstan, 6
Petroleum
    Marginal efficiency ME, 58
    Prices, 95
    Fluctuation, 95
Photovoltaic, 255
PLATT's price quote system, 113
PLS. *See* Partial Least Squares, 190
PLS regression, CO_2 emissions, 191
PMRS. *See* Pattern modeling and recognition system, 150
Policy implications, 18, 41
Pollution, 3
    Environmental, 1
Greenhouse gas, 3
Population, income levels, 147
Population growth, 76
Post-Kyoto, 173
PRC State Council Development Center, 75
Premcor Refining Group Inc, 6
Prices block, 124
Primary energy production, 10, 281
Primary energy-related carbon intensity, 176
Primary industry, 37, 45, 130
Production block, 124

Q

Qaidam Basin, 25
Qinghai, 8, 10, 20, 28
Qiong Basin, 25
Quality coal resource reserves, 3
Quantity and structure, energy, 13
Quick growth, energy consumption, 13
Quotas, greenhouse gas emission, 7
Quotient, carbon intensity, 179

R

R&D expenditure, 247
R&D funding, 255
R&D intensity matrix, 263
RACI. *See* Refiner acquisition cost, imported oil
Radiation. *See* Solar energy
Real crude oil price, 258
Real GDP, 34, 125
Refiner acquisition cost of imported crude oil, 95
Reform and opening-up
    China, 29
Regional coal resources, 19
Regional economic indexes, 27
Regional energy expense, 18
Regional industrial structure, 28
Regional net tax payment, 212
Regional population, 76
Renewable energy, 3, 30
    Law, 30
    Projects, 13, 275
Renewable Energy Law, 30, 277
Republic of Korea, 282
Residents' real income, 127
Revenue and consumption block, 124
Ride Wind Program, 13
Rio de Janeiro, Brazil, 255
Rio Earth Summit, 255
Risk Assessment Index System, 234

Root Mean Square Error (RMSE), 151
Rotterdam, 122, 158
Russia, 1, 102, 147, 230, 240

S

Saudi oil, 112
Scenario analysis, 71, 82
  Regional natural gas, 85
SDA. See Structure decomposition analysis
Secondary industry, 19, 37, 41, 130
  Structure, 49
Shanghai Futures Exchange, 113
Shanghai Security Newspaper, 148
Shenzhen, 221
Short Rules for Clean Development Mechanism Project Management, 174
Siemens, 243
SIMA. See Statistical Information Management and Analysis
Singapore, 32, 122
Sino-Kazakhstan oil pipeline, 282
Sinopec, 158, 160, 282
Sino-Russia AN-DA pipeline, 242
Small-scale collieries, closing, 47
SO_2. See Sulfur dioxide
Social accounting matrix (SAM), 127
Solar energy, 3, 9, 244
  Water heating, 13
Solar radiation, 10
Songliao Basin, 23
Southeast Asia, 166, 233
Special Plan for Medium and Long-Term Energy Conservation, 284
SPR. See Strategic petroleum reserve
State-owned coal mines, 70
State Council, 75, 163
State Planning Commission, 12
Statistical Information Management and Analysis (SIMA), 186
STIRPAT, 184, See Stochastic Impacts by Regression on Population, Affluence, and Technology
STIRPAT model, 184
Stochastic Impacts by Regression on Population, Affluence, and Technology, 184
Strategic petroleum reserve (SPR), 217, 282,
Structural decomposition analysis, 43
Sub-industries outputs, 54
Sudan, 167
Sulfur dioxide, 12
Sulfur-Tax, 125

Sulphur hexafluoride (SF_6), 171
Switzerland, 257
SWOT analysis, 272

T

Taishet-Nakhodka oil pipeline, 232
Tapirs for light crude, 104
tce. See Tons coal equivalent
Techno-economic paradigms, 249
Technological advances, 125
Technology scenarios, 77
Ten Forecasts in Chinese Economy, 149
Tertiary industry, 19, 131
Thermal efficiency, 200
Third Appraisal of Natural Gas Resources, 25
Tibet, 8
Time series, 18
Toe. See Tons of oil equivalent
Tons of oil equivalent, 3
Total generating capacity, 281
Trace test statistic, 108
Transformation from coal to petroleum, electricity, 67
Transition economy, 196
Transportation sector, 125, 136

U

Unidirectional causality, 41
United Nations Economic Commission for Europe (ECE), 222
United Nations Environment Program (UNEP), 255
United Nations Framework Convention on Climate Change (UNFCCC), 255
Unocal Corporation, 6
Urban residents' lifestyle, impact of, 171
Urbanization, 1, 191, 200
  Forecast, 76
Urbanization scenarios, 76
U.S. EIA website, 102
Utilization efficiency of energy, 17

V

Valero Energy Corporation, 6
Variable Importance in Projection, 191
Variance decomposition (VDC) analysis, 111
VECM. See Asymmetric Vector Error Correction Model 109
Vector Error Correction Model, 109, See
Venezuela, 5, 148
VIP. See Variable Importance in Projection
VIP, GDP, 191

Index

VIP, global population, 194
VIP, urbanization, 192

W

Wald test, 112
Walking out strategies, 164
Wavelet analysis, 96, 145
Wavelet decomposition, 146
Welfare, residents, 114
West Africa, 235
West Development Policy, 214
West Texas Intermediate, 142
Westinghouse, 254
Wind power, 12, 177
Workgroup of the Academy of Macroeconomic Research of the State Development Planning Commission, 77
Workgroup of the State Statistics Bureau, 76
World Bank, 95, 176,
World coal trade, 6
World crude oil imports, 5
World crude oil trade, 5
World energy production, 4
    Crude oil, 4
World gross domestic production, 3

World Meteorological Organization (WMO), 255
World oil
    Reserves and imports (2003), 233
World oil stocks, 153
World oil trade (2004), 232
World Renewable Energy Reserves, 3
World Trade Organization (WTO), 162
WTI, 105, *See* West Texas Intermediate
WTO. *See* World Trade Organization

X

Xing'an Mountains, 9
Xinjiang, 8, 20

Y

Yukos, 148

Z

Zhejiang Province, 222, 282
Zhenhai, 222
Zhenhai base, 233
Zone theory models, 145